T0272625

Logic Pro®

3rd Edition

by Graham English

Logic Pro® For Dummies®, 3rd Edition

Published by: **John Wiley & Sons, Inc.,** 111 River Street, Hoboken, NJ 07030-5774, www.wiley.com

Copyright © 2023 by John Wiley & Sons, Inc., Hoboken, New Jersey

Published simultaneously in Canada

No part of this publication may be reproduced, stored in a retrieval system or transmitted in any form or by any means, electronic, mechanical, photocopying, recording, scanning or otherwise, except as permitted under Sections 107 or 108 of the 1976 United States Copyright Act, without the prior written permission of the Publisher. Requests to the Publisher for permission should be addressed to the Permissions Department, John Wiley & Sons, Inc., 111 River Street, Hoboken, NJ 07030, (201) 748-6011, fax (201) 748-6008, or online at http://www.wiley.com/go/permissions.

Trademarks: Wiley, For Dummies, the Dummies Man logo, Dummies.com, Making Everything Easier, and related trade dress are trademarks or registered trademarks of John Wiley & Sons, Inc. and may not be used without written permission. Logic Pro is a registered trademark of Apple, Inc. All other trademarks are the property of their respective owners. John Wiley & Sons, Inc. is not associated with any product or vendor mentioned in this book. *Logic Pro For Dummies®*, 3rd Edition is an independent publication and has not been authorized, sponsored, or otherwise approved by Apple, Inc.

LIMIT OF LIABILITY/DISCLAIMER OF WARRANTY: WHILE THE PUBLISHER AND AUTHORS HAVE USED THEIR BEST EFFORTS IN PREPARING THIS WORK, THEY MAKE NO REPRESENTATIONS OR WARRANTIES WITH RESPECT TO THE ACCURACY OR COMPLETENESS OF THE CONTENTS OF THIS WORK AND SPECIFICALLY DISCLAIM ALL WARRANTIES, INCLUDING WITHOUT LIMITATION ANY IMPLIED WARRANTIES OF MERCHANTABILITY OR FITNESS FOR A PARTICULAR PURPOSE. NO WARRANTY MAY BE CREATED OR EXTENDED BY SALES REPRESENTATIVES, WRITTEN SALES MATERIALS OR PROMOTIONAL STATEMENTS FOR THIS WORK. THE FACT THAT AN ORGANIZATION, WEBSITE, OR PRODUCT IS REFERRED TO IN THIS WORK AS A CITATION AND/OR POTENTIAL SOURCE OF FURTHER INFORMATION DOES NOT MEAN THAT THE PUBLISHER AND AUTHORS ENDORSE THE INFORMATION OR SERVICES THE ORGANIZATION, WEBSITE, OR PRODUCT MAY PROVIDE OR RECOMMENDATIONS IT MAY MAKE. THIS WORK IS SOLD WITH THE UNDERSTANDING THAT THE PUBLISHER IS NOT ENGAGED IN RENDERING PROFESSIONAL SERVICES. THE ADVICE AND STRATEGIES CONTAINED HEREIN MAY NOT BE SUITABLE FOR YOUR SITUATION. YOU SHOULD CONSULT WITH A SPECIALIST WHERE APPROPRIATE. FURTHER, READERS SHOULD BE AWARE THAT WEBSITES LISTED IN THIS WORK MAY HAVE CHANGED OR DISAPPEARED BETWEEN WHEN THIS WORK WAS WRITTEN AND WHEN IT IS READ. NEITHER THE PUBLISHER NOR AUTHORS SHALL BE LIABLE FOR ANY LOSS OF PROFIT OR ANY OTHER COMMERCIAL DAMAGES, INCLUDING BUT NOT LIMITED TO SPECIAL, INCIDENTAL, CONSEQUENTIAL, OR OTHER DAMAGES.

For general information on our other products and services, please contact our Customer Care Department within the U.S. at 877-762-2974, outside the U.S. at 317-572-3993, or fax 317-572-4002. For technical support, please visit https://hub.wiley.com/community/support/dummies.

Wiley publishes in a variety of print and electronic formats and by print-on-demand. Some material included with standard print versions of this book may not be included in e-books or in print-on-demand. If this book refers to media such as a CD or DVD that is not included in the version you purchased, you may download this material at http://booksupport.wiley.com. For more information about Wiley products, visit www.wiley.com.

Library of Congress Control Number: 2023935197

ISBN 978-1-394-16210-9 (pbk); ISBN 978-1-394-16211-6 (ebk); ISBN 978-1-394-16212-3 (ebk)

SKY10070503_032124

Contents at a Glance

Table of Contents

Introduction

At its near-permanent spot in the top-ten grossing apps in the entire Mac App Store, Logic Pro has proven itself to be in high demand. You shouldn't expect anything less than stellar software from Apple. And there's a good reason why Logic Pro is professionally competitive. Apple designs intuitive software that music producers love at the best possible value. And unlike other digital audio workstations that have moved to a subscription pricing model, Logic Pro remains a relatively inexpensive one-time purchase with free updates included.

In line with Apple's mission, I wrote *Logic Pro For Dummies* to add value to your Mac and Logic Pro. You learn how to record, arrange, edit, mix, and share your music, becoming a self-sufficient musician with your computer and Logic Pro. This book will guide you to make more music.

About This Book

If I could give people one superpower, I would give them instant musical talent. My world would be a curious musical, filled with willing musical partners. *Logic Pro For Dummies* is my honest attempt to make musical partners out of every reader, including you.

This book is designed to get you making music fast. You don't even need to know how to play an instrument to make music with Logic Pro because it includes additional content you can use in your projects. Regardless of your current capabilities, the step-by-step instruction in this book guides you through everything you need to know to make music quickly.

I'm happy you came to me to learn Logic Pro because I have been coaching Logic Pro users since 2007 and know people's common frustrations and mistakes. I want you to feel confident using the software so you can complete more projects and share your music — with others and with me. This book gives you the most important information you need to quickly meet your musical goals and turn your ideas into completed projects.

Logic Pro For Dummies is organized for easy access. It's your productivity advisor and your reference for quickly finding the information you need. And because many people learn more quickly by watching someone else, I provide free videos and project templates to accompany the book at `https://logicstudiotraining.com/bookextras`.

Throughout the book, I use certain conventions to show you what to do. For example, when you choose items from menus, I use the command arrow, such as Choose File ⇨ Edit. Links to websites are presented like this: `https://logicstudiotraining.com`. If you purchased the e-book, links are live and will take you directly to the web page. Finally, Logic Pro uses the term *key command* for any combination of keys that can act as a shortcut to a function; when I refer to the Command key, I use the ⌘ symbol.

Foolish Assumptions

As I said, I want to give you instant talent, but I have to make some assumptions about you, my friendly reader. I'm pretty sure you have the music bug. But you may want to only record audio with Logic Pro, such as voice-overs, podcasts, or live seminars. This book covers those topics, but I'm also writing for the musician in you.

I believe you bought this book not only to learn how to use software but also to create music. Logic Pro is the tool, and your music is the reason it exists.

I also assume that you're not making as much music as you could be making. I know I'm not. I'm sure we could all be bringing more music into the world, and I often aggressively push for it. I love to train musicians because they are great listeners. Great listeners make great leaders, and if I didn't push for more great leaders, I would feel that I wasted a golden opportunity to inspire you to greatness. The more music you make, the better listener you become.

Even if you're a beginner, I assume you'll be able to make music that sounds great with Logic Pro. It might be a foolish assumption, but given everything Logic Pro can do for you, I don't think so. Finally, I make the safe assumption that you'll enjoy your time with Logic Pro.

Icons Used in This Book

You'll see helpful icons throughout this book. Scan for them, and you'll find useful information that will help pull everything together and even broaden your perspective. Readers love to scan, and I love to write for scanners.

The Tip icon is usually designed to give you an "aha" moment. Tips go beyond step-by-step instruction into strategies and techniques to make better sounding music. Pay close attention to the tips!

The Remember icon points out information that you need to keep in mind as you use Logic Pro. In some cases, you'll be given key commands that are important to remember. Other times, you'll see a short refresher on information relevant to the topic and covered elsewhere in the book. Whenever you see this icon, it's important to at least store the information in your short-term memory. After all, the book remembers everything for you in the long term.

The Technical Stuff icon points out information that can be skipped or treated as extra credit. The information in these sections shouldn't be beyond your understanding, but you don't need to know how the engine works to drive a car.

The Warning icon is reserved for potential mistakes that could cause you to sound bad. That's the last thing I want, and fortunately, sounding bad is hard to achieve with Logic Pro. So when you see the Warning icon, please read it!

Beyond the Book

As mentioned, I deliver content outside this book through videos and project files. Where appropriate, I've added a link to a web page with further instructions. These videos should help you visualize the book's content, and the project files are excellent resources for getting started.

In addition, *For Dummies* books include one of my favorite tools of all time, the cheat sheet. I make cheat sheets for a hobby, and I'm excited to give you what I've got. To get to the cheat sheet, go to `www.dummies.com` and type *Logic Pro For Dummies cheat sheet* in the Search box.

Where to Go from Here

Although I wrote the book to be somewhat linear and to follow a logical progression, you can start anywhere you want. Because I reference chapters throughout the book, you should be able to open any chapter and follow along.

If you're new to Logic Pro, you'll at least want to skim the first four chapters. These chapters make up Part 1 and will get you started using Logic Pro and understanding how it works. Part 2 shows you how to record audio, load and play software instruments, and add prerecorded media to your project. If you're upgrading from previous versions of Logic Pro, you might skip to Part 3 and learn about the new software instruments or head over to Part 4, where you learn how to use the exciting new editing features such as flex pitch.

Part 5 is dedicated to mixing audio so that the final result sounds good and is ready to share with the world. From the beginning of the book to the end, you have a powerful music production blueprint. I hope you get what you need. If you should have a question, you can find me online at `https://logicstudiotraining.com` or `https://grahamenglish.com`.

1 Leaping into Logic Pro

IN THIS PART . . .

Develop a productive workflow and mindset.

Discover timesaving tips to help you finish Logic Pro projects and share your projects for collaboration and backup.

Navigate the software interface, play and control your project, and explore the tools.

Understand how tracks and regions work in Logic Pro, adjust your tempo and time signature, save track settings for instant recall, and edit and loop regions.

IN THIS CHAPTER

» Understanding the benefits of creating with Logic Pro

» Getting into the Logic Pro mindset

» Developing a productive workflow

» Setting up your Logic Pro studio

Chapter 1
Getting Logic Pro Up and Sprinting

The joke used to be that Logic Pro wasn't logical. I would argue that it was logical but not intuitive. Nowadays, you can't make that joke without dating yourself. Apple, known for making the complicated simple, bought Logic Pro from Emagic in 2002 and continues to make the product better and better.

You'll find that creating music with Logic Pro can be a straightforward and rewarding experience. One caveat: As you explore Logic Pro, remember your desired outcome. With so many bright and shiny objects in this deep and powerful app, getting distracted is easy. But if you keep your musical and learning goals in mind, you'll discover why Logic Pro is behind so many Billboard hits.

Take command. Logic Pro listens.

In this chapter, you discover why Logic Pro users are proud, productive, and ready to play. You'll understand how to plan your creations, get the most value from your time with Logic Pro, set up your studio, and much more.

REMEMBERING THE LOGIC PRO JOURNEY

Logic Pro has come a long way since its inception. In the mid-80s, the German company C-LAB created Supertrack for the Commodore 64 computer. This product evolved into the Creator software program and eventually became Notator Logic, which ran on the Atari system in the early 90s. Here are some significant milestones on the path to Logic Pro:

- 1994: Audio recording capabilities were added to Notator Logic.
- 2000: Virtual instruments were added to Logic 4.
- 2002: Apple purchased Logic.
- 2004: Logic 6 became Mac-only.
- 2007: Several audio applications, including Logic Pro 8 and MainStage, were bundled as Logic Studio. New features such as Quick Swipe Comping and the Delay Designer plug-in were introduced, and the copy protection USB dongle was eliminated.
- 2009: Logic Pro 9 introduces features such as flex time editing, Amp Designer, and Pedalboard plug-ins.
- 2010: Logic Pro 9 goes 64-bit.
- 2013: Logic Pro X is released, with a redesigned look, flex pitch editing, new editors, the Drummer software instrument, the Bass Amp Designer plug-in, virtual vintage instruments, MIDI plug-ins, track stacks, smart controls, tighter integration with GarageBand, the Logic Remote iPad app, and much more.
- 2020: Apple drops *X* from the Logic Pro name while continuing to develop several incremental updates worthy of significant releases and without any additional cost to users. Notable features include live loops, step sequencer, remix FX, drum synth, and a new refined design with drag-and-drop workflows. The EXS24 sampler software instrument was phased out in favor of the new sampler and quick sampler. And spatial audio mixing with Dolby Atmos enables you to mix and monitor 3D immersive audio with dynamic head tracking using select headphones.

Embracing Logic Pro

Why would you want to choose Logic Pro when so many different digital audio workstations (DAWs) are on the market? Here's a list of reasons why you don't need to look any further than Logic Pro:

- Logic Pro is designed by Apple, so hardware and software compatibility are simple and usually hassle-free. Logic Pro doesn't lag behind Mac operating system updates and takes advantage of the hardware. However, if you rely on third-party plug-ins, upgrade cautiously. If you're wondering whether it's safe to upgrade, stop by `https://logicstudiotraining.com` and ask me.
- Logic Pro has thousands of sampled instruments and effects presets, saving you thousands of dollars in additional expenses. You get a complete studio — including a virtual drummer who won't show up late or scuff your floors.
- Logic Pro excels at both recording and editing audio and MIDI. Some DAWs do one or the other well, but Logic Pro does both with superb sound quality and ease of use.
- Logic Pro is compatible with most audio and MIDI hardware. I rarely get asked hardware questions from my clients because the product just works.
- Logic Pro enables you to provide professional notation for lead sheets and full scores. When you need to hand out chord charts to the band or provide fully engraved charts with advanced markup to the orchestra, Logic Pro has you covered. Logic Pro can also create guitar tablature and add lyrics to your score.
- Logic Pro supports many hardware control surfaces, so you can control knobs, faders, buttons, and other parameters in Logic Pro right from your hardware. An inexpensive MIDI controller can be turned into a remote control for advanced control (or when the hand you use for your mouse or trackpad begins to ache from overuse).
- Logic Pro is a powerful mixing console. If your computer can handle it, you can have 255 audio tracks, 255 software instrument tracks, and 255 auxiliary tracks. You weren't worried about running out of tracks, were you? And because you don't have 255 hands, you can automate parameter changes on all those tracks.
- Logic Pro is a 64-bit application that gives you increased power. Older 32-bit apps allow the use of only 4MB of RAM, but Logic Pro can access all the memory your computer has installed. You can run more plug-ins and more software instruments without a hiccup.

I've only touched the surface of what Logic Pro can do. Surround sound, virtual vintage instruments, drum machines, guitar amps and pedals, pitch and time editing, and MIDI effects are a fraction of what you have available as a Logic Pro user. Congratulations on making such a smart choice to embrace Logic Pro. Welcome to the club!

Adjusting Accessibility Settings

Apple products include accessibility features to assist users. To find out how to use these features, visit the following:

- » Apple accessibility: `www.apple.com/accessibility`
- » Accessibility support: `support.apple.com/accessibility`

If you want to use Logic Pro with VoiceOver, you can make plug-in windows open in Controls view, which will list the plug-in parameters instead of displaying a graphical user interface. Choose Logic Pro ⇨ Settings ⇨ General. Select the Accessibility tab and then select the Open in Controls View by Default check box.

Transitioning from Other Software

I won't bad-mouth or slam other DAWs — it's a good rule, especially when collaborating with musicians who use different software. I'm a fan of GarageBand and use it sometimes because it integrates so well with Logic Pro. If you've used GarageBand, you'll find the Logic Pro interface familiar and welcoming.

Logic Pro has many of the same features as GarageBand, including a professional mixer and finer control over audio and MIDI regions. GarageBand 10 users are familiar with smart controls, but with Logic Pro, you can do a lot more with them, such as control and edit more parameters. Best of all, Logic Pro can open GarageBand projects, including GarageBand projects created in iOS, the operating system that powers the iPhone, the iPad, and the iPod touch. Starting projects on one of those devices and continuing working on them in GarageBand or Logic Pro creates a powerful workflow.

If you're coming from another software environment, maintain a beginner's mind as you explore Logic Pro. The workflow is probably similar to what you're accustomed to. It won't take long to understand that a Logic Pro project has tracks containing MIDI or audio regions and that you can arrange those regions in the main window or edit them in the various editors. Did I just give the whole book away right there? Not quite, but understanding Logic Pro is almost that simple.

Just remember, if your objective is clear, Logic Pro can help you reach it.

Creating with Logic Pro

With Logic Pro, you have a full band at your fingertips. From country to dance music, songwriting to film scoring, capturing MIDI performances to professional audio recording, Logic Pro will excite your passions and showcase your strengths. Lead the way, and Logic Pro will follow.

Thinking like a pro

Whether or not you're in it for the money, having a professional mindset can make your experience with Logic Pro more productive and enjoyable.

My advice to new and seasoned Logic Pro users alike is to set a goal and work steadily towards that goal. Here's what I've found, and I hope you agree. Creativity loves speed. Slow and tedious labor kills creativity faster than anything. So set a clear objective and move quickly towards completion.

Planning your creative process

You can always wait for inspiration to strike you — but you could end up waiting a long time. And when inspiration does strike, it often doesn't stick around long. For all the moments when you're not feeling inspired, having a plan for your creative process can help.

What's your purpose for the project? Are you learning something or creating something? If you're learning, what's your learning goal and how much time will you give it? You could easily get lost trying to learn everything Logic Pro can do and never create anything. That will suck the joy out of using Logic Pro, and you'll find yourself opening it less and less. So give yourself learning limits and give yourself creative projects. Connecting to the purpose of your project will keep you motivated and moving in a clear and forward direction.

Here are some example projects to get you learning and creating:

- **Learn a tool.** Logic Pro has lots of tools and editors to help you achieve your creative goals. Spend 5 to 15 minutes learning a single tool, function, or editor. You will have many opportunities to do this throughout the book.
- **Write a song.** Lots of my students find it easier to separate writing a song from recording a song. Writing a song means starting from scratch and trying out ideas. You'll end up doing some recording, but your purpose is not to create "keeper" tracks but to experiment and organize your ideas.

- **Record a song.** If you've already written a song, record it. Logic Pro has an intuitive interface, and you'll be able to polish your tracks until they're as shiny as a platinum record.
- **Sequence an 8-bar loop.** Not all projects have to be big and grand. A simple 8-bar piece of music can become a loop that you can use in another project or license to another artist.
- **Compose a score.** Use the orchestral instruments or synthesizers to create a score for a video or just for fun. If you have any home videos on your computer, you can import them to Logic Pro and give them a soundtrack. Play the video at the next holiday reunion, and your film composition is sure to impress friends and family.
- **Design a sound.** Spend 30 minutes with any of the Logic Pro virtual instruments and come up with your own sounds. Several synths even come with a randomize button to keep the sound fresh. Ever wondered what a six-foot guitar made of cardboard would sound like? Me neither, but you can make it happen with the Sculpture software instrument in Logic Pro.
- **Mix a song.** You can use a project you've recorded or Apple loops to practice your mixing chops and share your project with the world. Be sure to consult my mixing guide at `https://logicstudiotraining.com` to make your time in audio engineering land productive and free of earaches.

These examples are just a sampling of the types of projects you can start. The main point is to set a clear objective so you can achieve your goal. Set yourself up to win, and you'll stay motivated.

Getting to the finish line

Most Logic Pro users come to me for help not with getting started but with getting things done. When inspiration fades, so does motivation, unless you have a strategy for getting to the finish line.

Here are some tips for completing projects:

- **Set time limits.** Give yourself the shortest time frame for completing a task. Parkinson's law states that work expands to fill the time available to complete it. Set aside 10 hours to complete a job, and it will take 10 hours. Set aside 30 minutes, and the job will take 30 minutes. For larger projects, put a deadline on the calendar and stick to it. This suggestion might sound rigid, but the only thing you have to lose is your uncompleted projects.
- **Make projects attainable.** Dream big but be realistic. If you've never done a particular task, give yourself time to learn and improve. You might want to

write a chart-buster, but begin by writing a simple song with a clear structure. Then you can build upon your new skills and improve with each project.

- **Break the project down.** Typically, an album is made up of many songs, each song is made up of many instruments recorded on separate tracks, and each track is made up of many takes that are edited and turned into a final take. When you list all the tasks required to complete your project, you'll have an easier time completing each task.
- **Keep it simple.** The fewer parts you commit to your project, the easier it will be to complete. Many pop songs have 32 tracks on the low end and more than 100 tracks on the not-so-high end, but that doesn't mean you have to do the same. Lots of great songs have only four instruments, including the lead vocal. Try to simplify your project; you'll find that completing your project is much easier when it isn't complicated.

The more you create, the better you'll get. The more projects you complete, the more confidence you'll have. With confidence and chops, you'll tackle more ambitious projects and find yourself working with higher-level musicians.

Now that you've read about the Logic Pro producer's mindset, it's time to get your gear ready to handle your mad genius.

Connecting Your Logic Pro Studio

You can do a lot just with the Logic Pro software. You can play the software instruments with your computer keyboard by using musical typing (described in Chapter 7). You can import media from a variety of sources, such as Apple's Music app or iMovie (as detailed in Chapter 8). You can use headphones or your computer speakers to listen to your project. However, you'll want to connect some peripheral devices to take advantage of all that Logic Pro offers.

Consider adding some or all of the following devices to your studio:

- **Audio interface:** You'll need to get audio into and out of your computer. Your computer probably has a built-in mic, but you might also want to capture audio from a variety of sources, such as a keyboard, guitar, and microphone. You may want to be able to record more than one instrument at a time. Audio interfaces allow you to get professional sounding audio into your Logic Pro project. You can get inexpensive, good-quality input devices through major retailers and the Apple store or by searching for used options. USB and Thunderbolt connectivity are standard on the latest Apple computers.

- **Speakers:** In the pro audio community, your speakers are known as *monitors.* In a pinch, your computer speakers will do just fine. I use them to do a chunk of my mixing because many people listen to music on their laptop's speakers. (The same advice goes for mixing with common headphones, like Apple's AirPods. Even studios with expensive monitors listen to mixes in cars — or trucks if they're mixing country music.) For a more accurate audio picture, listen to your project on a set of monitors that produces the entire frequency range. There's a good chance that your audio interface will have a stereo monitor output for connecting a pair of speakers.
- **Headphones:** When you record vocals or instruments using microphones, you need a quiet environment. Headphones allow you to hear what you're recording without speakers, so you capture just the instrument — not any other audio coming from your speakers. Also, live instruments can be loud, and headphones allow the performer to hear all the tracks well while recording. Headphones are also useful during mixing when you need to hear the details of your audio.
- **MIDI controller:** MIDI (Musical Instrument Digital Interface) allows devices to talk to each other and lets you control all the wonderful software instruments that come with Logic Pro. A MIDI controller can be a keyboard, drum pads, or any other device that transmits MIDI. MIDI controllers connect to your computer by USB, Bluetooth, or a separate MIDI interface for devices with 5-pin DIN connectors.
- **iPad and iPhone:** The Logic Remote iOS app can control software instruments and the mixer, execute key commands, and more. Best of all, the app is free! It connects to Logic Pro through your Wi-Fi network. Have an instrument you want to record, but it's sitting 20 feet away from the computer? No problem: Use your iOS device to control Logic Pro remotely. I cover Logic Remote in Chapter 20.

Setting up your computer

If you haven't already installed Logic Pro on your computer, get it from the Apple App Store. Open the App Store in your Applications folder and search for Logic Pro. Purchase and install it (an installation wizard will guide you) — but be patient while downloading because the program size is about 1.1GB.

To download the additional Logic Pro content, launch Logic Pro. Choose Logic Pro ⇨ Sound Library ⇨ Download All Available Sounds or Download Essential Sounds. If you're updating the sound library, you may need to select the libraries available for download individually. Click the Continue button to download and install the content.

TIP

I recommend finding the hard drive space to download the additional content because the instruments, samples, and loops you get are fantastic and fun to play.

Here are some tips to make your experience with Logic Pro as smooth as possible:

- **Pay attention to Time Machine backups.** If you use Time Machine to back up your computer, Time Machine might access the hard drive while you're accessing the hard drive with Logic Pro. As a result, you might get an error if you're working on a project with a high track count or lots of samples streamed from the disk. Turn off Time Machine temporarily if you encounter any issues.
- **Shut down other apps if you run out of power or if Logic Pro gets glitchy.** When I arrange or edit my projects, I often leave other apps open. When I'm recording or in the final process of mixing, however, I close all other apps so that the hard drive is free and extra power is available for processor-intensive Logic Pro effects.

REMEMBER

- **Enable advanced features.** Logic Pro has several advanced features that aren't available by default. You should enable them so that you can follow along with the examples in this book. In the main menu, choose Logic Pro ⇨ Settings ⇨ Advanced and then select the Enable Complete Features check box, as shown in Figure 1-1.

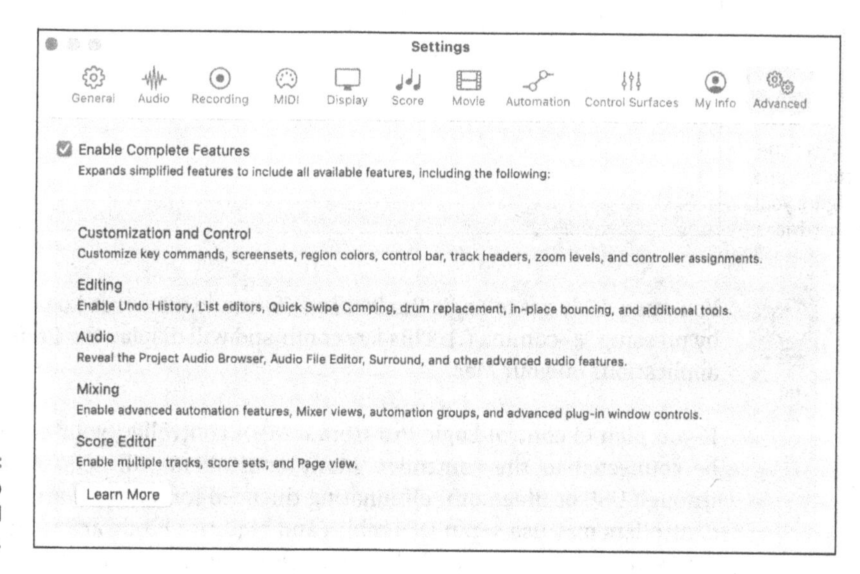

FIGURE 1-1: The Logic Pro Advanced Settings pane.

Now that you've installed Logic Pro and made a few tweaks, you're ready to connect your hardware and start making music.

Connecting your hardware

If you plan on recording audio, you'll need a way to get audio into Logic Pro. Your Mac probably has a built-in line in or microphone. These may work in a pinch, but professional-quality recordings need professional hardware. Most professional hardware is compatible with Logic Pro, so you should have a simple plug-and-play experience.

After you connect your audio hardware, you tell Logic Pro how to use it by choosing Logic Pro ⇨ Settings ⇨ Audio. Select the Devices tab, as shown in Figure 1-2, and use the Output Device and Input Device drop-down menus to choose your audio hardware.

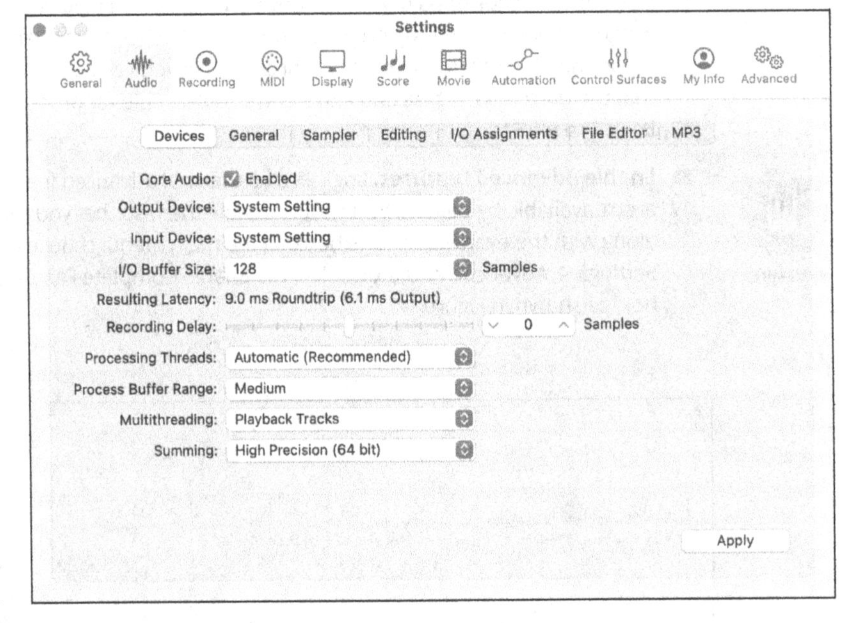

FIGURE 1-2: In the Audio Settings pane, choose and configure your audio hardware.

TIP

You can quickly get to Logic Pro Settings (formerly known as Logic Pro Preferences) by pressing ⌘+comma (,). This key command will display the Preferences in most applications on your Mac.

If you plan to control Logic Pro from a MIDI controller, your computer needs to be connected to the controller. Many controllers will send and receive MIDI through USB or Bluetooth, eliminating the need for a third-party interface. Other controllers may use 5-pin DIN cables and require a hardware MIDI interface to go between the computer and the controller. Check the literature; Logic Pro will probably be mentioned.

After you've connected your MIDI controller, Logic Pro will automatically listen for your controller's signals on the currently selected track. Playing software instruments with a MIDI keyboard controller is usually a plug-and-play experience without any need for customization.

With all your hardware connected, Logic Pro is ready and waiting for your commands.

Building common setups

To help you visualize what a complete hardware and software setup looks like, I've designed some possible systems. You can create music with a computer and the Logic Pro software alone. However, if your goals are more ambitious and you want to get good sounds into and out of Logic Pro, consider your hardware and the acoustics of your listening environment. Idealized scenarios aren't necessary for good quality or enjoyment. With a Mac computer and Logic Pro, your baseline for quality and fun is already high.

It's easy to believe that you need the best equipment (and a lot of it) to create anything worthy of attention. Don't believe the hype, and don't get GAS (gear acquisition syndrome). Spend your money wisely and spend your time creating, not buying. The setups in this section can be built inexpensively.

In a recording setup, such as the one shown in Figure 1-3, you need to get audio from instruments or microphones into Logic Pro. You also need to hear what you're recording through monitors or headphones. Your audio interface is the intermediary between your computer and the peripherals.

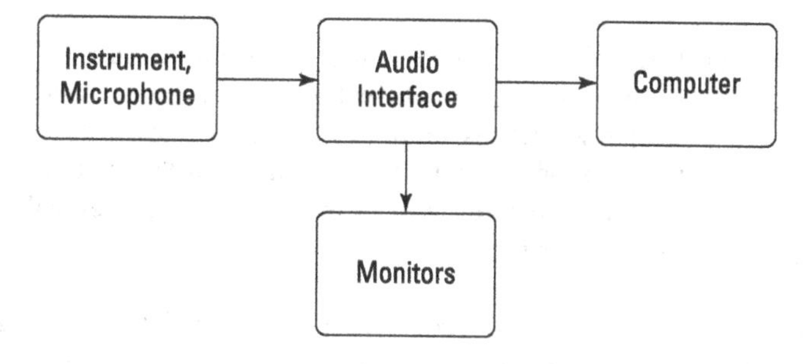

FIGURE 1-3: A typical recording setup.

In a MIDI studio, such as the one in Figure 1-4, instruments such as synthesizers, drum machines, samplers, and even alternate controllers (for example, a guitar MIDI system) connect to a MIDI interface, which transmits the MIDI messages from the various instruments to the computer. An audio interface is still needed to transmit audio in and out of your system. However, some devices handle the audio and MIDI together.

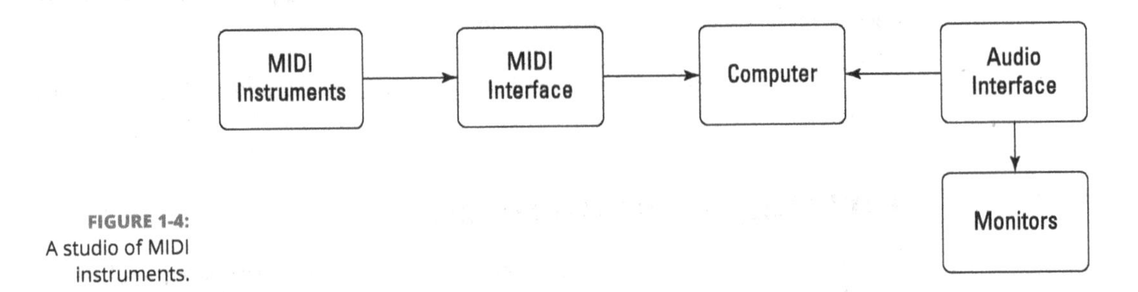

FIGURE 1-4: A studio of MIDI instruments.

If all you do is mix other people's music, you may never need to do any recording of your own. In this case, you need an audio interface to get audio from Logic Pro into a pair of monitors, as shown in Figure 1-5. You may, however, use a controller to mimic a mixing console with faders, knobs, buttons, and other useful features that control Logic Pro remotely.

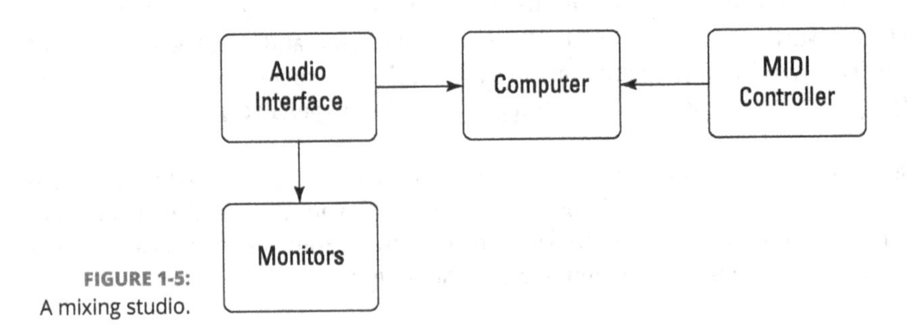

FIGURE 1-5: A mixing studio.

I love to travel and never go anywhere without a laptop and an instrument. I travel with guitars, drum machines, small keyboards, and sometimes more than one of each. I like to keep the setup as simple as possible while affording me the flexibility I love. In a mobile rig, such as the one shown in Figure 1-6, headphones will replace speakers, and portable interfaces and microphones will be used to get audio in and out of Logic Pro.

TIP

If you plan on performing live, consider buying MainStage, the companion app to Logic Pro. MainStage shares all the instruments and presets with Logic Pro but is designed for live performance. It has a gorgeous full-screen interface that's easy to read on stage. MainStage doesn't include features you don't need live, such as notation or audio and MIDI editors.

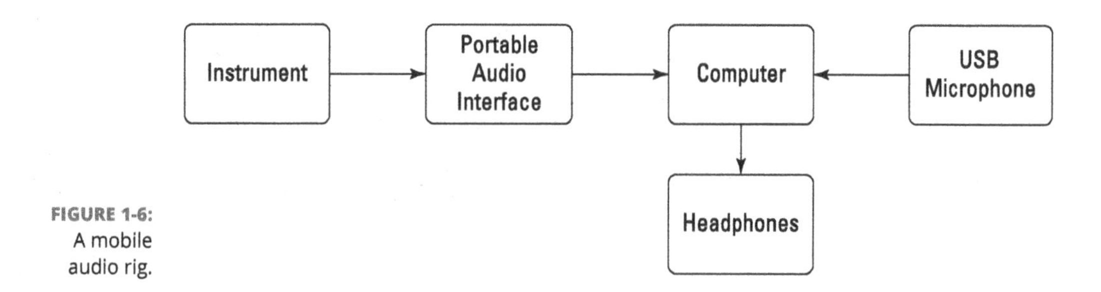

FIGURE 1-6: A mobile audio rig.

There's never been a better time to be a Logic Pro user. You get an amazing studio at an amazing price and can do amazing things with it. I hope you're as excited as I am to dig into what Logic Pro can do.

IN THIS CHAPTER

» Starting and finishing projects

» Discovering some timesaving workflow tips

» Backing up and securing your projects

» Sharing your projects for collaboration

Chapter 2
Examining Logic Pro Projects

Logic Pro projects are similar to any computer file type, except they're larger in scope than files such as text documents. You might be shocked to see an entire chapter about a file type, but there's so much more you can do with Logic Pro projects than other file types. You'll soon understand how important they are to the creative process.

Projects are flexible. When disk space is a concern, for example, you can save only the assets you want to keep. You can create project templates to speed up your workflow and set up Logic Pro exactly how you want to work. Each project contains global settings of the entire project as well as different snapshots of the project, such as different arrangements, mixes, or treatments. For example, you can create an alternate version of your project if the producer calls for a version without a vocal (for when the performer needs to sing live on TV to a backing track).

In this chapter, I cover naming conventions to keep all your projects organized, tips and tricks to speed up your workflow, strategies for archiving and backing up your work, and much more.

Starting Your Project

A *project* is the file type that you work with in Logic Pro. The file extension of a Logic Pro project is .logicx. The project file contains MIDI events, parameter settings, and information about the audio and video in your project.

To get the big picture of your project and how it relates to Logic Pro, the hierarchy goes like this:

> Project ⇨ Tracks ⇨ Regions ⇨ Events

Your project contains tracks. Your tracks contain regions. Your regions contain events. The File menu is where you do most of your project-level work.

To start an empty default Logic Pro project, choose File ⇨ New or press Shift+⌘+N. A New Tracks dialog window opens, as shown in Figure 2-1. At the top of the screen, choose the type of track you want to begin working with and click Create.

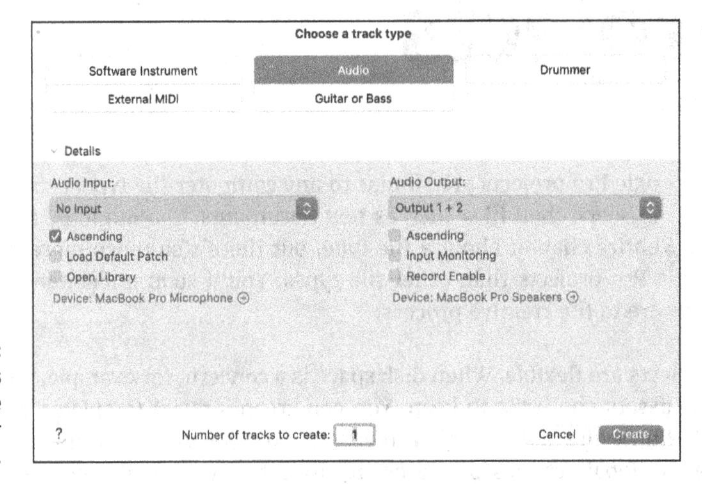

FIGURE 2-1: The New Tracks dialog is where you choose your first track.

A more advanced and customized way to start a new project is to choose File ⇨ New from Template (⌘+N). The Project Chooser window opens, as shown in Figure 2-2. You can select a premade project template, a recent project, or your own customized project template. I show you how to create a customized orchestral template in Chapter 12. Click the Details disclosure triangle (at the bottom left) to display even more options for your new project, such as the tempo, time and key signatures, and audio input and output.

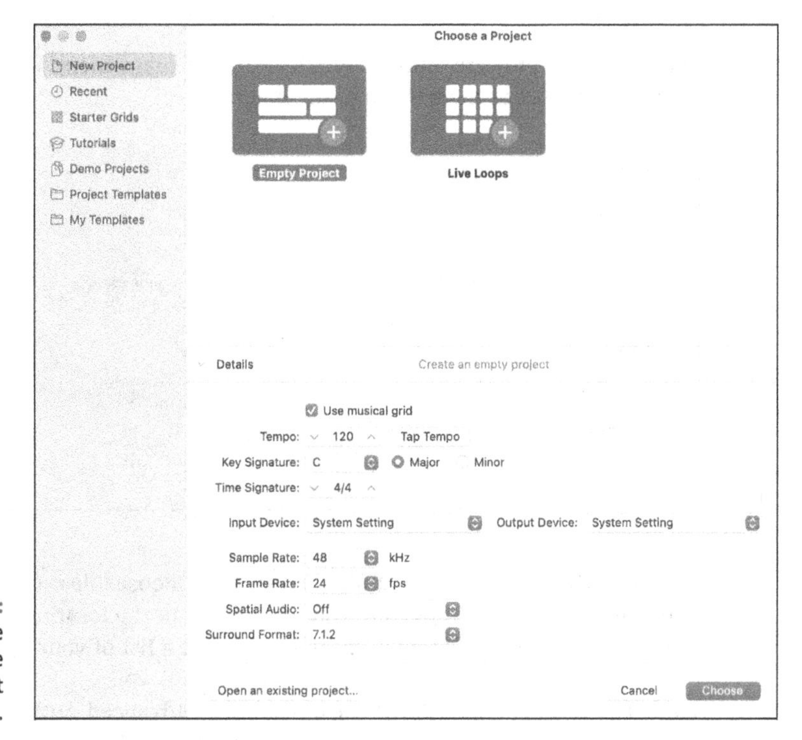

FIGURE 2-2: Choose customizable project templates here.

TIP

You can change any project option, but you should stick to a single sample rate. The default sample rate, 44.1 kHz, is the standard rate for an audio CD. If your goal is to use your audio in a video project, consider using a 48 kHz sample rate, which is the video standard. Using higher sample rates depends on your hardware capabilities and project needs.

After you start your project, you're ready to begin recording, arranging, editing, and mixing. It's a dream come true.

TIP

Choose a custom startup action to tell Logic Pro what to do when you launch it. If you're the prolific type, you can create a new project every time you launch the app. If you, like me, are a mere mortal, you might want to open the most recent project on startup. Choose Logic Pro ➪ Settings ➪ General and select the Project Handling tab, as shown in Figure 2-3. Then select your startup action.

Opening a project

You can open a project in several ways. You can double-click a project file in Finder, which will launch Logic Pro and open the project. If another project is open, Logic Pro will ask if you want to close the project. More than one project can be open simultaneously, so closing the current project is unnecessary. To switch between open projects, choose Window on the main menu and then select the project in the list at the bottom of the menu.

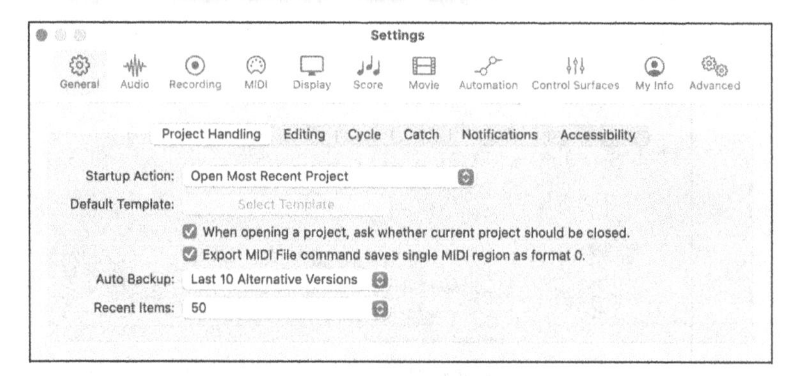

FIGURE 2-3: Customize your startup action.

You can open a project also from the File menu. Choose File ⇨ Open or press ⌘ +O, and a dialog will open, allowing you to navigate to the location of your project in Finder. Choose File ⇨ Open Recent instead to see a list of your recent projects.

Logic Pro can also open MIDI files, AAF files (Advanced Authoring Format files used by digital audio workstations such as Pro Tools), XML files (used by Final Cut Pro), and GarageBand projects. It can also open projects created with earlier versions of Logic Pro.

TIP

If this is your first time launching Logic Pro, you might want to explore a demo project by a popular artist. Choose File ⇨ New from Template (⌘+N) and select Demo Projects in the Project Chooser sidebar. Then select a demo project and press Choose.

Saving a project

When you create your project, it is autosaved in the Logic folder under the temporary name Untitled.logicx. (You can get to the Logic folder in Finder by navigating to Users ⇨ USERNAME ⇨ Music ⇨ Logic.)

To save your project manually, choose File ⇨ Save. In the Save dialog that appears, name your project and choose a location or keep the default location, which is the Logic folder. You can choose to organize your project as a package or a folder. A *package* saves your project as a single file that includes all project assets. A *folder* saves the project file and saves its assets in subfolders.

You can also choose to copy specific file types into your project. It's a good idea to copy your audio files into the project, but you might not want to copy samples due to their potentially large file size. The benefit to saving a project without assets is that you conserve hard drive space. The downside is that it can be easy to mistakenly delete assets that the project depends on. Hard drive space is inexpensive, so it makes sense to include all assets in your project folder. By doing so, organizing, moving, and archiving projects will be easier.

TIP

I find saving projects as packages is the simplest approach. You can view the contents of packages (all package file types, including Pages, Keynote, and Numbers files) by Ctrl-clicking the package in Finder and choosing Show Package Contents. All your audio files and assets will be in the Finder window that opens.

If you want to save the project with a different name or in a different location, choose Save As on the File menu. If you want to create a copy of the project, choose Save a Copy As on the File menu.

Closing a project

When you're ready to close your project, choose File ➪ Close Project. If you've made any changes since you last saved your project, Logic Pro asks if you want to save the project. If you don't want to keep those changes, select Don't Save from the Save dialog.

Don't confuse the Close Project command with the Close command. Both are on the File menu. The Close command simply closes the currently focused window. However, if your project has only one window open, which is often the case, using the Close command will also close your current project.

Naming and renaming a project

It's a good idea to have a naming convention and stick with it. When I'm saving a project that doesn't yet have a title, my file-naming convention is to use the date, key signature, harmonic mode, and tempo, followed by any other useful descriptors such as the musical genre, song section, and version numbers. For example:

```
2022-06-03 Cmin Dorian 120bpm EDM verse 01
```

Naming your file this way enables you to match projects based on mode and tempo. If you eventually come up with a title for your project, use the Save As command (described earlier) or just rename the project file in Finder. Or you can use the Rename function by choosing File ➪ Project Management ➪ Rename.

Augmenting Your Project

Projects are so basic to your workflow that you may take them for granted after a while. But you can do several cool things at the project level that will make your time with Logic Pro more productive.

Saving time with project templates

When you create a project (see earlier in this chapter), you see Project Chooser, where you can begin a project from a premade template. These default templates are excellent starting places. You can also create your own project templates.

How can you save time with templates? Suppose you're recording several songs with a band and each song has a similar setup, or you're a film composer and use identical orchestra setups for all your scores. In either case, you could create a project template once and use it over and over again. (You find out how to create an orchestral template in Chapter 12.)

To save a template, set up your project how you want it and then choose File ➪ Save as Template. Project templates are saved in a special folder located at Users ➪ USERNAME ➪ Music ➪ Audio Music Apps ➪ Project Templates.

I love using templates, and I've saved dozens of them. I have genre-based templates, templates that include my favorite third-party software, and templates that I've created from analyzing popular hits (and not-so-popular guilty pleasures). Templates are excellent productivity tools.

Autosaving your hard work

In high school, I spent an entire evening sequencing a popular song in one of the first Apple MIDI sequencers. I spent hours hunched over a computer keyboard instead of doing my homework, completely focused, with no bathroom breaks, until the electricity went out. I hadn't saved the project even once.

Even though I was crushed, I jumped back in, and sequencing was much easier the second time. I also developed a habit of pressing ⌘ +S, which I still have to this day. You might even find my left hand "air-saving" while I'm away from the computer. It's a habit I'm happy to have.

Fortunately for you, Logic Pro autosaves your work. If Logic Pro should crash, when you reopen the project, it will ask you to choose an autosaved version or the last manually saved version. However, even though the program autosaves, get in the habit of saving your work after every important change you make.

Recovering from problems with project backups

What would you do if your computer was stolen or ruined? Barring the financial considerations of buying a new computer, could you recover quickly? If I were to lend you my own computer, could you rebound and save the game? If you couldn't, please pay attention, for the sake of your music.

I'm a backup fanatic. I back up all my computers using Apple's Time Machine software and a few rotating external drives. I also back up my entire computer offline using Apple iCloud and Amazon S3 cloud storage. But wait, there's more. I sync my current projects using Dropbox so I can work on them on multiple computers.

TIP

A smart best practice for backing up your data is to use the three-two-one rule. Back up at least three copies of your work, in two different formats, with one copy offsite. Back up your work often and make it part of your regular work routine.

Logic Pro also creates project backups every time you save your project. As long as you have Show Advanced Tools selected in the Advanced Settings pane (see Chapter 1), you can revert to an earlier saved version of your project. Every time you save your project, a backup is made. You can revert to these backups by choosing File ➪ Revert To. A list of your time-stamped project backups allows you to go back in time to a previously saved project. This feature saves you when you try things out that you don't like or make mistakes while working.

Creating options with project alternatives

You can create alternative projects within a project. This feature saves you from creating new projects or copies of projects every time you want to try something new. The downside of making copies of projects is that if you don't do it properly and you delete audio from one project that's being used in another project, you might lose that file in the first project. If everything is self-contained, you can try things out until your mad genius is content. Also, if you create two projects and want to A/B them, switching back and forth means all the plug-ins have to be reloaded each time, which can take enough time that your ear can't tell the difference between the two.

To create a project alternative, choose File ➪ Project Alternatives ➪ New Alternative. You can rename and remove the alternatives by choosing File ➪ Project Alternatives ➪ Edit Alternative. You can export an alternative by choosing File ➪ Project Alternatives ➪ Export Alternative as Project. As shown in Figure 2-4, each project alternative is time-stamped, which helps you know which project is the most recent and which is the original.

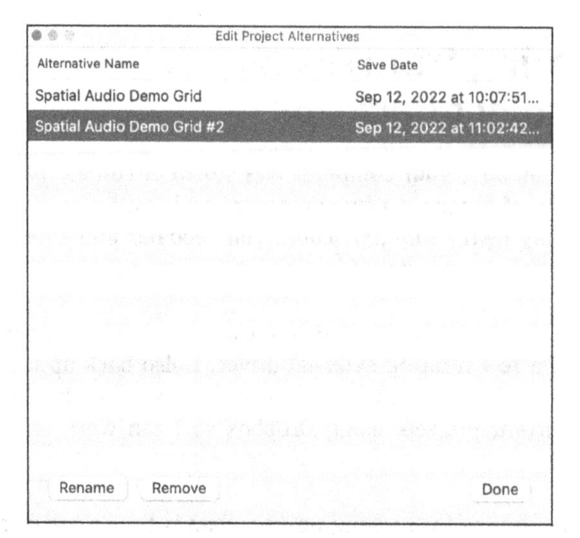

FIGURE 2-4: Rename and remove project alternatives.

Following are some examples of project alternatives you might want to try:

- **Arrangements:** Create project alternatives with different lengths for advertising, such as 10, 15, 30, and 60 seconds. Create an alternative with a short intro geared towards music industry professionals. Or try different instrumentations. Or, if your band has a prima donna who needs to be louder than everyone else, let that band member hear the louder version while the rest of the band hears the version you so expertly crafted. Or perhaps you want to hear how the project would sound with a song section in a different place.
- **Singers:** If you plan on pitching your songs to artists and publishers, record both male and female lead vocal tracks. With project alternatives, you can create a male and female version of the song and increase your chances of success.
- **Mixes:** You can create a project alternative with a loud vocal and one with a soft vocal. Or create an alternative without the vocal, in case the singer needs to perform the song to a backing track. (See Chapter 19 for information on the different alternate mixes you should produce.)

Customizing Your Project Settings

Similar to the Logic Pro Settings, your project has global settings you can adjust. You get to the project settings shown in Figure 2-5 by choosing File ➪ Project Settings.

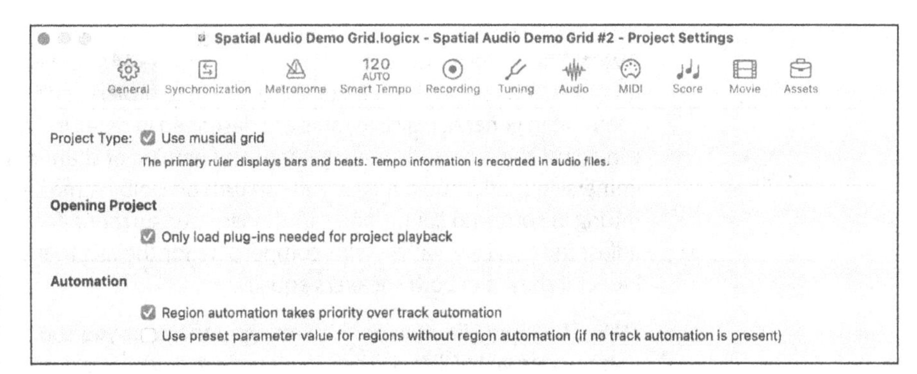

FIGURE 2-5: Set global settings for your project.

Here's a description of the Settings panes you can choose at the top of the Project Settings window:

- **General:** Set the project grid and automation settings. Select the Use Musical Grid check box to display bars and beats in the primary ruler and store tempo information in the audio files you create. Storing the tempo in files is helpful if you'll use the assets in other projects that recognize the data. Deselect Use Musical Grid if you want to display time instead of bars and beats in the primary ruler and don't want to record tempo information in the audio files. You can also adjust your automation settings, as described in Chapter 18.
- **Synchronization:** Sync your project to an external device or control your project from an external device.
- **Metronome:** This window is where you control the internal metronome sound generator, uniquely called Klopfgeist. You can also control an external click via MIDI if you prefer to use a different sound. When you don't want to hear the metronome while recording, select Only During Count-in, which is useful when the sound gets in the way of the prerecorded media you're listening to.
- **Smart tempo:** Adapt the project tempo to imported or new audio recordings.
- **Recording:** Adjust the number of bars or seconds that play before you begin recording. The MIDI settings give you several options for recordings that overlap. You can create take folders, join the regions, or create new tracks. (These settings are covered in depth in Chapter 7.) You can also set the audio recording path in this window.
- **Tuning:** If you play with a lot of live instruments, such as the piano, you might need to detune the software instruments that accompany the acoustic instruments to make their pitch match. You can also experiment with several alternate scale types, though most people stick with Equal Tempered.

» **Audio:** The sample rate is important to set here — 44.1 kHz is standard for CD quality audio, and 44.8 kHz is standard for video projects and becoming more common in general. (Sample rates are described in detail in Chapter 5.) You can adjust the automatic management and naming of channel strips. For immersive spatial audio mixes, you can turn on Dolby Atmos. If you plan on mixing in surround sound, you can choose your surround format. You can adjust the Pan Law, which helps compensate for the fact that sounds get louder if they're in both speakers equally.

» **MIDI:** The Input Filter tab on this Settings pane gives you some useful options if your external MIDI device sends a lot of extra data you don't need. For example, *aftertouch* (pressure applied to a keyboard key while it is being held down) and system exclusive data can add a lot of data, and if you're not using those features, turning them off makes editing in the list editor much easier because it's not filled with extraneous data.

» **Score:** If the MIDI tracks in your project will be printed for musicians to play, you can adjust the settings here. This pane has many settings, and most are for professional notation. If you plan on printing a lead sheet, quick parts, or guitar tablature, you may need to check out these settings. Otherwise, the default settings are often all you need.

» **Movie:** If you import a movie into your project to compose a film score, Movie Position is a useful setting because movies don't often start at 0. You can also adjust the movie volume if it's getting in the way of your work.

» **Assets:** When you select every option in this pane, all your assets are copied (not moved, but copied) into the project folder or package. Selecting every option is the safest way to go because all your assets are in one place, but you might not want to go this route if you're concerned about your hard drive space.

TIP

After you adjust your project settings the way you like, save your project as a template (choose File ➪ Save as Template). That way, when you start a new project, you won't have to repeat your work. Project templates save you time and give your clicking finger relief so you can use it for more creative pursuits.

Importing settings and content from other projects

If you want to import track content (the audio and MIDI regions) and channel strip settings from another project, choose File ➪ Import ➪ Logic Projects. In the dialog that appears, select the project from which you want to import and then click Import.

The all files browser opens on the right side of the main window and displays the track import view, as shown in Figure 2-6. Track content, plug-ins, effects sends (see Chapter 16), I/O (input/output) settings, and automation are shown for every track.

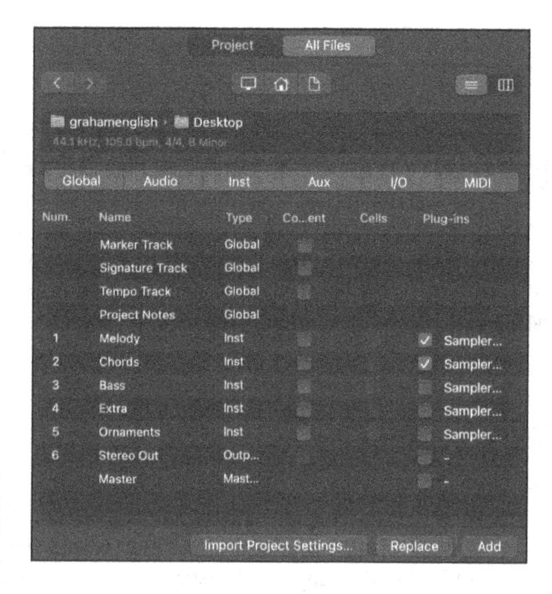

FIGURE 2-6: The track import view of the all files browser.

Decide what you want to import by selecting the check boxes. You can also bring in markers and other global track content. When you're ready to import, click Add to import the content into the current project or click Replace to import the data and replace the content of the currently selected track. Replace mode works on only a single track at a time.

What if you've already started a project but have another project that's set up the way you want? Importing settings from other projects into your current project is a breeze. After choosing File ⇨ Import ⇨ Logic Projects, press the Import Project Settings button at the bottom of the all files browser, and the window in Figure 2-7 appears, displaying the settings you can import into your project. Select the project settings you want to copy and click Import.

Another way to get to the projects import function is to use the all files browser and navigate to a Logic project. Choose View ⇨ Show Browsers or press the F key. Click the All Files tab of the browser, select the project from which you want to import, and double-click the project or click the Import button at the bottom of the browser. The browser displays the track import view, and you can choose what you want to import into your current project.

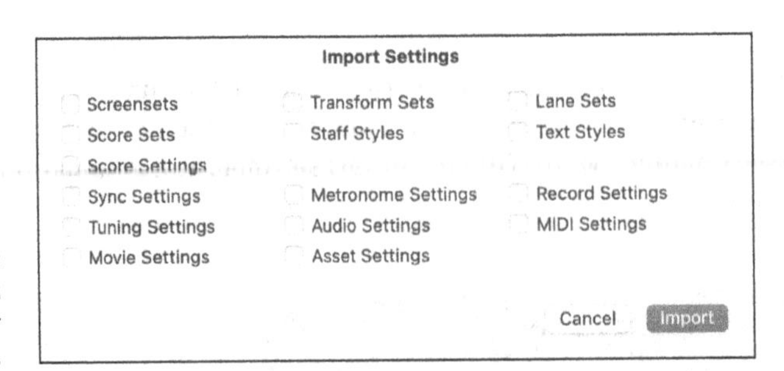

FIGURE 2-7: Copy settings from other projects.

Exporting your project for collaboration

You might want to export your project for several reasons. Perhaps you want to collaborate with other artists or work on your project in a different software application. You can also export portions of your project for use in other projects. To export regions, MIDI selections, tracks, and the entire project begin by choosing File ➪ Export.

Exporting regions and cells

To export a region from the tracks area or a cell from the live loops grid and add it to your Apple loop library, choose File ➪ Export ➪ Region/Cell to Loop Library. The dialog shown in Figure 2-8 appears. Name your file; choose the loop type, the scale, genre, and key; and add other tags and instrument descriptors. Click Create to export your Apple loop and add it to the loop library.

If you simply want to export the region to your hard drive as an audio file, choose File ➪ Export ➪ Regions as Audio File. In the dialog that appears, select the file location, audio file format, and bit depth.

Exporting MIDI selections

You can export a selection of MIDI tracks as a MIDI file by choosing File ➪ Export ➪ Selection as MIDI File. Selecting more than one MIDI region will result in a single MIDI file.

Exporting tracks

To export your tracks for use in a different audio application, such as Pro Tools, choose File ➪ Export ➪ All Tracks as Audio Files. In the dialog that appears, select the audio format, the bit depth, and other options that will determine how the tracks are processed before they are exported. If your exported audio tracks will be mixed in another application, it's best to select the Bypass Effect Plug-ins and Include Volume/Pan Automation options.

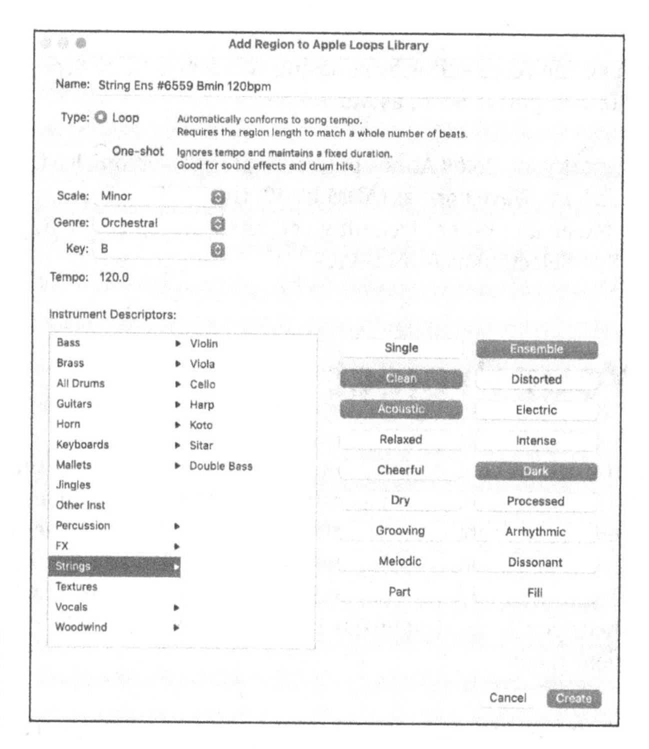

FIGURE 2-8: The Add Region to Apple Loops Library window.

TIP

If you're collaborating with a Pro Tools user, export a MIDI track the length of your project. That way, if you have marker data, it will be exported along with the tempo information, and your Pro Tools user will have a marker and tempo map to import with the audio files. The Pro Tools user will thank you and treat you like a hero! You can't export an empty MIDI region, so add any MIDI event to the region (see Chapter 7) before you export the track.

Exporting projects

Exporting a project as an AAF (Advanced Authoring Format) file is another option for collaborating with Pro Tools users. Choose File ➪ Export ➪ Project as AAF File. All the used audio regions will be exported, including their track and position references and volume automation.

To export a project as a Final Cut Pro XML file, choose File ➪ Export ➪ Project to Final Cut Pro XML. Software instruments and automation data are exported as audio, but MIDI tracks are ignored.

You can export your MIDI score as a MusicXML file if you want to edit the notation in a program such as Sibelius, Finale, or a music optical character recognition (OCR) application such as SmartScore or PhotoScore. Select the MIDI you want to export, open the score editor by choosing Window ➪ Open Score Editor, and then choose File ➪ Export ➪ Score as MusicXML.

You can export your Dolby Atmos project or project section also in Audio Definition Model Broadcast Wave Format (ADM BWF), the file format for spatial audio mixes on Apple Music and other streaming services. Choose File ➪ Export ➪ Project as ADM BWF or Selection as ADM BWF.

Tidying Up Your Project

When you work on your projects, you'll probably try lots of things, record takes that don't make the mix, and basically add a bunch of audio and MIDI data that doesn't need to be there when it comes time to share or archive your project. You should give your project a spring cleaning to get it ready for the next season in its life. If you plan to send your project to a collaborator or archive it because it's finished, you'll appreciate the following useful tools in the File ➪ Project Management menu:

- **Clean Up:** Deletes unused files, backups, and media browser files, as shown in Figure 2-9. This function is safe to use because it deletes only unused data. However, it deletes media you may have shared with the media browser, so sharing with the media browser isn't the best strategy if you want your media to be available for other projects or other applications in the future.
- **Consolidate:** Creates a copy of all the assets you select and includes them in the project, as shown in Figure 2-10. If you haven't been including assets in your project up until now, this is your chance to pull everything into your project. This function is also useful when you want to share the project for collaboration because the Logic Pro user you're sharing with might not have the same samples or content installed.

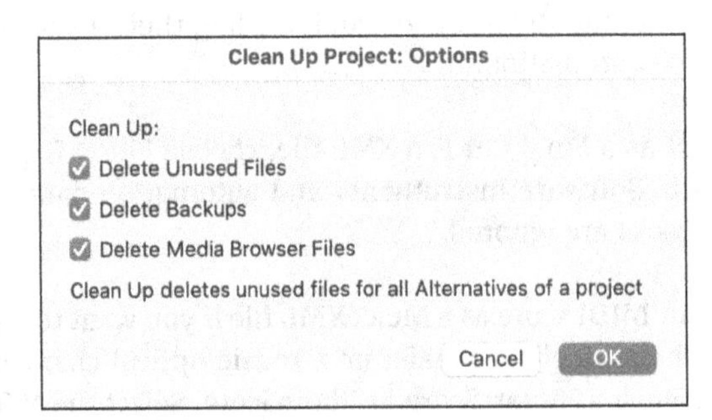

FIGURE 2-9: Delete unused project data.

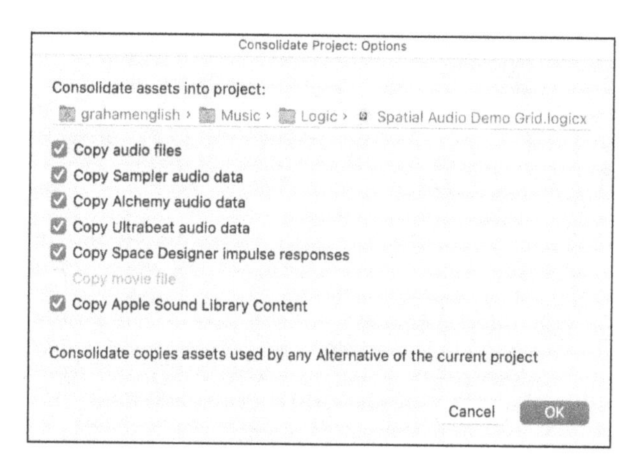

FIGURE 2-10: The consolidate function copies and includes all used assets in the project.

- **Rename:** Opens a dialog asking you to select your new project name. (This function is similar to the File ➪ Save As command.)
- **Show in Finder:** Opens a Finder window with your project selected. Use this function when you need to get to your project file quickly.

Now that your project is nice and tidy, it's important to consider protecting it for the future. A good backup strategy can save you from downtime and avoidable, life-shortening stress. Hard drives fail, so back up your project files by maintaining multiple copies of your projects in separate locations and in the cloud. Also, because Logic Pro will probably not be the final version of this fantastic software, export your tracks as audio files so you can import them into a version of Logic Pro down the line. Your future self will thank you.

IN THIS CHAPTER

» Navigating the windows

» Playing and controlling the project

» Exploring the tools and controls

» Saving time and speeding up your workflow

Chapter 3
Exploring the Main Window and Tracks Area

Logic Pro has a simple single-window design that helps you be more productive and puts everything you need only a click away. You can do most of what you need right in the main window. But you can also open windows separately and push windows to different displays. The program is as flexible as it is simple.

In this chapter, you discover how to navigate the Logic Pro interface. You also learn some time-saving tricks and smart ways to use key commands to accomplish the bulk of your work. Navigating Logic Pro with speed and purpose will put your music out in the world and build your project catalog. You'll be zipping around Logic Pro in no time.

Navigating Logic Pro

To get the most out of Logic Pro, you should know the name and purpose of each area of the main window. Plus, if you ever need to contact the folks in product support, you'll be able to communicate the problem precisely.

Getting comfortable in the main window

The *main window*, shown in Figure 3-1, was called the arrange window in earlier versions of Logic Pro. The main window title makes more sense because you can use it to do much more than just arrange. The name also stresses the importance of this Logic Pro key element. As the highly effective Stephen Covey once said, "The main thing is to keep the main thing the main thing."

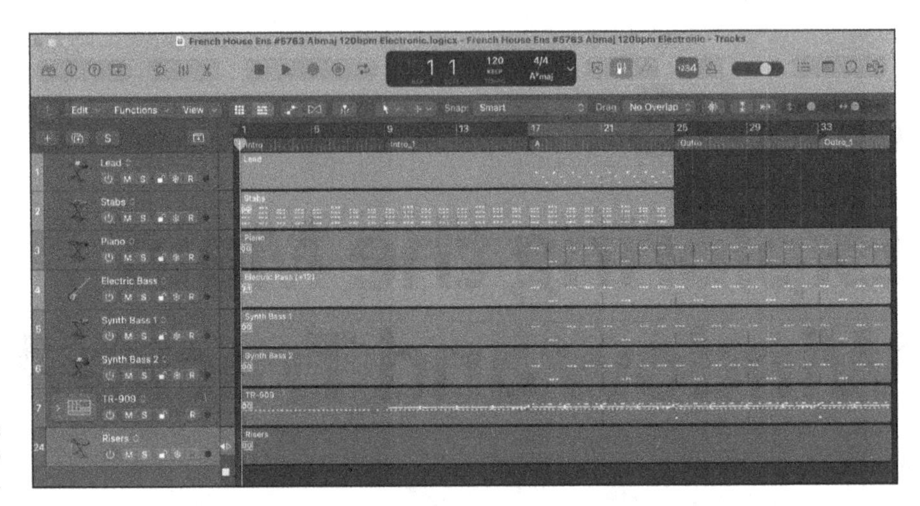

FIGURE 3-1: The Logic Pro main window.

The main window includes the control bar along the top and the tracks area or live loops grid, or a split screen of both in the area below. It's highly customizable and gives you all the tools you need to begin making music. To display the main window, choose Window ⇨ Open Main Window.

Working in the tracks area

The tracks area, shown in Figure 3-2, contains your project's tracks of audio and MIDI regions. When you create a project, Logic Pro asks you what type of track you would like to create. The tracks you create are added to the vertical track list to the left of the tracks area. To create more tracks, click the new tracks (plus sign) icon at the top of the track list or choose Tracks ⇨ New Tracks. I cover the track list in more detail later in this chapter.

TIP

Download a Logic Pro project template with several tracks and regions so you can follow along with the examples in this chapter. Visit `https://logicstudiotraining.com/bookextras`.

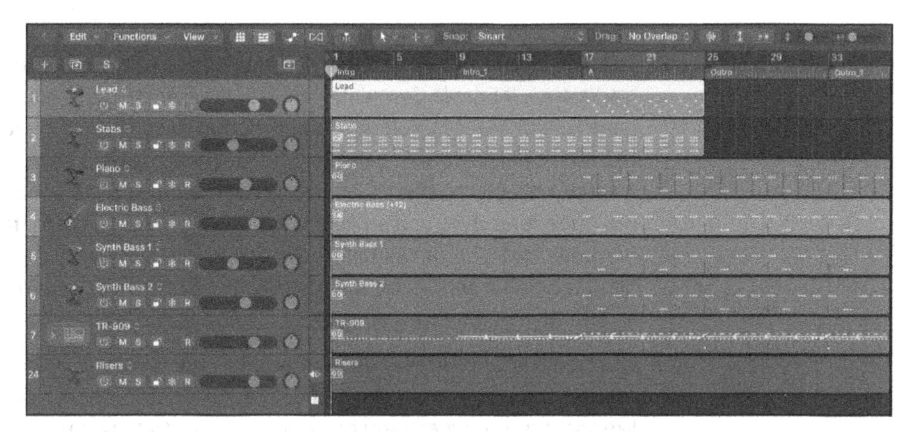

FIGURE 3-2: The tracks area in the main window.

The toolbar above the tracks area, shown in Figure 3-3, contains several menus to help you work.

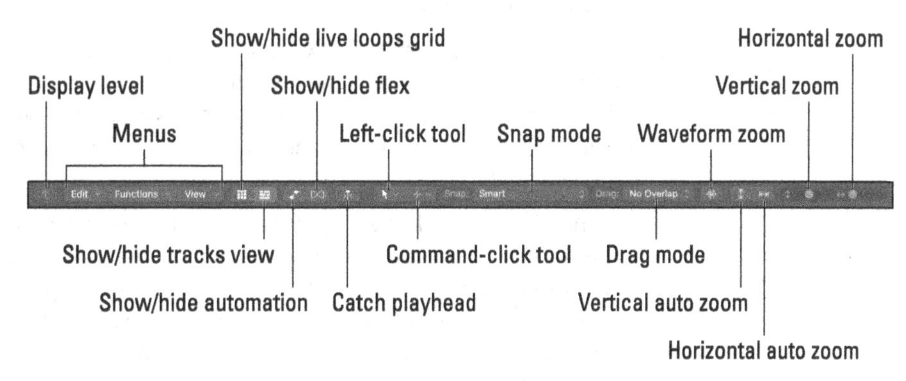

FIGURE 3-3: The tracks area toolbar.

Here's an overview of what's available in the menus:

- The Edit menu gives you several functions you can apply to your regions, including essential functions such as copy and paste and region-specific functions such as splitting, joining, and repeating. The options in this menu are almost identical to the options in the Edit submenu in the main application menu, so you have two places to execute commands. This feature is useful when you don't have the main window open and are working in an editor in a separate window. For details on editing audio and MIDI, see Chapters 14 and 15.
- The Functions menu gives you even more actions to take on your regions, such as naming and coloring your tracks and regions, MIDI quantizing (snapping your MIDI events more precisely to the time grid) and transforming, and removing silence from audio regions. When you want to do something to a region, check out the Functions and Edit menus first.

- The View menu gives you more viewing options and useful tweaks to the tracks area. If Enable Complete Features is selected in the Advanced Settings pane (see Chapter 1), you have a Link option to control the relationship between open windows. Selecting the secondary ruler is useful when you want to view your project ruler in both clock time and bars and beats. The marquee ruler gives you a visual indicator of selections you've made with the marquee tool. You can eliminate the grid in the tracks area if you're working in free time or if it gets in the way visually. The Scroll in Play option continually scrolls the tracks area as you play.

TIP

I turn off the Scroll in Play option because I like my tracks area to snap when it reaches the end instead of continually scrolling. The tracks area behaves more like sheet music, where you have to turn a page when you get to the end.

- To the right of the View menu are icons to display or hide the live loops grid, tracks view, automation, and flex modes (more on flex editing in Chapter 14) and to catch the playhead — that is, the tracks area will follow *(catch)* the playhead instead of remaining still.
- In the center of the toolbar are two tool menus. The left tool menu is the currently selected tool that's visible when your cursor is in the tracks area. The right tool menu selects the ⌘-tool that becomes available when your cursor is in the tracks area and you press the ⌘ key. Both tool menus are described in detail later in the chapter.
- The snap function makes dragging regions easier. You can choose the snap mode from the drop-down menu to the right of the tool menus. Smart mode is usually all you need until you're trying to do something specific. At the top of the drop-down menu, you can choose a finer snap value if you need to move a region more precisely. If you snap a region to a relative value and the region doesn't start exactly on the beat, it will move relative to its current position. That's the default value and usually what you want until you need to move something more precisely.
- To the right of the snap modes is your Drag mode drop-down list. The different Drag modes allow you to overlap, not overlap, crossfade, or shuffle regions in a track. Overlap mode preserves the region borders of the selected region when you drag it on top of another region. No Overlap shortens the right boundary of the region on the left. X-Fade creates a crossfade the length of the overlapped area.

  The shuffle modes move the regions in the direction of the particular shuffle mode selection; in addition, resizing a region resizes all the regions, and deleting a region moves the regions by the length of the deleted region. Shuffle is a useful function for editing voice-overs or audio interviews.
- Finally, on the far right of the toolbar are a zoom button that zooms your audio waveforms and a pair of buttons and sliders that control the vertical and horizontal track zoom.

Using the live loops grid

The live loops grid, shown in Figure 3-4, is a powerful feature for composing, improvising, and performing live. Each cell in the grid contains a loop or musical phrase that will stay in tempo with the project. You can record new loops, add effects, and control the playback of individual cells or groups of cells in real time. Cells can contain audio or MIDI, and you can record a live loops performance in the tracks area.

FIGURE 3-4: The live loops grid in the main window.

Each column in the grid is called a *scene*. You can think of scenes as song sections containing all the cells that will play together. The cells in each row are assigned to the same audio track as the adjacent track in the tracks area. Scenes are queued and played by clicking or right-clicking the scene triggers in the bottommost row of the grid.

To show the live loops grid, click the Show/Hide Live Loops button in the tracks area menu bar (refer to Figure 3-3). You can toggle between tracks view and the live loops grid by pressing Option-V or view them both by pressing Option-B.

The live loops grid is separated from the tracks area by a divider column. The divider column includes cell status indicators that tell you if the cell is playing or queued to play. Click the divider column to stop the cell from playing, and Option-click it to pause and restart the cell. Clicking while no cells are playing activates regions in the tracks area.

REMEMBER

Live loops take precedence over the tracks area. When you start playing a scene in the loops grid, the tracks that have cells will override the regions in the tracks area. However, you can alternate playing tracks with regions in the tracks area and cells in the loop grid by clicking the divider column.

TIP

You can also use your iPhone or iPad to control live loops by using the Logic Remote app, described in Chapter 21. Live loops also sound amazing with the DJ-style multi-effects plug-in called Remix FX.

Controlling the control bar

The control bar, shown in Figure 3-5, is located at the top of the main window. It contains view icons that show and hide windows, transport controls for playback and recording, an LCD display area for viewing important information about your project, and icons for different behavior modes and specific functions.

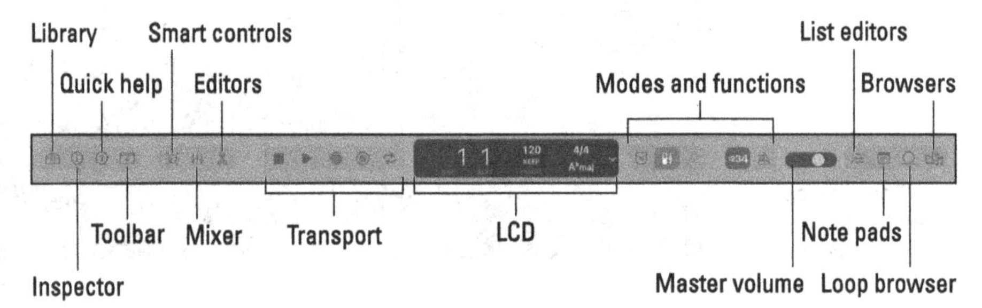

FIGURE 3-5: The control bar.

Your control bar may look different than the one in Figure 3-5. You can customize the control bar by Control-clicking an empty area and choosing the Customize Control Bar and Display option. On my MacBook Pro, I can't see all the default control bar items, so I always customize it to show only what I need.

TIP

After you customize the control bar the way you like, Control-click it and choose Save as Defaults and choose Apply Defaults to recall it later. You can also import your customized control bar from another project by importing the project's screensets. You learn more about screensets later in this chapter.

When you become comfortable using key commands to navigate Logic Pro, you may never need to click the control bar. But it still provides an excellent heads-up display, indicating whether certain functions are engaged.

The leftmost group of view icons displays the library, the inspector, Quick Help, and the toolbar. Here's a brief description of what these icons allow you to do:

- The library is where you load and save settings for recall later. It's like having a modular studio where you can save and load entire mixer setups, guitar amp settings, or even groups of software instruments. The library is one of the most powerful features in Logic Pro, and you learn about it throughout the book. The key command to open the Library is Y.
- The inspector icon opens a dynamic editing window for the object or objects currently selected in the tracks area. The inspector is more like three inspectors in one. It inspects the selected track, the channel strip where you adjust the track's sound, and the selected region or regions. The key command to open the inspector is I.
- The Quick Help icon opens a floating window with details about whatever your cursor is hovering over. Quick Help is a good learning tool for new Logic Pro users. The Quick Help icon doesn't have a key command, but you can create one for it. Details on creating key commands are provided later in the "Navigating with Key Commands" section.
- The toolbar icon puts a customizable row of functions along the bottom of your control bar. You can customize the toolbar by Control-clicking it and choosing Customize Toolbar. When you're new to Logic Pro, displaying this toolbar is helpful. When you've memorized the key commands or need more screen real estate, however, you might want to hide the toolbar. You can do so by pressing Control-Option-⌘-T.
- The smart controls icon opens an editor in the main window below the tracks area. The content of the smart controls editor depends on the selected track. Smart controls decide which parameters you need the most, and they do an almost perfect job. The key command to open the smart controls is B.
- The mixer icon opens the pristine-sounding or dirty-sounding mixer, depending on how you decide to mix it. The mixer is so versatile and great that it gets plenty of attention throughout this book, including an entire chapter on mixing (Chapter 16). Like the smart controls, the mixer appears at the bottom of the tracks area. The key command to open the mixer is X.
- The editors icon displays the group of editors associated with the selected region type. The editors open at the bottom of the tracks area; which editors you see depends on whether you have an audio, a MIDI, or a drummer region selected. All editors are covered in detail in Chapters 14 and 15. The key command to open the editors is E.
- The *transport* area is a collection of icons that provide basic controls for playing, recording, pausing, stopping, forwarding, rewinding, and generally navigating your project.

TIP

Because all these functions are easily performed with key commands, I display only the most important controls on my transport. I always display the stop, play, and record icons because when you click and hold down on them, you see additional options that you often need to adjust throughout your work.

- The LCD display indicates the current location of the playhead. You can see your project in beats, time, or a customized display. To change the display mode, click the display mode pop-up menu on the right side of the display and make a selection. A useful feature of the custom display options is to open a giant beats or time display in a separate floating window for viewing from afar.

  You can enter data directly into the LCD display by double-clicking or click-dragging (for tempo and location). You edit other key project parameters, such as key and time signatures, by clicking the display and manually entering the data.

TIP

During hard-drive- and processor-intensive work, such as recording and mixing, I like to show the performance meter on the display so I know how far I'm pushing my computer.

- To the right of the LCD display are several mode and function icons. You can customize the control bar to display your most used modes and functions.

TIP

A few icons, such as cycle mode, are helpful to have available even if you know the corresponding key command because their pressed state lets you know quickly whether the mode is enabled. And some icons, such as the tuner, don't have corresponding key commands, so it's handy to have them on the control bar.

- The master volume slider might be familiar to GarageBand users. Because GarageBand doesn't have a mixer, the master volume slider is the primary way to turn your project volume up or down. In Logic Pro, a master fader in the mixer serves the same function as the master volume slider on the control bar, so the slider isn't necessary.
- On the far right of the control bar is another group of view icons. The list editors icon opens a window on the right side of the tracks area with four tabs: Event, Marker, Tempo, and Signature. Each tab allows editing access to the smallest details of your project data. The Event tab updates its display depending on what you have selected. The Marker, Tempo, and Signature tabs show events that affect your project globally. The key command to open the list editors window is D.
- The note pads icon opens a window on the right side of the tracks area with two notes tabs: Project and Track. I keep copious notes throughout my project, so I often open these tabs. The Project notes tab is a great place to

write song lyrics. You could use the Track notes tab as a change log so you don't forget what you've tried out or how you've edited your track. The key command to open the note pads is Option-⌘-P.

» The loop browser icon opens a list of prerecorded audio and software instrument loops in a window on the right side of the tracks area. You can filter your loops by using the descriptive keyword buttons or search for them directly in the search field. Selecting a loop automatically auditions it. When you find one you like, drag it to the bottom of the tracks area, where you see the *Drag Apple Loops here to create tracks* text. The key command to open the loop browser is O.

» The browsers icon opens a window on the right side of the tracks area with two tabs: Project and All Files. The project browser shows you all the audio in your project. The all files browser works similarly to Finder; you can navigate to any location on your computer to import media. The key command to open the browser is F.

As you can see, the control bar gives you lots of, well, control. Even though most of these modes and functions are available as key commands, having them in your line of sight can help your workflow. For starters, you can click them. But more than that, they make excellent visual reminders of what you can do with Logic Pro. And if you're up to the challenge, you can use them to help you memorize their key commands.

Polishing in the editors area

The control bar editor icons and corresponding key commands aren't the only ways to open the various editors. Double-click an audio, a MIDI, or a drummer region, and the corresponding editor will open at the bottom of the tracks area. An audio region defaults to the audio track editor. A MIDI region defaults to the piano roll editor, as shown in Figure 3-6. A drummer region defaults to the drummer editor.

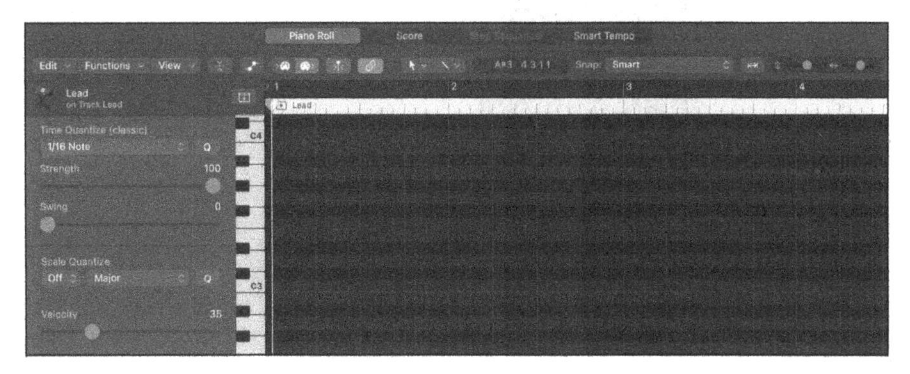

FIGURE 3-6: The editors area.

In the audio and MIDI editor windows, you'll see tabs for additional editors. The MIDI editor displays tabs for the piano roll editor, score editor, step sequencer, and smart tempo. The audio editor shows a tab for the audio track editor, audio file editor, and smart tempo.

Like the tracks area, the editors area has a toolbar with edit, functions, and view menus; icons; tool menus; and snap and zoom settings. (You find out what the editors can do in Chapters 14 and 15.) When you learn how to navigate the tracks area, you'll have a pretty easy time navigating the editors.

Investigating the inspector

The inspector, shown in Figure 3-7, is an essential tool, and you'll quickly memorize its key command, I. (When you've memorized the key command, consider deleting the inspector icon on the control bar to free up space.) The inspector can take up a lot of space in the main window, but it's integral to editing your tracks and shaping your sound.

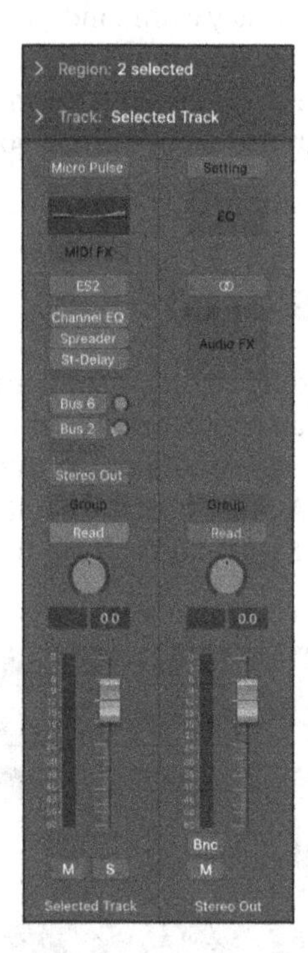

FIGURE 3-7: The inspector.

At the bottom of the inspector are two channel strips. The strip on the left corresponds to the currently selected track. The strip on the right is dynamic and is discussed further in Chapter 16. Above the channel strips are two panes that you can open and close:

- The track inspector pane gives you details about the entire track. You can change the track icon here, which I love to do. You discover what the track inspector can do in Chapter 4.
- The region inspector shows you details about the currently selected region or regions. The details in both the region and track inspector panes depend on what kind of track or region is selected.

Understanding the difference between the track and region inspector panes will save you from a lot of confusion as you work. You discover more about tracks and regions in Chapter 4.

Taking Inventory of Your Track List

Every track you create is added to the track list and given a track header, as shown in Figure 3-8. You can reorder tracks by dragging the track headers to new locations in the list. To select more than one track at a time, ⌘ -click the track headers. You can delete tracks by choosing Track ⇨ Delete Track or by Control-clicking the track header and choosing Delete Track. You can also navigate your track list using the up and down arrow keys.

FIGURE 3-8: A track header.

Making headway with track headers

Track headers are customizable and resizable. Choose Track ⇨ Configure Track Header (or press Option-T) to display the Track Header Configuration dialog shown in Figure 3-9. Select the additional items you want to see in your track headers. You can also Control-click any track header to pull up the Track Header Configuration dialog. You can resize the track header vertically or horizontally by placing your cursor at the top, bottom, or right edge of the track header and dragging when your cursor changes to the resize pointer.

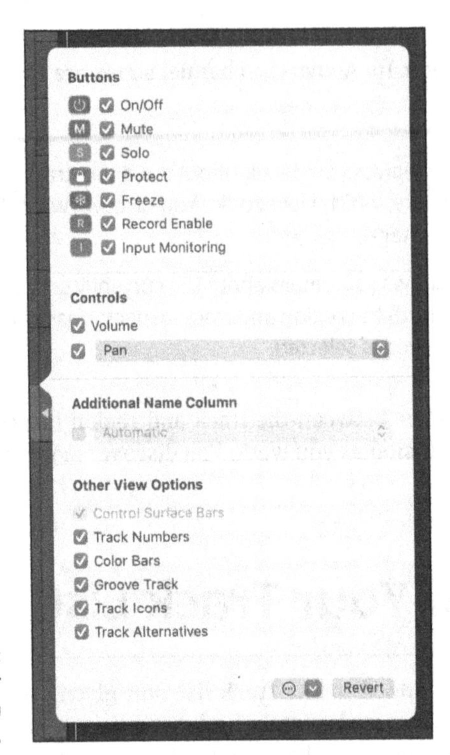

FIGURE 3-9: Track header configuration dialog.

By default, all the buttons and controls aren't shown on the track headers. Here's a brief description of the track header configuration options:

- **On/Off:** Use the on/off icon to turn the track on and off. When a track is turned off, it is silenced. To turn off tracks, you must select Enable Complete Features in the Advanced Settings pane (see Chapter 1).
- **Mute and Solo:** The mute icon silences the track. You can mute all tracks by pressing ⌘ while you click the mute icon. The solo icon mutes every track except the soloed track. You can mute or solo multiple tracks at one time by clicking and holding down on the icon and dragging your cursor up or down the track list.
- **Protect:** This icon prevents changes to the track. With the protect icon enabled, you won't be able to record or edit the track. Protecting a track is helpful to ensure that it remains exactly as it is, without accidental changes.
- **Freeze:** When you freeze a track, you reduce processing power on the track by temporarily turning the track and all its audio effects into an audio file. This feature is helpful for software instruments and audio tracks with many plug-in effects that require a lot of processing power. With freeze enabled, all plug-ins (including software instruments) are temporarily deactivated, and the track is turned into an audio file that includes all effects.

- **Record Enable:** Click the record enable icon to enable your track for recording.
- **Input Monitoring:** This icon allows you to monitor audio tracks that aren't enabled for recording. Use this icon whenever you need to set audio levels for recording or to practice a part you plan to record.
- **Volume and Pan:** If you're used to working in GarageBand, having the volume and pan control in the track header might help you get accustomed to Logic Pro. I usually need more workspace in the tracks area, so I don't display this control. Plus, I find the mixer controls better suited for the job. But when I'm on a bigger computer screen, I like to fill my track headers. The volume slider does double-duty as a level meter. You can change the pan control to an effects send control. (You learn more about send effects in Chapter 17.)
- **Additional Name Column:** Add a name column to the right of the track name. You can also customize what the column displays by using the drop-down menu in the track header configuration dialog.
- **Control Surface Bars:** Select this option to add a thin bar on the far left of the track header that indicates whether the track is being controlled by a control surface. A *control surface,* such as your iPad, is a hardware device that allows you to control a digital audio workstation such as Logic Pro. Many MIDI controllers can also be used as control surfaces, enabling you to utilize hardware to control the onscreen faders, knobs, buttons, and displays. To read an in-depth manual on control surfaces, choose Help ➪ Logic Pro Control Surfaces Support.
- **Track Numbers:** Select the Track Numbers option to display the track number on the left of the track header.
- **Color Bars:** Selecting this check box colorizes the left side of the track header and visually organizes your tracks. You find out more about track colors in the next section.
- **Groove Track:** You can set one track in your project as the groove track and select other tracks to follow the timing of the groove track. When a groove track is selected, a star will appear to the right of the track number and all other tracks will have check boxes that you can select to make a track follow the groove track. For the details on groove tracks, see Chapter 13.
- **Track Icons:** If you Control-click the track icon in the track header, you can choose a new icon from the icon pop-up menu. Track icons are useful visual indicators and look cool, too.
- **Track Alternatives:** Adds the Track Alternatives menu to the track header. You can create alternate versions of a track by choosing New from the menu.

TIP

You can quickly rename a track in the track header by pressing Shift-Return and typing the new name. You can get through the entire track list by pressing Tab between each new name.

Making it pretty with track colors

Track colors not only make your tracks pretty but also help you identify tracks and groups of tracks quickly. In the track header configuration dialog (refer to Figure 3-9), you can display the color bars to aid visual recognition. When you create new regions on a track, they are also colorized in the same color. You can even change the color of regions independently of the track color. However, selecting a track automatically selects all the regions on the track, so if you colorize a track while all the regions are selected, those regions will also change color.

To change a track color, Control-click a track and choose Assign Track Color. You'll be shown a beautiful palette of 96 color swatches that you can use to colorize your selected tracks and regions. By default, MIDI tracks are colorized green and audio tracks are blue. I always group my tracks by color. Drums get their own color, lead vocals get their own color, background vocals get a different color — you get the idea. Group your groups with color.

Zooming Tracks

You'll probably need to zoom in and out of tracks a lot, especially when you're editing. Fortunately, you can zoom tracks in several ways. The first trick is turning on Zoom Focused Track by pressing Control-Z or choosing View ⇨ Zoom Focused Track from the tracks area toolbar. With Zoom Focused Track on, as in Figure 3-10, the currently selected track will automatically zoom horizontally. You see more of the track contents and a quick indicator of which track is selected and has focus. The key command makes it easy to toggle between the two zoom states.

At the top right of the tracks area toolbar are two zoom sliders, vertical to the left and horizontal to the right. Drag the sliders to adjust the zoom level. You can also use key commands to zoom; I suggest you memorize them because they're so easy to use. To zoom in on all your tracks vertically, press ⌘-down arrow; to zoom out, press ⌘-up arrow. Likewise, to zoom in horizontally, press ⌘-right arrow; to zoom out, press ⌘-left arrow.

If you want to zoom in on a specific area of your tracks, you can use the zoom tool. It's always available when your cursor is in the tracks area: Simply press Control-Option while dragging over the area you want to zoom. Your cursor will temporarily turn into the zoom tool, and the area you select will automatically zoom when you release the cursor. To get back to the previous level of zoom, press Control-Option while clicking anywhere in the tracks area.

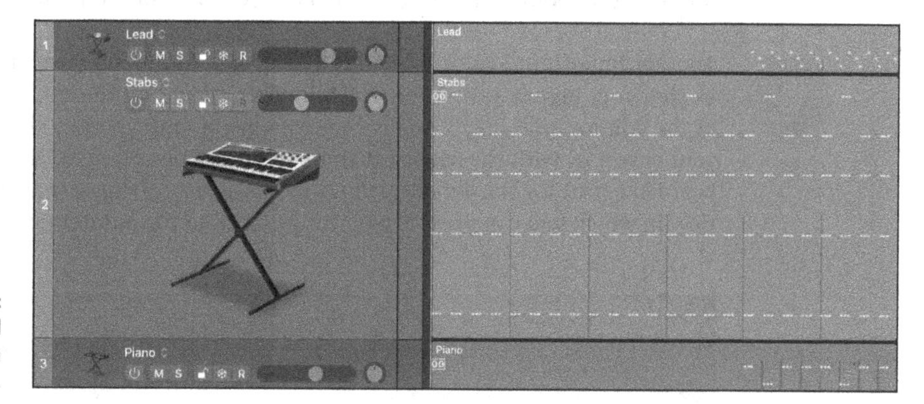

FIGURE 3-10: A zoom-focused track on a selected track.

TIP

You can recall multiple levels of zoom by using the zoom tool. This means you can zoom in on a large portion of your project, zoom in on a smaller section, zoom in on a single part of a region, and then recall each one in sequence just by pressing Control-Option and clicking the tracks area.

Another useful key command to memorize is Z, which toggles the Zoom to Fit Selection or All Contents command. If a region is not selected, this zoom command zooms out to fit all the content in the tracks area. The tracks and regions are smaller, and you can view all your content. If you have any regions selected, the same command will zoom in on those regions.

One more zoom to consider is the Waveform Vertical Zoom. Press ⌘-minus (-) or ⌘-plus (+) to zoom only your audio waveforms. This zoom feature makes your audio waveforms larger in the regions without making the regions themselves bigger. I often use this feature for audio editing and voice-over work when waveforms aren't tall because they're not recorded at high volumes. If you forget the key command, use the waveform vertical zoom icon to the left of the zoom slider in the tracks area toolbar (refer to Figure 3-10).

TIP

The default state of zoom should be to see the entire project. You learned how to zoom out to see your entire project by pressing Z with no regions selected. From that position, you can easily see where you want to focus and get there quickly by Option-Command-dragging over the area. When you're finished, zoom back out and decide what to do next. This method of zooming creates an efficient workflow.

You can zoom in and out of your project in many other ways, as you discover later in the chapter when you read about creating your own key commands.

Opening Your Logic Pro Toolbox

Tools imply work. But Logic Pro is about having fun. So think of your toolbox as a fun box. The toolbar in the tracks area has several tools you can play with.

TIP

Another important key command is T. This key command opens the tools menu, as shown in Figure 3-11. In several windows, including the tracks area and most of the editors, pressing T opens the tools menu, and you can choose a tool with your cursor or with the keyboard shortcuts listed on the right. Note that the keyboard shortcut for the default pointer tool is also T, giving you an efficient workflow in which you can press T twice to get back to the pointer quickly.

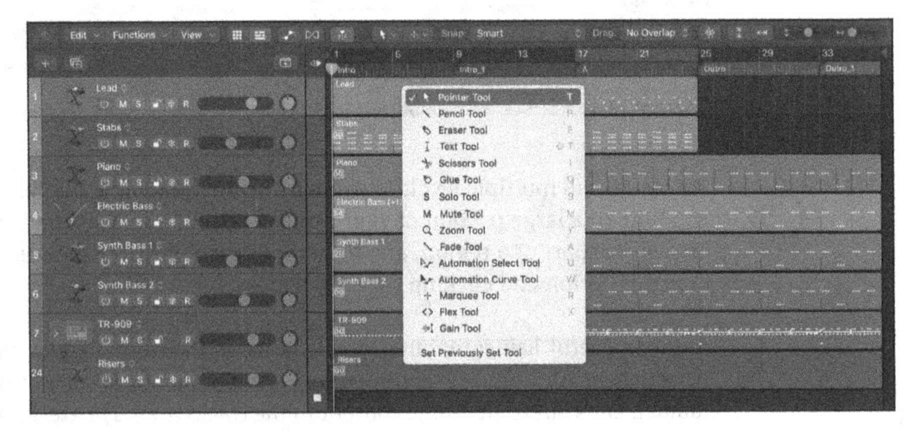

FIGURE 3-11: The tool menu in the tracks area.

Here's what's on the tool menu:

- **Pointer:** The pointer is your default tool for selecting and moving things. Using the pointer tool, you can copy items by Option-dragging them. Grabbing the corners and edges of regions can temporarily cause the pointer to take a descriptive shape as an indicator of additional pointer functions. Place your cursor over the upper half of the right side of a region to turn the cursor into the loop tool. With the loop tool active, dragging the region corner to the right loops the region. Place your cursor over the lower half of the region's right side to change the length of the region as you drag the corner. You'll get the hang of the pointer quickly because it will be your most-used tool.
- **Pencil:** The pencil tool is similar to the pointer tool in that it can also loop, drag, alter the length, and even select regions and other events. The pencil is unique because it creates regions when you click in empty track areas. Note that if the project hasn't been saved and you click an empty audio track with the pencil tool, you'll be asked about opening an audio file.

- **Eraser:** The eraser tool deletes regions and events from the tracks area. If multiple regions or events are selected and you click one of them with the eraser tool, all selected items will be deleted. This tool doesn't get much use because pressing Delete has the same effect. However, if you are going to delete several items in a row, clicking with the eraser tool is faster than selecting each item one by one and pressing Delete after each. And if you're trying out for the Logic Pro Editing Olympics, every keystroke counts.
- **Text:** With the text tool selected, you can rename regions and other events.
- **Scissors:** Use the scissors when you want to split items. The scissors tool has a special Option-click behavior that can split a region into portions of equal length. You can also click-drag the scissors tool over a region to find the right place to make your split.
- **Glue:** The glue tool joins selected items. You can also click-drag over items to select them before joining them.
- **Solo:** Use the solo tool when you want to listen to only a single region. With the solo tool, click and hold down on a region to hear it.
- **Mute:** The mute tool mutes or unmutes the items and other selected items it touches. You can select multiple items and mute or unmute them all at once or simply click any region to mute or unmute it.

TIP

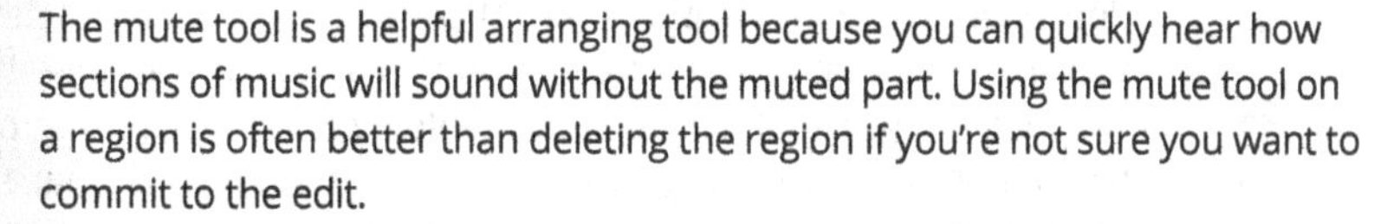

The mute tool is a helpful arranging tool because you can quickly hear how sections of music will sound without the muted part. Using the mute tool on a region is often better than deleting the region if you're not sure you want to commit to the edit.

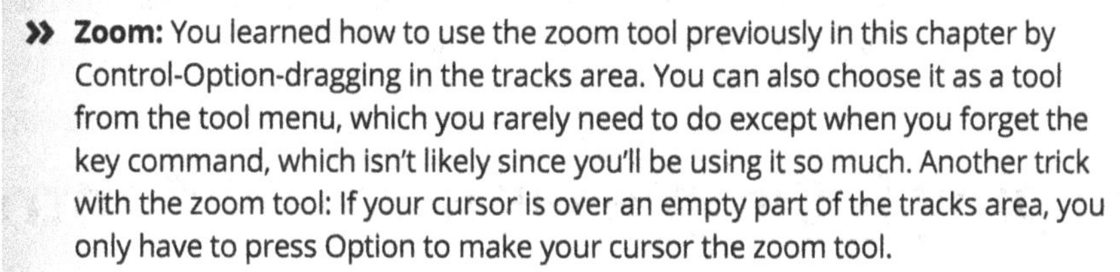

- **Zoom:** You learned how to use the zoom tool previously in this chapter by Control-Option-dragging in the tracks area. You can also choose it as a tool from the tool menu, which you rarely need to do except when you forget the key command, which isn't likely since you'll be using it so much. Another trick with the zoom tool: If your cursor is over an empty part of the tracks area, you only have to press Option to make your cursor the zoom tool.
- **Fade:** With the fade tool, you can fade in and fade out the volume of your audio regions by dragging over the start or end of the region, respectively. You may need to zoom in horizontally to see the fade that's applied to the region. You can edit the fade length by dragging the start or end point with the fade tool. You can also adjust the fade curve by dragging up or down within the start and end points.
- **Automation Select:** I discuss automation in detail in Chapter 18. When automation is active, the automation select tool allows you to select automation data for editing. Enable Complete Features in the Advanced Settings pane must be selected to enable automation tools.

» **Automation Curve:** You can bend an automation curve by dragging it with the automation curve tool. (You find out more about automation in Chapter 18.)

» **Marquee:** Use the marquee tool to select and edit regions and parts of regions. You drag the marquee tool over the objects you want to select or edit. After you've made a selection with the marquee tool, clicking Play on your transport will start your project at the beginning of your selection, and playback will stop at the end of the selection. The marquee selection can also be used for punch recording, which you learn about in Chapter 6. The marquee tool is flexible, and you'll use it throughout this book.

» **Flex:** The flex tool will blow your mind when you learn more about it in Chapter 14. With the flex tool selected, you can grab an audio region's waveform to manipulate it and change its rhythm. The flex tool will save you from throwing away recordings that contain mistakes because you can fix them. It's like having a time machine.

» **Gain:** The gain tool is used to make changes in the gain of audio regions. You can click and drag the region up or down, respectively, to increase or decrease its gain. Using a marquee selection, you can make gain changes only to the selected part of the region.

You have two tools available at all times. The first tool, chosen by the left tool menu on the menu bar, is the tool that's currently available. The second tool, chosen by the right tool menu, is available by pressing ⌘. You can select any tool to be your ⌘-click tool. If you're slicing a lot of regions, make the scissors your ⌘-click tool. Different workflows require different tools, and the ⌘-click tool will help you accomplish your work quickly.

Keeping It Simple with Smart Controls

Smart controls are a gorgeous interface that instantly gives you a customized and dynamic set of controls for shaping the sound of your track. Smart controls give you the best tools for the job. They don't give you every available parameter — only the most important ones.

Smart controls are dynamic. The controls you're given depend on the selected track and the software instrument or plug-in effects on that track. For example, if you have a compressor and EQ plug-in on a track, the smart controls will give you a combination of the most important controls of each plug-in. If a software instrument track is selected, the smart controls will also include parameters that affect the instrument's sound.

Last but not least, smart controls look cool and make you want to play with them.

Opening smart controls

Click the smart controls icon in the control bar or press the key command B to open the smart controls at the bottom of the tracks area, as shown in Figure 3-12. I memorized the smart controls key command by remembering the word *best*, as in *best controls*. You can also choose View ➪ Show Smart Controls. To open smart controls in a new window, choose Window ➪ Open Smart Controls.

FIGURE 3-12: The smart controls.

At the top of the smart controls is a menu bar. In the center of the menu bar are two buttons to quickly switch between the track's smart controls and dedicated channel EQ. If no EQ is inserted on the track and you press the EQ button, a plus sign icon in the center of the control area will allow you to insert an EQ into the track's channel strip instantly.

If the selected track is a software instrument track, an arpeggiator icon appears on the right side of the menu bar, as shown in the margin. An *arpeggiator* turns the chords you play into *arpeggios*, or one note played after another instead of simultaneously.

Click the icon to turn on the arpeggiator, and a pop-up menu will appear so you can choose a preset or adjust the settings. The arpeggiator is a popular synth effect across many genres. (The arpeggiator and other MIDI effects are covered in more detail in Chapter 17.)

REMEMBER

To enable the smart controls icons and features described in this chapter, Enable Complete Features must be selected in the Advanced Settings pane. Choose Logic Pro ➪ Settings ➪ Advanced and select the Enable Complete Features option.

On the left side of the smart controls menu bar are the smart controls inspector icon and a Compare button. The icon opens the inspector on the left side of the Smart Controls window, as shown in Figure 3-13. The Compare button compares the edited smart controls with the saved version.

FIGURE 3-13: The smart controls inspector.

Editing smart control layouts

Smart controls automatically give you quick control over the most important parameters, so you won't need to edit them often. Suppose you want to control a parameter not included in the automatic smart control layout. In that case, you can edit the sound directly in either the software instrument interface or the Effects plug-ins inserted in the track. (You learn how to adjust the plug-ins directly in Chapter 17.)

Think of smart controls as shortcuts to the sound parameters that you use most often. For example, if your track contains a software instrument, such as an electric piano, the smart controls will give you the knobs that are frequently needed to adjust the sound of an electric piano. If you were to also add an effect to the track, the smart controls will readjust based on the new setup. Smart controls will continue to adjust as you add or remove effects to a track — that's why they're smart.

But what if you want to manually customize the smart controls for a particular purpose? Fortunately, smart controls are as flexible as they are intelligent. They also have a menu of gorgeous layouts designed to emulate the look and feel of gear you may be familiar with, such as classic guitar amps and instruments.

To change the layout of the smart controls, follow these steps:

1. **Click the inspector icon.**

   The smart controls inspector opens to the left of the screen controls. At the top of the inspector is the name of the current layout (refer to Figure 3-13). The default layout is Automatic Smart Controls.

2. **Click the name of the current layout at the top of the inspector.**

   A pop-up menu appears.

3. **Make a selection from the menu of layouts.**

   The smart controls are updated.

Smart control layouts are predefined. You can't add knobs, buttons, or other controls to the layout, though you can map the relationship between the controls and your track's parameters, as you discover next.

Manually mapping smart controls

When you open the smart controls and choose Automatic Smart Controls as the layout, all screen controls are automatically mapped to the track parameters. If the controls aren't mapped to the parameters you want or some of the controls remain unmapped, you can both manually and automatically map the smart controls.

To map smart controls automatically:

1. **Open the smart controls inspector by clicking the inspector icon in the smart controls menu bar.**
2. **Open the Parameter Mapping area by clicking the disclosure triangle next to Parameter Mapping (refer to Figure 3-13).**
3. **Click the Parameter Mapping pop-up menu and make a selection as follows:**
   - *Choose Map All Controls* when you want to completely reset the controls and have them automatically mapped. Automatically mapping your controls is a great starting place and usually gives you control over the parameters you need.
   - Choose *Map All Unmapped Controls* when you want to map only the controls that are currently labeled Unmapped.

You can manually map smart controls in two ways. The first way is to map controls by using the Learn button:

1. **Open the smart controls inspector.**
2. **In the layout, select the control that you want to map to a parameter.**

   Depending on the current layout, your controls could include knobs, faders, switches, buttons, and other interfaces. You should select the controller type that's similar to the parameter you want to control. For example, a switch or button will control a parameter with an on/off state, and a knob will control a parameter with a range.

3. **Click the Learn button next to the Parameter Mapping menu in the inspector.**

   The Learn button turns red to indicate that learn mode is active.

4. **Click the plug-in or channel strip parameter you want to control.**

   For example, you might click the volume slider on the track. (Channel strip parameters and plug-ins are described in Chapters 16 and 17, respectively.)

5. **Click the Learn button again to finish mapping controls.**

   The selected control now adjusts your track volume.

The first method is excellent if you know which parameters you want to adjust. The second way allows you to map controls by browsing the Parameter Mapping pop-up menu.

1. **Open the smart controls inspector.**
2. **Select the control in the layout you want to map to a parameter.**
3. **Click the parameter mapping disclosure triangle (see the margin) to display the mapping area.**
4. **Click the parameter mapping pop-up menu and choose the parameter you want to control.**

TIP

You aren't limited to mapping a control to a single parameter. You can add additional parameters to the control by clicking the parameter name in the inspector and choosing Add Mapping. In addition to adding mappings, you can copy and delete mappings from the same pop-up menu.

Editing smart control parameters

After you've mapped a control to parameters, you can adjust how the control modifies the parameters. For example, you might want a volume knob never to go all the way down and all the way up. You might want the control to modify a specific range. Open the Parameter Mapping area in the inspector by clicking the disclosure triangle, and you'll see values that you can edit below the parameter name:

- **Range Min:** This value sets the minimum range of the parameter.
- **Max:** This value sets the maximum range of the parameter.

» **Invert:** This check box switches the minimum and maximum values.

» **Scaling:** Click the Open button to open a graph to adjust the minimum and maximum range values, as shown in Figure 3-14.

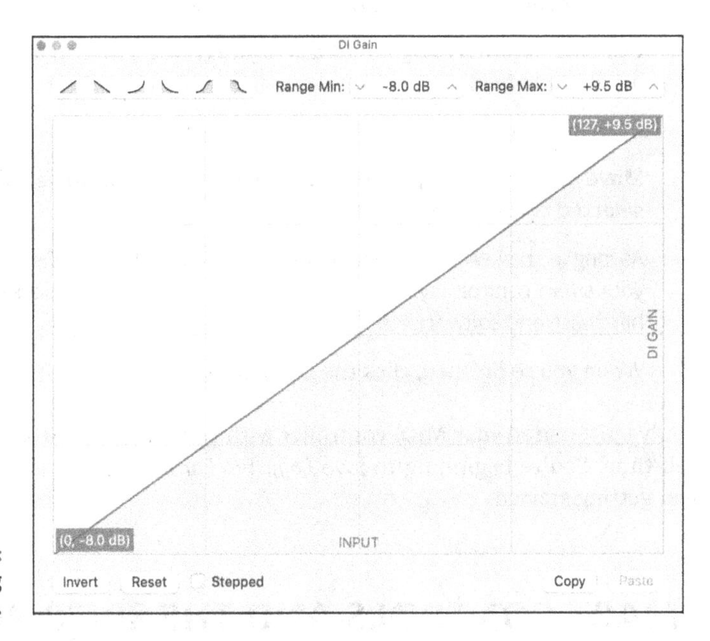

FIGURE 3-14: The scaling graph window.

TIP

If you have more than one parameter mapped, you can change the order of the parameters by dragging the left side of the parameter name up or down the list.

REMEMBER

Don't forget to use the Compare button, which lets you audition changes before fully committing to them.

Controlling the controls with your controller

A significant benefit of smart controls is how much time they save you. Instead of having to open plug-ins and instrument interfaces and find the parameters you need to change, smart controls give you the most used controls in a simple and beautiful interface.

As Steve Jobs would say in every keynote, "But wait, there's more." Your MIDI controller can control all smart controls. The setup is superfast:

1. **Click the inspector icon to display the smart controls inspector.**
2. **In the layout, click the control that you want to control.**
3. **Click the external assignment disclosure triangle to display the assignment name field.**
4. **Click the Learn button.**
5. **Move the control on your MIDI device that you want to pair with the selected control on the layout.**

   As long as the Learn button is enabled, you can continue to select controls in your smart controls layout and move controls on your MIDI device to pair the hardware and software.
6. **When you're finished, click the Learn button.**

You've just paired your MIDI controller with your smart controls and saved yourself time. You're beginning to love *Logic Pro For Dummies*, aren't you? And we're just getting started.

SMART CONTROLS AND THE 80/20 RULE

Vilfredo Pareto, an Italian economist, discovered that 80 percent of the income in Italy (and every country he studied afterward) was in the hands of 20 percent of the population. Since his discovery, the 80/20 rule has been applied to almost everything: 80 percent of your business comes from 20 percent of your clients, 80 percent of your results come from 20 percent of your efforts, and 80 percent of your guitar picks will go missing and never be found again.

And now, I will make the connection between Logic Pro and the 80/20 rule: Smart controls give you 20 percent of the controls you will use 80 percent of the time.

Use the 80/20 rule as a lever to get more accomplished and with better results. I use it to select my opportunities, schedule my time, and guide my songwriting. And when you find out that your tools are purposefully designed to give you 80/20 leverage, you can believe that the Logic Pro designers want you to make more music.

Smart controls are your 80/20 lever.

Navigating with Key Commands

The absolute fastest way to navigate Logic Pro is with key commands. I've talked about key commands quite a bit and don't intend to stop. I sometimes talk about key commands on intimate dates and important holiday gatherings. I find them to be festive and captivating. And even if my date doesn't agree that key commands are great conversation starters, I'm sure you will.

Learn any new key commands lately? Open the Key Commands window, shown in Figure 3-15, by pressing Option-K or by choosing Logic Pro ⇨ Key Commands ⇨ Edit. In the search field, type the name of the command you're looking for, even if you have to guess at the keywords, and your search results will be displayed in the key commands list. If you press Tab, your cursor will move to the key commands list, and pressing any key combination will instantly take you to the associated command.

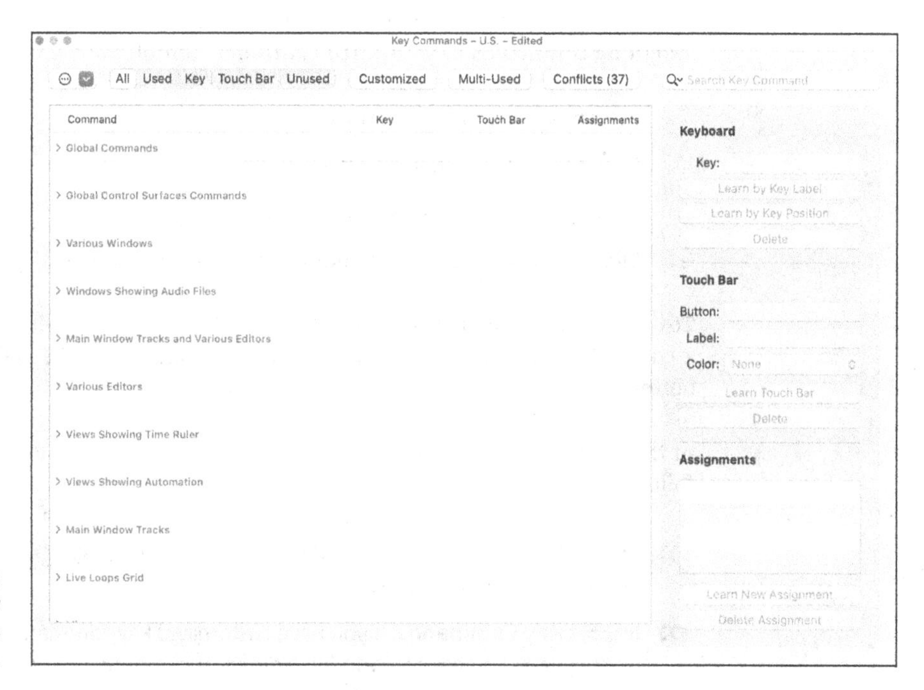

FIGURE 3-15: The Key Commands window.

REMEMBER

The key command to open the key commands is Option-K. If you can't remember which modifier key to use, it doesn't hurt to try all the modifier keys until you get it right. Anything you do to your project, you can undo.

From the Key Commands window, you can create your own key commands:

1. **Search or browse for the command.**
2. **Select the command and then click the Learn by Key Label button.**

   If you have a keyboard with a numeric keypad and you want to distinguish between number keys on the alphanumeric keyboard or numeric keyboard, press the Learn by Key Position button instead. A full-size keyboard is an excellent tool for music production. The additional keypad can store a lot of key commands.
3. **Press the key and modifier key or keys.**

   If the key command is already in use, an alert will ask you to cancel the operation or replace the key command.
4. **Click the Learn by Key Label button again to finish.**

What's the next best thing to being able to assign thought commands? How about assigning commands to your MIDI controller? Simply follow these steps:

1. **Search or Browse for the command.**
2. **Click the Learn New Assignment button.**
3. **Press a button on your MIDI controller.**
4. **Click the Learn New Assignment button again to finish.**

To delete a key command, select the command and click the Delete button. To delete a controller assignment, select the command and click the Delete Assignment button.

At the top of the Key Commands window is an Options drop-down menu with the following options:

- **Presets:** Choose a key command preset for another language and, in some instances, other presets installed on your computer.
- **Import Key Commands:** If you have customized key commands that you use regularly and have to work on a different computer, you can simply import your key commands from this menu. Be a good guest and back up the owner's key commands first.
- **Save/Save As:** Save your key commands to an external file for backup and importing into other systems.
- **Copy Key Commands to Clipboard:** This menu item is useful if you want to print your key commands for reference, all 30 or so pages of them.

- **Expand All/Collapse All:** So many key commands are available that it's necessary to group them into categories. Being able to expand and collapse the key command menus will help you browse all your choices.
- **Scroll to Selection:** If you have a key command selected at the bottom of the list, but you've traveled to the top of the list looking for another key command, you can quickly jump to your selection using this command.
- **Initialize all Key Commands:** This item resets all key commands to their original state, erasing all key command modifications you may have made. Fortunately, the original set of key commands is a great place to start.

To the right of the Options menu is an additional drop-down menu to show all, used, or unused key commands. Some functions are available only by using a key command. If you open the Key Commands window once a session and learn a new key command, it will be a valuable session.

Saving Workspaces with Screensets

As you've discovered so far, Logic Pro has many windows, inspectors, icons, and interfaces. You could probably imagine having two or three 30-inch displays with everything open all at once. So what do you do when you get your screen just the way you like it? You create a *screenset*, a snapshot of your current screen layout.

I love screensets. Whenever I move windows around and get them the way I like for a particular job, I save it as a screenset. If I need to do a different job with a different focus, I create a new screenset. I give you some ideas about how to use screensets later, but first let me show you precisely what screensets are and how they help you make more music.

You're always using a screenset. The numbered menu to the right of the Window menu shows you what screenset is currently selected. Screensets store window size and placement, your control bar customization, your zoom level, and much more. You can import another project's screensets by using the import project settings (see Chapter 2). You can assign screensets to all the number keys except 0, which makes them easy to navigate. You can also store double-digit screensets. Press Control for only the first digit of double-digit screensets (but you still can't use 0).

To create a screenset, do one of the following:

- Press any number key from 1–9. If a screenset doesn't already exist, one is created. To create screensets higher than 9, press Control with the first digit.

- Click the Screenset menu (the number in the main menu to the right of the Window menu) and choose Duplicate to make a copy of your current screenset. Name your screenset in the dialog that appears and press OK.

After you have a screenset exactly how you like it, you can lock it from the Screenset menu. You can also delete and rename screensets from the Screenset menu.

Screensets are easy to recall because all you have to do is use your number keys. Most projects don't need more than nine screensets, but it's nice to know you can have as many as you want in case you need them for a specific workflow.

Following are some ideas for using screensets:

- Open the score in a separate window from the main window to reference the music notation while you record or edit.
- Open the Mixer and Floating Transport windows when you're focusing on mixing.
- Audio and MIDI editing might need windows positioned just right for a good workflow.
- Software instruments and their interfaces can get their own screenset for quick sound editing or playing.
- Open the main window with no inspectors or editors so you can see your entire arrangement quickly. Don't forget the Z key command, which zooms everything to fit in the main window.
- If you work on a laptop and a desktop computer like I do, you can create a set of screensets optimized for the display size and save them as a template for either starting projects or importing the screensets into current projects.

Here's how I use screensets. I reserve screenset number 9 for project notes and number 8 for track notes. Both screensets have the Notes window open and set to the correct tab so I can quickly jot down ideas and references and keep a change log. Screenset number 1 is reserved as an ad hoc workspace where I can set up windows for specific workflows and then duplicate the screenset to its own name and number. Whenever I get the windows just right for what I'm about to do, I duplicate it as its own screenset and give it a descriptive name.

You've learned why Logic Pro is a music producer's powerhouse and timesaver. It's capable not only of sounding great but also of bringing out the great in you.

IN THIS CHAPTER

» Understanding how tracks and regions work

» Adjusting your tempo and time signature

» Saving track settings for instant recall

» Editing and looping regions

Chapter 4

Embracing Tracks and Regions

Tracks and regions are the basic building blocks of your project. You can have up to 255 audio tracks and 255 software instrument tracks in a project, so it's safe to say that you won't ever run out of tracks.

Each track gives you independent control over the sound and placement of a single sound source. These days, multitrack recording might not seem like such a big deal because your iPhone has similar capabilities, but it wasn't that long ago when multitrack recording didn't exist. Now we have nondestructive recording and editing with 200 steps of undo history. It's tempting to say with a straight face that you can record better than The Beatles.

Regions are objects on your tracks that you create, edit, and manipulate until your ear is content. Regions are versatile containers for your creative ideas.

In this chapter, you learn about several track types and region types and gain some basic region editing skills.

Knowing Your Track Types

As described in Chapter 3, the tracks you create are added to the vertical track list to the left of the tracks area, as shown in Figure 4-1. You create more tracks by clicking the add tracks (plus sign) icon at the top of the track list or choosing Tracks ➪ New Tracks. You can also create specific track types with key commands and the Tracks menu.

FIGURE 4-1: The track list.

In this section, you discover what the different track types do and how to create them.

Audio track

An *audio track*, like the one shown in Figure 4-2, can contain audio regions, audio Apple loops, and imported audio files. You use an audio track when you want to record a live instrument or a microphone, as described in Chapter 6. You can also import prerecorded audio files and loops into your project, as described in Chapter 8.

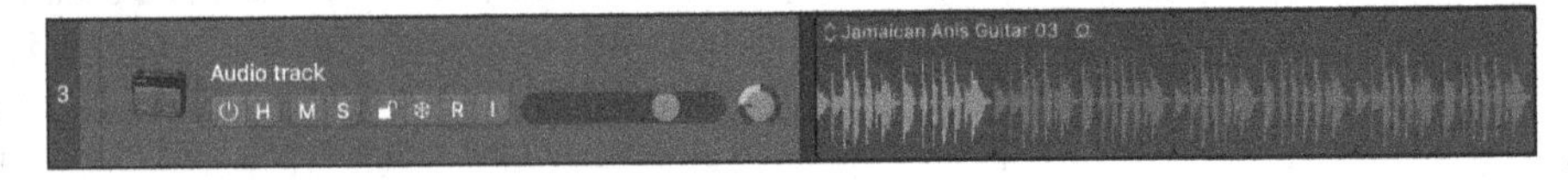

FIGURE 4-2: An audio track.

To create an audio track, do one of the following:

- Press Option-⌘-A.
- Choose Track ➪ New Audio Track.

You can edit audio tracks by using the audio track editor or the audio file editor. For more on editing audio, see Chapter 14.

TIP

You can spot an audio track by the audio waveforms in the track's regions, as shown in Figure 4-2.

Software instrument track

A *software instrument track*, shown in Figure 4-3, can contain MIDI regions, MIDI Apple loops, and imported MIDI files. You use a software instrument track when you want to record one of Logic Pro's software instruments or a third-party instrument you've installed.

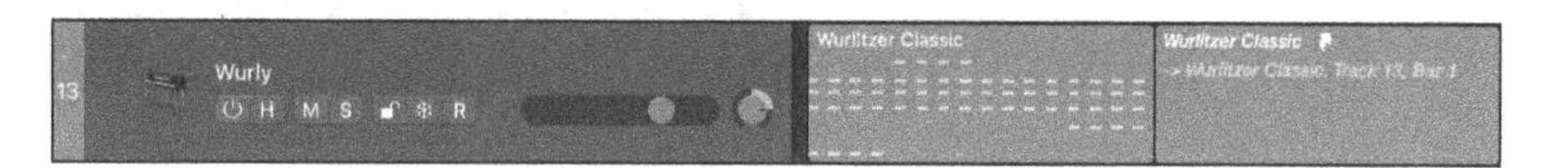

FIGURE 4-3: A software instrument track.

To create a new software instrument track, do one of the following:

- » Press Option-⌘-S.
- » Choose Track ⇨ New Software Instrument Track.

TIP

Logic Pro is compatible with third-party Audio Unit (AU) instruments. Audio Units are Apple's own plug-in format. The marketplace for Audio Unit instruments and effects is massive. Lots of great developers are out there to help you bring your imagination into reality.

You record MIDI data on software instrument tracks, which are covered in Chapter 6. You can edit software instrument tracks by using the piano roll editor, score editor, and step editor, which you will learn about in Chapter 15.

Drummer track

A *drummer track* is used when you want to add a virtual drummer to your project. The drummer track, shown in Figure 4-4, won't let you play its drums (sounds like a lot of drummers that I know); the only way to control a drummer track is in the dedicated drummer editor, which you'll learn about in Chapter 9.

To create a drummer track, choose Track ⇨ New Drummer Track. Logic Pro doesn't have a built-in key command for this option, so why not create one now? I chose Shift-Option-⌘-D for my key command. (For information on creating key commands, see Chapter 3.)

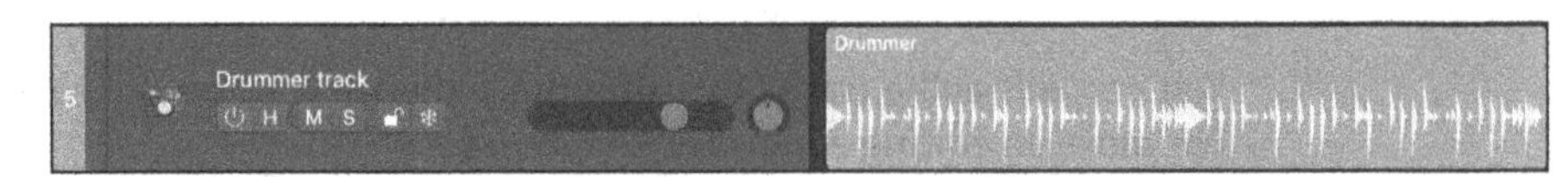

FIGURE 4-4: A drummer track.

External MIDI track

An *external MIDI track,* shown in Figure 4-5, can contain MIDI regions. Use an external MIDI track when sending MIDI data to an external device such as a synthesizer or a drum machine. External MIDI tracks make no sound, so you must send the MIDI data out to your MIDI device and receive the audio signal from your MIDI device on a separate audio track.

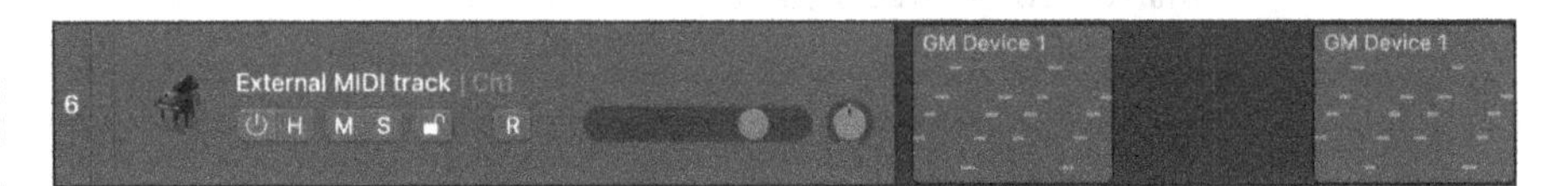

FIGURE 4-5: An external MIDI track.

To create a new external MIDI track, do one of the following:

- Press Option-⌘-X.
- Choose Track ➪ New External MIDI Track.

External MIDI tracks look and behave like software instrument tracks in the tracks area because they both contain MIDI data. But if you open the inspector (press I), you'll see that the channel strip of an external MIDI track looks very different than a software instrument track. The external MIDI track has no audio capabilities and no plug-ins. A MIDI channel strip, covered in Chapter 16, is used to route MIDI data through a MIDI port on a MIDI channel. Using different MIDI ports and MIDI channels allows you to build a project that can communicate with many different external instruments, up to 16 channels per instrument.

REMEMBER

External MIDI tracks only send MIDI data out of Logic Pro. To receive the audio signal from the external MIDI device, you must create a separate audio track to monitor or record the audio.

Track stacks

Track stacks, an innovation in Logic Pro, help you organize your tracks by placing them as subtracks in the main track. You expand and collapse the track stack using the disclosure triangle, as shown in Figure 4-6.

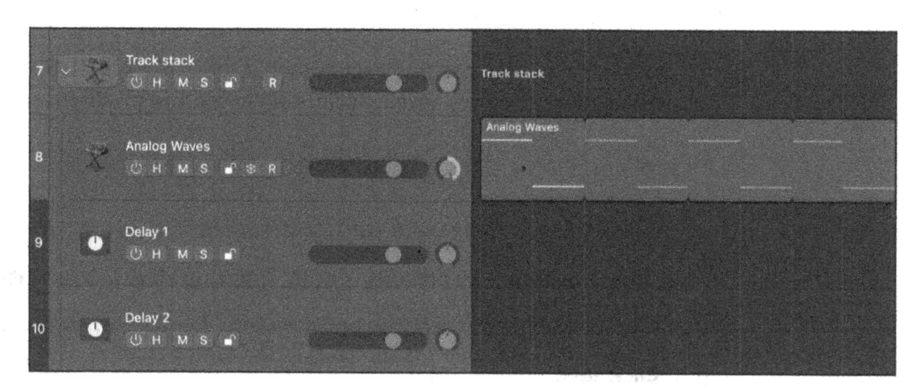

FIGURE 4-6: A track stack.

Two types of track stacks are available:

- **Folder stack:** A folder stack can control the volume of all the tracks in the track stack. All the tracks can be automated, soloed, muted, and grouped as a whole. But folder stacks can't be saved as patches, and you can't add effects to the entire group or control their audio output as a group.
- **Summing stack:** This stack routes audio, saves all tracks and their settings and can be saved as a patch for later recall. All tracks in a summing stack send their audio outputs to a collective auxiliary track. You find out more about audio routing and auxiliary tracks in Chapter 16.

To create a track stack, follow these steps:

1. **Select the tracks you want to include in the track stack.**

   To select more than one track at a time, hold down the ⌘ key while making your selections.

2. **Press Shift-⌘-D or choose Track ➪ Create Track Stack.**
3. **Select the type of track stack you want to create.**

   Your choices are Folder Stack or Summing Stack.

4. **Click Create.**

   The tracks are now grouped in a track stack.

Track stacks can be nested inside each other. If you want to remove the tracks from a track stack, select the track stack and choose Track ➪ Flatten Stack or press Shift-⌘-U.

As mentioned previously in this section, summing track stacks can be saved as a patch for instant recall. For example, you can build an entire orchestral project with dozens of software instrument tracks (as described in Chapter 12) and save the tracks as a patch. Your orchestral track stack is then available for all your projects. Like smart controls, track stacks deliver a more productive workflow.

To save a summing track stack as a patch, follow these steps:

1. **Select the summing track stack in the track list.**
2. **Open the library by pressing Y or by choosing View ⇨ Show Library.**
3. **Click the Save button at the bottom of the library and name your patch.**
4. **Click Save.**

   Your patch is now saved in the library.

To load a patch on a selected track, simply select the patch in the library.

Folder track

A *folder track* is similar to a track stack, but folder tracks are focused more on regions. Track stacks organize tracks; folders organize regions. A folder track, shown in Figure 4-7, can contain multiple tracks and their regions but doesn't have its own channel strip.

FIGURE 4-7: A folder track.

You might use a folder track to:

- » Store tracks you're no longer using
- » Organize the regions of song sections
- » Organize arrangements and parts in a song
- » Sequence longer material, such as albums and musicals

To create a folder track, do the following:

1. **Select the regions in the tracks area.**
2. **In the tracks area toolbar, choose Functions ⇨ Folder ⇨ Pack Folder.**

To view the contents of a folder track, double-click the folder. To exit the folder, click the display level arrow on the far left of the tracks area menu, as shown in Figure 4-8.

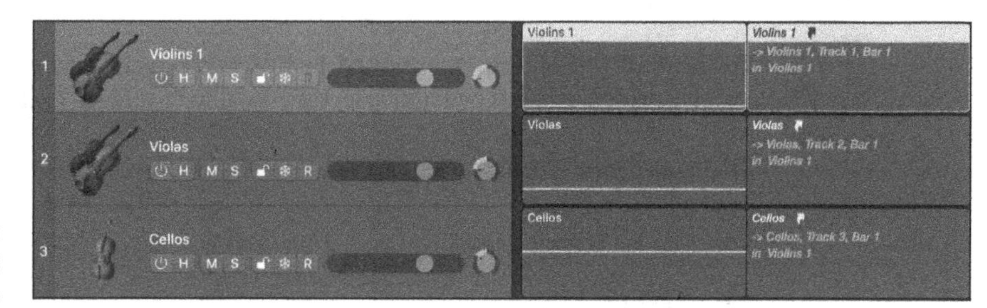

FIGURE 4-8: Folder track contents.

TIP

After you've created a folder track and put some MIDI tracks in it, you can move that folder to an empty software instrument track, and the MIDI in the folder will play the software instrument.

You can unpack a folder by selecting the folder and choosing Functions ⇨ Folder. Then choose Unpack Folder to New Tracks if you want to create tracks in the tracks area of the current project level, or choose Unpack Folder to Existing Tracks to use the existing tracks of the current project level.

As you just discovered, tracks do more than just hold audio and MIDI. They are creative instruments that you can play and explore.

Around the Global Tracks

Global tracks, shown in Figure 4-9, contain lanes and data, but they're global to the project. Global settings such as tempo and time signature changes affect the entire project, and you set them in your global tracks.

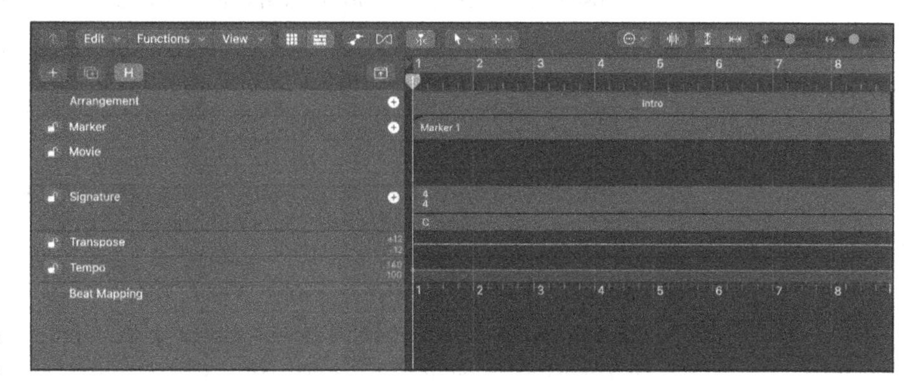

FIGURE 4-9: Global tracks.

To have all the options mentioned in this book available to you, choose Logic Pro ⇨ Settings ⇨ Advanced and select the Enable Complete Features check box.

To display your global tracks, do one of the following:

- Choose Track ⇨ Global Tracks ⇨ Show Global Tracks.
- Press G.
- Click the global tracks icon at the top right of the track headers menu.

You might not see all available global tracks on the screen. To choose which global tracks are displayed, do the following:

1. **Choose Tracks ⇨ Global Tracks ⇨ Configure Global Tracks or press Option-G.**
2. **Select the global tracks you want to show.**

Here's a brief description of the different global tracks:

- **Arrangement track:** The arrangement track creates special markers to help you organize your project based on common song sections, such as chorus, verse, and intro.
- **Marker track:** The marker track lets you mark spots on your project. You can name your markers, put descriptive text on them, and navigate to them by using key commands.
- **Movie track:** The movie track shows thumbnails of imported movies. To import a movie into your project, choose File ⇨ Movie ⇨ Open Movie. You can choose to extract the audio from the movie into a track in your project.
- **Signature track:** The signature track controls both the key signature and the time signature of your project. You can change either at any point in your project by selecting the pencil tool on the tool menu and clicking where you want the change to occur.
- **Transposition track:** The transposition track transposes MIDI and Apple loops but not audio files. Click the track with the pencil tool to create a transposition node. Drag the node up or down, and MIDI regions, MIDI Apple loops, and audio Apple loops will follow the transposition track.
- **Tempo track:** The tempo track defines your project tempo. Double-click anywhere in the tempo track lane to create a new tempo node. Adjust the tempo by dragging the node up or down.

- **Beat-mapping track:** The beat-mapping track aligns the project tempo to the content of your tracks. It's useful for projects that don't have a consistently perfect tempo.

Global tracks are covered in even more detail in Chapter 13, where you learn about arranging.

Sorting and Hiding Tracks

You can move tracks by dragging the track headers up and down. You can also sort tracks automatically according to the type or whether or not they are being used by choosing Tracks ➪ Sort Tracks By.

Hiding tracks is useful when you want to clean up your project. You might not be ready to delete a track that's no longer being used, but you want it out of the way. Here's how you can hide tracks:

1. **Click the green hide icon on the track header menu (or press H if the icon isn't visible), as shown in Figure 4-10.**

   Hide icons will also appear on each track.

2. **Click the hide icon on the track you want to hide.**

   The icon turns green.

3. **Click the hide icon in the track header again.**

   The icon turns orange, indicating that tracks are hidden. The track is hidden from the track list, and the channel strip disappears from the mixer.

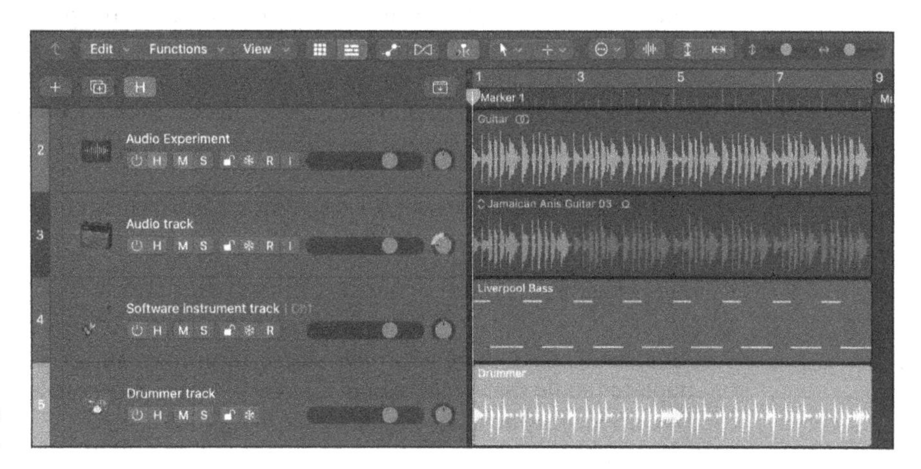

FIGURE 4-10: Hiding tracks.

To unhide your tracks, click the hide icon in the track header. The tracks are again displayed in the track list and mixer.

Creating Options with Track Alternatives

You can create alternative versions of a track within a single track. In Chapter 14, you learn how to edit multiple recordings into a single composite recording, called *quick swipe comping*. For example, you can record a singer multiple times and use only the best parts of each performance. With track alternatives, you can create multiple different versions of a track, giving you the freedom to experiment with different edits.

To create track alternatives, you need to add the Track Alternatives menu to the track header, as described in Chapter 3. To create a track alternative, choose New from the Track Alternatives menu in the track header. A new empty track alternative is made active and added to the menu. If you want to create a track alternative using the content contained in the track, choose Duplicate from the Track Alternatives menu.

By default, your track alternatives are named A, B, C, and so on. You rename them by choosing Rename from the Track Alternatives menu. Choosing Rename by Region will use the currently selected region as your track alternative name. You can also display inactive track alternatives and delete inactive track alternatives.

Knowing the Region Types

The rounded rectangles in your tracks area are called regions. Think of *regions* as flexible placeholders for your audio and MIDI data. Regions can be empty, waiting for you to add content, or created as you record new content. Regions are references to your MIDI and audio data. They can even reference a reference, allowing you to change one region and have all other regions follow suit.

In this section, you discover the different types of regions and how to use them.

Audio region

An *audio region* can be your own audio recording, an audio Apple loop, or an imported audio file. Audio regions, shown in Figure 4-11, can be only audio tracks. You can identify an audio region by its audio waveforms.

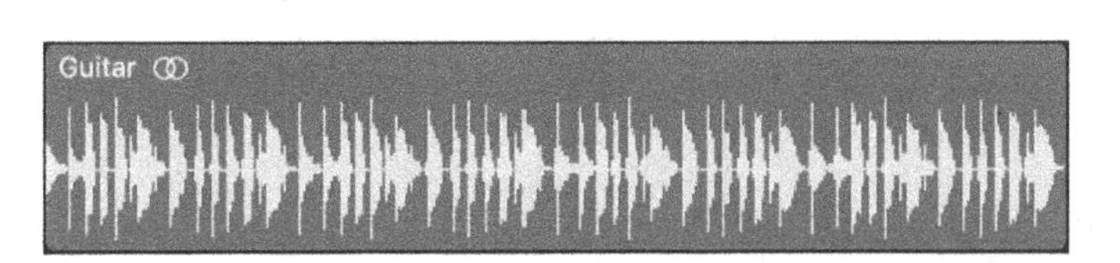

FIGURE 4-11: An audio region.

Audio regions contain references to audio files. An audio region isn't the file itself, which is important to remember. If you split an audio region in half, for example, the audio file isn't split in half. And if you delete a portion of an audio region, the audio file isn't deleted — it simply isn't referenced by the region anymore. If you want to get that portion of the audio back, all you have to do is drag the edge of the region. You learn more about basic region editing later in this chapter.

MIDI region

A *MIDI region* contains MIDI data. MIDI regions can be used on software instrument tracks or external MIDI tracks. You can identify MIDI regions by their thin rectangle note events, as shown in Figure 4-12.

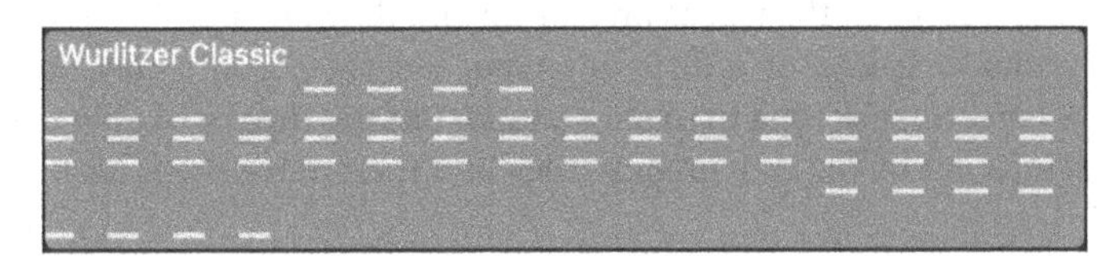

FIGURE 4-12: A MIDI region.

MIDI regions are more flexible than audio. You can definitely manipulate audio beyond recognition, but MIDI regions allow you to experiment and compose with complete freedom.

Drummer region

A *drummer region* can appear only on a drummer track. Drummer regions, shown in Figure 4-13, are like a MIDI-audio hybrid region. They look like audio, but they contain MIDI data. The difference between a drummer region and a MIDI region is that you can't edit the MIDI data directly on a drummer region. You have to use the drummer editor to edit the drummer region. However, after you're happy with how a drummer region sounds, you can export the region as MIDI to edit in a MIDI editor. For details, see Chapter 9.

Think of these regions as virtual drummers. You tell drummer regions what you want, such as playing simple and soft (versus complex and loud), by using the drummer editor. The drummer editor generates drumbeats within the drummer regions. Drummer is great for songwriting because you can instantly create varying drum patterns for all your song sections.

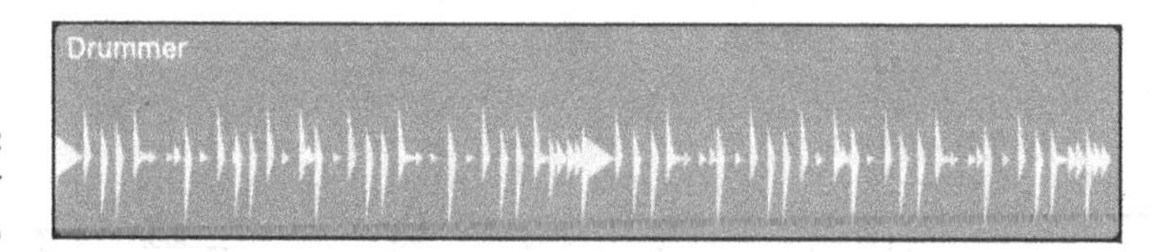

FIGURE 4-13: A drummer region.

Editing Regions

Most of your Logic Pro arranging and editing will probably consist of working with regions in the tracks area. In this section, you discover the basics of editing regions.

Dragging, moving, and resizing regions

Regions can be dragged to different locations on the timeline. They can be moved to different tracks entirely, and they can be resized.

To drag a region, use the pointer or pencil tool to select the region and drag it left and right in the tracks area or up and down to a different track. You can move regions to tracks of a different type, but they won't play. Regions must be on the correct track type to make a sound. But it's nice to be able to temporarily move regions out of the way on any track and then move them back again.

You can also move regions by using the Edit ➪ Move menu. Here are your options:

- **To Playhead:** Move all selected regions to the current playhead position.
- **To Recorded Position:** Move a region back to its original recorded position. This command works only on time-stamped audio regions.
- **To Beat:** Move a time-stamped audio region, so it aligns with the nearest beat.
- **First Transient to Nearest Beat:** Move the first transient of an audio region to the nearest beat. A *transient* is a loud and short sound in a waveform. This command is useful when the region starts on the beat, but the first audio waveform transient is after the start of the region.
- **To Focused Track:** Move regions to a selected track. The regions will keep their current time position.
- **Shuffle Left/Shuffle Right:** Align regions so that their start or end points are aligned with the neighboring region.
- **Nudge Left/Nudge Right:** Nudge the region left or right. First, set the nudge value in Edit ➪ Move ➪ Set Nudge Value To.

To resize a region, move your cursor to the lower left or right of the region until the cursor changes to the resize cursor, as shown in Figure 4-14. Then click and drag to resize the region. A help tag pops up to give you the details of your edit.

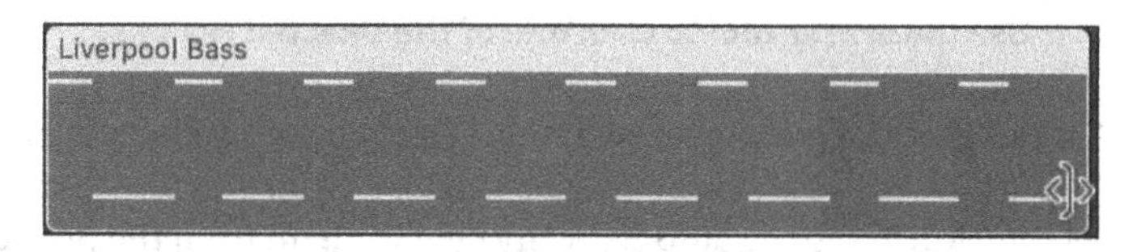

FIGURE 4-14: The resize cursor.

Several other ways to move and resize regions are available only as key commands. For example, press Control-\ (backslash) for the Set Optimal Region Sizes Rounded by Bar. This command is useful when you want your selected region's start and end points to align with the bar.

To find more key commands that move and resize regions, choose Logic Pro ➪ Key Commands ➪ Edit. Then search using the keywords *region, move,* and *length* or browse the categories.

TIP

If you have overlapping regions, you can choose Track ➪ Other ➪ New Track for Overlapped Regions to put the regions in their own track lane.

Splitting, joining, and deleting regions

Sometimes you want to make two regions out of a single region. This is called splitting regions. You can split regions in several ways:

- Use the scissors tool to split a region wherever you click.
- Choose Edit ➪ Split ➪ Regions at Playhead.
- Press Control while clicking a region to display an edit menu with several split commands.

You can join regions to make a single region. This is useful when you've done a lot of editing and want to simplify your edits into a single region. To join regions:

1. **Select the regions you want to join.**
2. **Use the glue tool to click one of the selected regions.**

You can join regions also by pressing ⌘ -J or by choosing Edit ➪ Join ➪ Regions.

You can delete a region from the tracks area or the project entirely. To delete regions:

1. **Select the regions you want to delete.**
2. **Use the eraser tool to click one of the selected regions.**

You can delete regions also with the Delete key or by choosing Edit ➪ Delete.

Deleting MIDI regions automatically deletes them from the project. Deleting audio regions removes the audio from the tracks area but not from the project. You can find the audio files that have been deleted from the tracks area in your project audio browser, as detailed in Chapter 8.

Snapping regions to a grid

When you move regions, they snap to the grid based on the snap settings in the tracks area menu bar. The following snap values are available:

- **Smart:** The Smart setting snaps regions to the nearest value on the grid and depends on the current ruler division and level of zoom. This setting is usually all you need except when you want to get specific.
- **Bar:** This setting snaps regions to the nearest bar.
- **Beat:** This setting snaps regions to the nearest beat.
- **Division:** This setting snaps regions to the nearest division based on the project time signature.
- **Ticks:** This setting snaps regions to the nearest clock tick, which is 1/3840 of a beat.
- **Frames/Quarter Frames:** These settings snap regions to the nearest SMPTE (Society of Motion Picture and Television Engineers) timecode frame.
- **Samples:** This setting snaps regions to the nearest sample, which depends on your project sample rate.

Looping and copying regions

Want to hear a musical part again? And again? And again? Copy or loop your regions.

Repetition is an important part of music composition, so Logic Pro enables you to repeat regions in the tracks area. To place a copy of a region at a new location,

Option-drag the region to the new location on the track. Here's another way to copy and paste regions:

1. **Select the region you want to copy.**
2. **Press ⌘ -C or choose Edit ➪ Copy.**
3. **Place the playhead where you want to paste the region.**
4. **Press ⌘ -V or choose Edit ➪ Paste.**

If you want the region to repeat continuously for any length of time, you can loop it. The benefit of looping is that if you edit the original region, all the loops are edited also. Loops reference the original region; they aren't copies of the region.

To create a loop, place the cursor in the upper-right corner of the region you want to loop, and the cursor turns into the loop cursor, as shown in Figure 4-15. Drag the cursor as far as you want the region to loop.

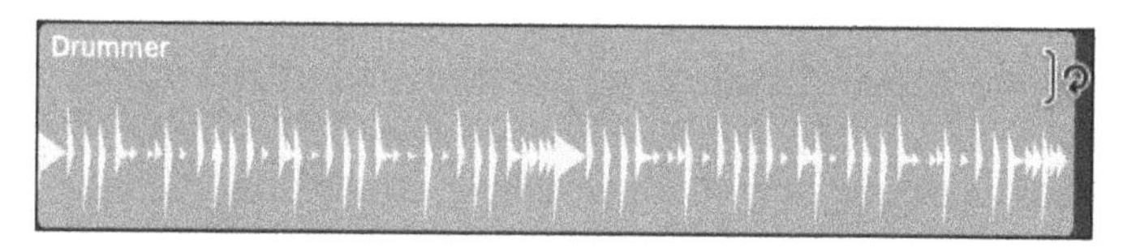

FIGURE 4-15: The loop cursor.

You can loop both audio and MIDI regions. Looping regions saves a lot of time when you're mocking up your arrangement.

Tracks and regions are fundamental to your workflow. With just a little experience, you'll understand how to get the most from them. And now you know some basic editing and how to save track settings for instant recall by using track stacks. With these fundamental skills, you're ready to dive into digital audio and MIDI and begin recording some music.

2

Digital Recording and Using Prerecorded Media

IN THIS PART . . .

Set up and connect your audio and MIDI hardware.

Record audio tracks, set up separate monitor mixes, and discover advanced recording techniques.

Record external MIDI instruments and software instrument tracks.

Import audio, MIDI, and video into your projects.

IN THIS CHAPTER

» Understanding the fundamentals of digital audio

» Getting high-quality audio from Logic Pro

» Setting up audio and MIDI settings

» Connecting your audio and MIDI hardware

Chapter 5

Introducing Digital Audio and MIDI

Even though Logic Pro does a great job of giving you exactly what you need, understanding the basics of digital audio and MIDI is still essential for making high-quality recordings. One of the first steps you take with a project is defining your project's audio settings. These settings can significantly affect the sound quality of your recordings. Your project's audio settings can also affect compatibility with your hardware and other audio systems.

In this chapter, you learn how to set up Logic Pro to get the best sound quality. You discover how to connect your audio and MIDI devices and avoid common pitfalls. You also choose the recording audio file type that's right for the job.

Understanding Digital Audio

So what's the big difference between digital and analog audio anyway? In *analog recording*, a representation (an analog) of the sound source is reproduced on a physical medium, such as records or tape. In *digital recording*, multiple measurements of the sound source are taken and stored digitally as binary code, or 1s and 0s. The process of measuring and recording digital audio is called *sampling*.

The *sample rate* is how often a slice of audio is turned into a digit each second. The higher the sample rate, the higher the audio fidelity. You want high-quality audio, even if your goal is to mangle and distort it. The noise you introduce into your audio should be a choice, not the result of a misunderstanding.

You also want a wide *dynamic range*, which is the ratio of loudest to quietest sound. Dynamic range is measured in decibels, or dB. CDs have a dynamic range of about 90 dB. Logic Pro is capable of 24-bit recording and has a dynamic range of around 125 dB. You'll be pleased that Logic Pro can record audio at and exceeding industry standards, depending on your hardware capabilities.

In this section, you find out how to set your project's sample rate and bit depth and choose the best audio file type for your project. This knowledge will ensure high sound quality and compatibility.

Acoustics 101

Ready to get out your calculus textbooks and start plotting sine waves? You aren't? Phew. Me neither. But it's important to define some audio terms, so you understand the choices you make as you record digital audio.

After you understand some basic acoustic theory, you'll be able to identify how your choices in Logic Pro affect what you hear. Without getting too technical and long-winded, let's define some aspects of sound:

- *Frequency* is the *number* of cycles a sound wave completes in one second, as shown in Figure 5-1. Frequency is measured in Hertz (Hz).
- *Wavelength* is the *distance* traveled over one cycle of a sound wave.
- *Period* is the *duration* of a sound wave cycle in time and is inversely proportional to frequency. The lower the frequency, the longer the period.

Here's why understanding some audio fundamentals is important. Audio has to travel through the atmosphere to get to you so you can perceive it. A low E string on a bass guitar has a frequency of 41.2 Hz and takes about 27 feet to complete a full cycle. Bass frequencies travel far, which is why you can hear the low boom of a loud car stereo coming a mile away. But bass doesn't always sound good in a small space because it can't complete a full wavelength without hitting a wall and bouncing around the room. And before you know it, all those bass frequencies pile on each other and multiply, causing room modes. *Room modes* are frequencies that can be too loud or too quiet based on how sound reacts to the room's dimensions.

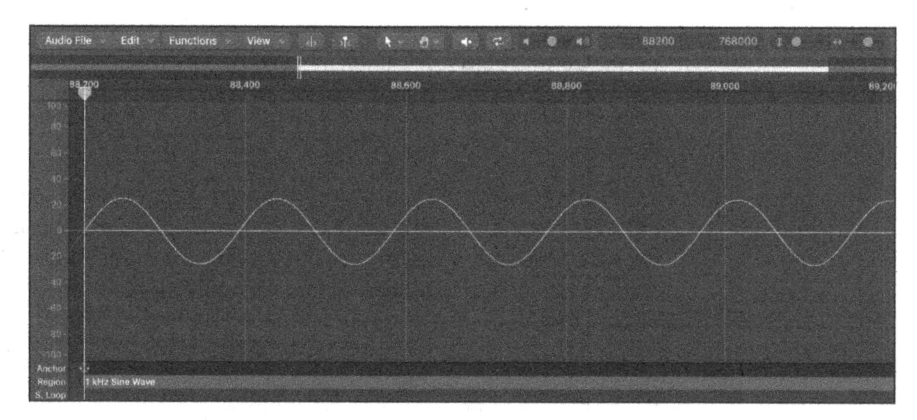

FIGURE 5-1: A sine wave in the audio file editor.

Knowing a little bit about audio will enable you to adjust your sound to improve its quality. Knowing what frequency is and that every sound is made up of multiple frequencies will help you as you record and mix your audio. In the upcoming sections, I will keep the technical information to a minimum and the discussion as practical as the subject will allow.

Setting your sample rate

Setting your project sample rate is one of the first things you should do before you begin recording audio. If your project consists only of software instruments and MIDI, you can change the sample rate at any time. But when audio is included in your project, changing the sample rate will require your audio to be converted to the new sample rate.

Logic Pro can convert your audio to any sample rate, but this processing introduces the potential for degradation. Strive to capture clear recordings that need little processing. Beginning your project with the correct sample rate will help you achieve this goal.

If you must change the sample rate, *downsampling* (converting the sample rate from high to low) is preferable to upsampling. So start your projects with as high a sample rate as your hardware will allow or the project requires. The downsides of very high sample rates are larger file sizes and additional hardware processing. Your computer can't hold as many tracks or Effects plug-ins on a project with a higher sample rate.

To set your sample rate:

1. **Choose File ⇨ Project Settings ⇨ Audio.**

   The Project Settings window opens to the Audio pane, as shown in Figure 5-2.

2. **In the Sample Rate drop-down list, select the sample rate.**

   Logic Pro supports the following sample rates: 44.1, 48, 88.2, 96, 176.4, and 192 kHz.

REMEMBER

   CD audio uses a sample rate of 44.1 kHz and is ideal for most situations. For video production, 48 kHz is common. Higher sample rates are usually reserved for audiophile recordings, such as classical music and DVD audio. However, higher sample rates may become more common as hardware and processing power improve.

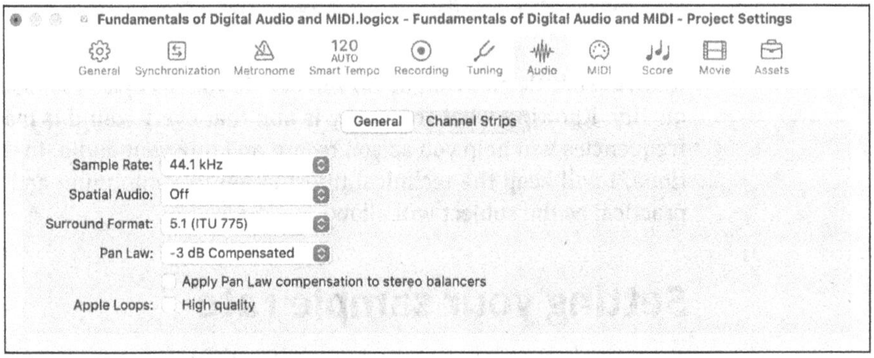

FIGURE 5-2: Audio project settings.

Determining your bit depth

Bit depth is the number of bits of information in each sample. A higher bit depth equals a higher resolution for each sample. The default Logic Pro setting is 24-bit recording. If you turn off 24-bit recording, Logic Pro will record in 16 bit. Similar to your sample rate, the downside of using a higher bit depth is that it takes up more drive space and uses more processing power.

To turn on or off 24-bit recording:

1. **Choose Logic Pro ➪ Settings ➪ Recording.**

   The Settings window opens to the Recording pane, as shown in Figure 5-3.

2. **Select or deselect the 24-Bit Recording option.**

   When 24-bit recording is not selected, Logic Pro records in 16 bit.

REMEMBER

Bit depth differs from *bit rate*, which is the number of bits of information processed each second, as opposed to each sample. When exporting your project to lossy audio formats such as MP3, higher bit rates equal higher quality sound. You choose a bit rate when you export your audio in Chapter 20.

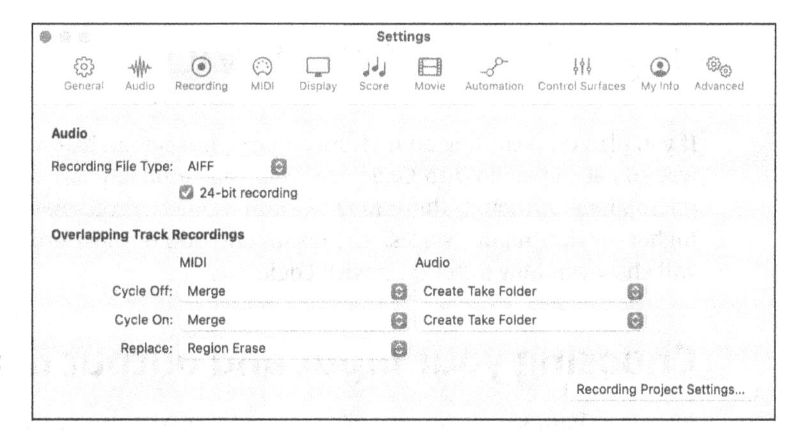

FIGURE 5-3: The Recording tab of Audio Settings.

Choosing audio file types

When Logic Pro records audio, it creates an audio file on your hard drive. Different file types have different advantages. All recording file formats are lossless and have the same sound quality. The major difference in the recording file type is the recording duration.

Logic Pro can record audio in the following file types:

- **AIFF:** 4 GB size limit. AIFF has a stereo-file time limit between 20 minutes and 3 hours and 15 minutes, depending on bit depth and sample rate. AIFF is the default Mac lossless audio file type. All Macs and most other operating systems can read AIFF files.
- **WAVE:** 4GB size limit. WAVE has a stereo-file time limit between 40 minutes and 13 hours and 30 minutes, depending on bit depth and sample rate. This file type is the default audio file type in Windows operating systems. To be nice to a Windows user, it can't hurt to use the WAVE file type. WAVE also has a longer recording duration, which gives it an advantage over AIFF.
- **CAF:** No size limit. CAF has a time limit of hundreds of years, if you can call that a limitation. Apple loops and sound effects installed with Logic Pro use the CAF file type. Although CAF files have certain advantages, they can't be played on all media players as AIFF and WAV files can.

To choose the recording file type for your audio recordings, follow these steps:

1. **Choose Logic Pro ⇨ Settings ⇨ Recording.**
2. **On the Recording File Type drop-down menu, make a selection.**

Connecting Your Audio Devices

If you plan on recording audio from a microphone or an instrument, you'll need a way to get the audio into Logic Pro. Your Mac probably has a built-in line-in or microphone. Although these may work in a pinch, professional recordings need higher-quality input devices. The documentation of most professional hardware will show you how to set it up with Logic Pro.

Choosing your input and output device

After your hardware is set up and connected to your computer, you must select the hardware in Logic Pro Settings. To choose your audio input and output devices:

1. **Choose Logic Pro ⇨ Settings ⇨ Audio.**
2. **Click the Devices tab.**
3. **On the Input Device and Output Device drop-down menus (see Figure 5-4), make your selections.**

TIP

You can choose separate input and output devices. You may want to record a guitar by using a mobile guitar interface and monitor it through the computer speakers or headphones. In this case, you select the audio interface as the input and select Built-in Output as the output.

4. **Click the Apply Changes button.**

Logic Pro will now begin using your selected input and output device for audio.

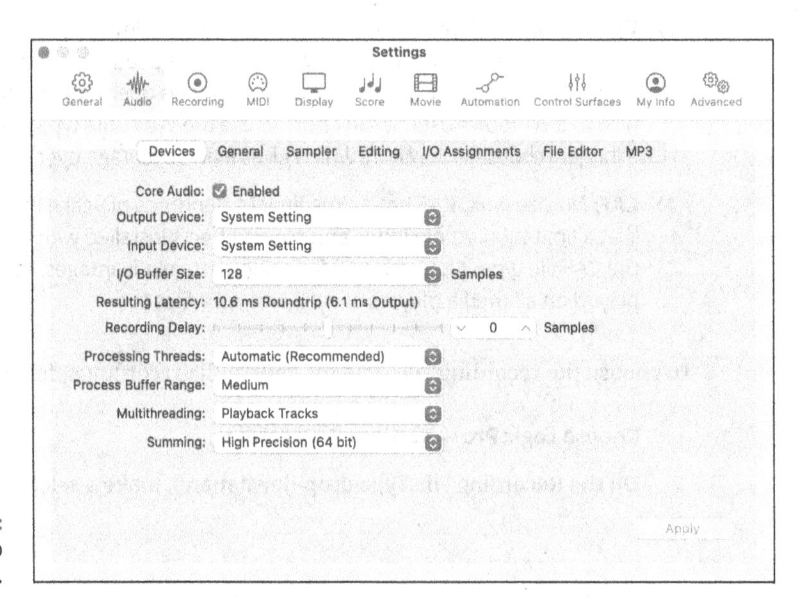

FIGURE 5-4: The Devices tab of Audio Settings.

Setting your I/O buffer size and reducing latency

Latency refers to the delay between your audio input and audio output. This delay is unavoidable in software because your sound source must be converted to digital audio and processed before it can be heard. In many cases, latency is negligible and won't be noticeable. However, latency can become an issue when a track has many plug-ins requiring significant processing.

TIP

It's a good idea to record most of your tracks before you begin adding lots of plug-ins. The more plug-ins you add, the more processing power required and the more latency you can introduce into the system. If you record without many plug-ins, latency won't be an issue.

When you record, latency can be a hindrance if you can't hear what you're playing exactly when you're playing it. Fortunately, you can adjust the latency as follows:

1. **Choose Logic Pro ⇨ Settings ⇨ Audio.**
2. **Click the Devices tab.**
3. **On the I/O Buffer Size drop-down menu (refer to Figure 5-4), make your selections.**

   Note that a smaller buffer size will reduce latency.

   REMEMBER

   A smaller buffer size requires more processing power. If you set the buffer size too low, you can introduce clicks and pops in your audio. Experiment with the buffer size until you find a setting that minimizes latency but doesn't introduce unwanted audio artifacts.
4. **Click the Apply Changes button.**

   Logic Pro will begin using your selected I/O buffer size.

Monitoring signals through your hardware or software

Some audio interfaces have built-in monitoring capabilities. This means you can listen to the source material you're recording through your hardware instead of the Logic Pro software. Monitoring through your hardware is helpful if the software introduces too much latency. Keep in mind that monitoring through hardware means you're hearing audio before it reaches Logic Pro, so you won't hear any plug-ins you've added to the track in your project.

By default, Logic Pro monitors audio through the software. To turn software monitoring off, follow these steps:

1. **Choose Logic Pro ⇨ Settings ⇨ Audio.**
2. **Click the General tab.**
3. **Deselect the Software Monitoring check box (see Figure 5-5).**

   If no hardware is connected, the check box will appear dimmed and will be unavailable.

   REMEMBER

   The Software Monitoring check box is available only if Enable Complete Features is selected in the Advanced pane of Logic Pro Settings, as discussed in Chapter 1.

4. **Click the Apply Changes button.**

   Audio will no longer be monitored through Logic Pro.

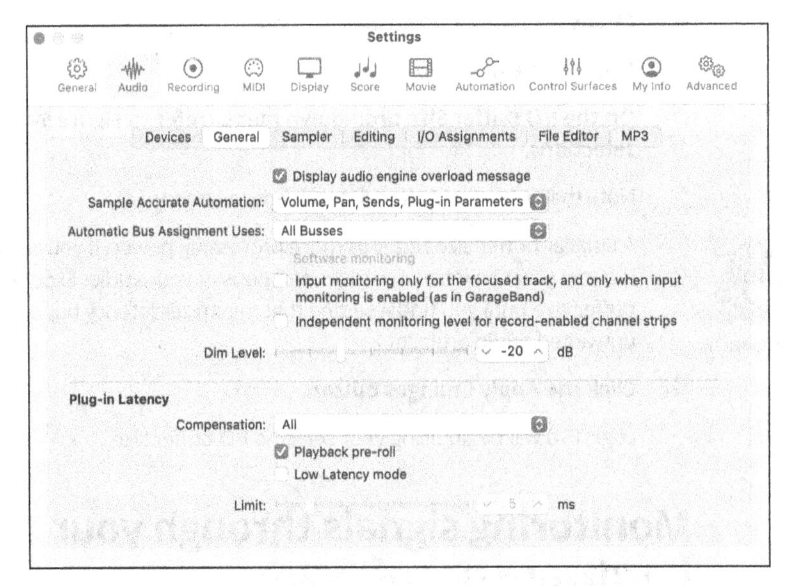

FIGURE 5-5: The General tab of Audio Settings.

Exploring audio settings

You've seen Logic Pro audio settings throughout this chapter. As you've discovered, you can customize the software to your audio needs in many ways. Next, I briefly cover some global audio settings that will be useful as you record, edit, and mix your audio in Logic Pro.

Open the Audio Settings window by choosing Logic Pro ➪ Settings ➪ Audio. Here's a brief description of the audio settings:

- **I/O Assignments tab:** The I/O Assignments tab has three tabs: Output, Bounce Extensions, and Input, as shown in Figure 5-6. The Output tab lets you send your audio to a different output pair if your hardware supports it. Select the Mirroring check box to send your output through stereo out in addition to a second output.
- **File Editor tab:** This tab, shown in Figure 5-7, lets you choose the settings for the audio file editor. (For details on this editor, see Chapter 14.) The audio file editor has its own undo history with a definable number of steps. You can also choose an external audio editor if you have one that you prefer.
- **MP3 tab:** The MP3 tab, shown in Figure 5-8, enables you to set the default bit rate of mono and stereo MP3 files. As discussed previously in this chapter, a higher bit rate will result in higher quality audio.

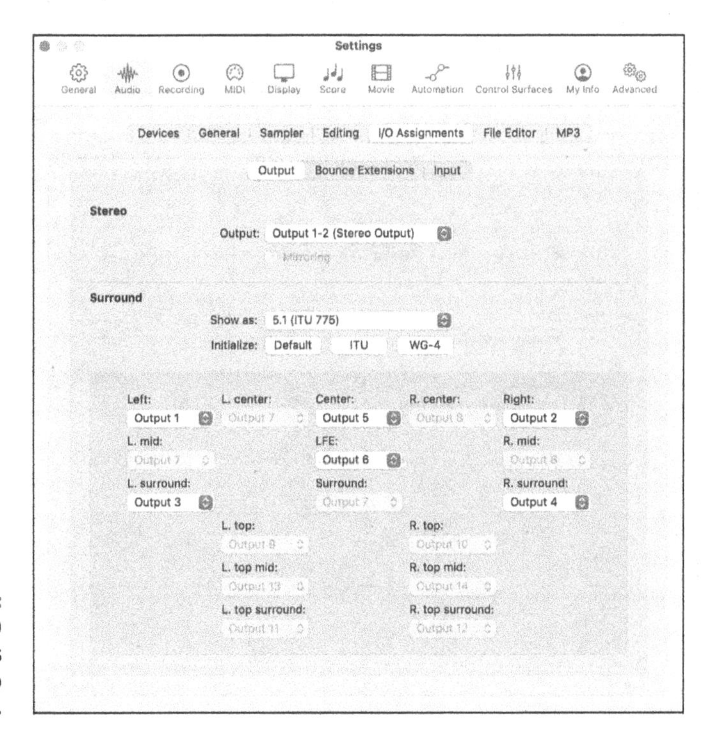

FIGURE 5-6: The I/O Assignments tab in Audio Settings.

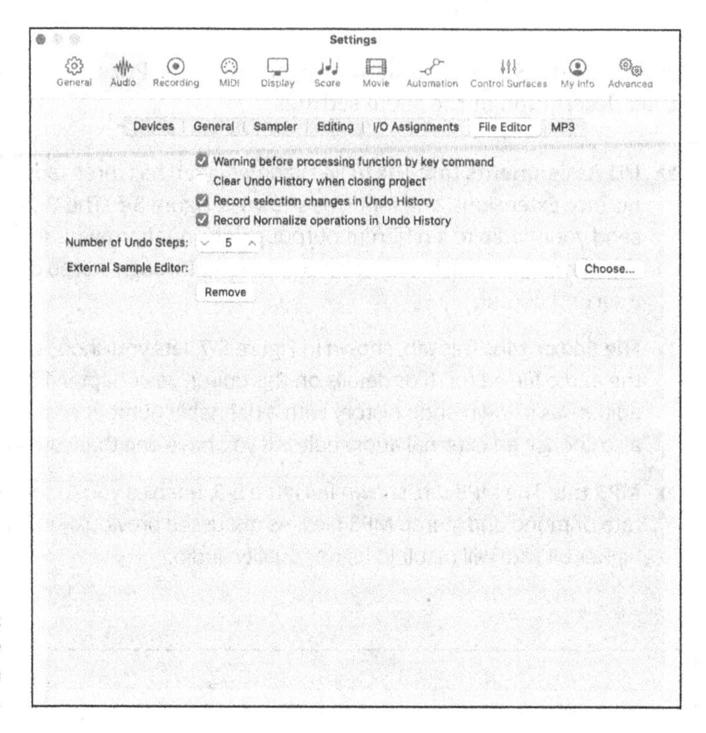

FIGURE 5-7: The Audio File Editor tab in Audio Settings.

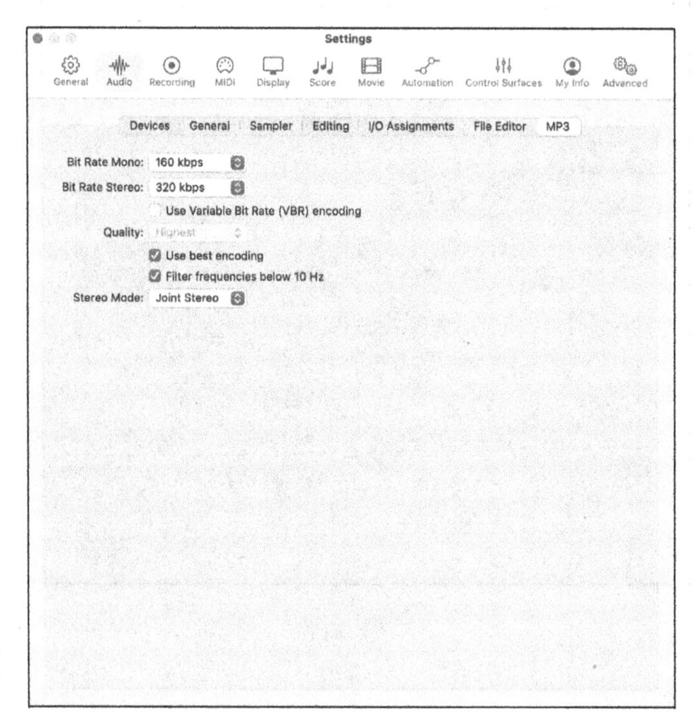

FIGURE 5-8: The MP3 tab in Audio Settings.

REMEMBER

In addition to global audio settings, you can make changes to the project's audio settings by choosing File ⇨ Project Settings ⇨ Audio (refer to Figure 5-2). These settings usually work just fine in their default state. The only setting you'll want to adjust when you begin your project is the sample rate, which you learned how to do previously in this chapter.

Understanding MIDI

The *MIDI* (Musical Instrument Digital Interface) protocol enables compatible devices to send and receive data. It was created to allow hardware to speak to each other. Logic Pro uses the MIDI protocol to play and record software instruments, automation, and external MIDI hardware. In this next section, you learn how to use MIDI and connect your MIDI devices to Logic Pro.

REMEMBER

MIDI does not contain audio. Your external MIDI devices still need their audio routed into your computer and Logic Pro to be heard. Follow the instructions for connecting your audio devices as described previously in the "Connecting Your Audio Devices" section.

Sending MIDI messages

MIDI messages are sent from one device and received by another, such as from a MIDI controller to Logic Pro. These messages can include information about pitch, velocity, sustain, and much more. Logic Pro translates the MIDI messages and sends them to a software instrument or external hardware. (You create software instrument tracks and external MIDI tracks in Chapter 4.)

You'll use MIDI to accomplish many of your musical goals in Logic Pro. As described in Chapter 3, you can even assign your MIDI controller to a Logic Pro command. It's not difficult to map a button on your MIDI keyboard to tell Logic Pro to play your project. But while Logic Pro is always using MIDI under the hood, you'll mainly be using MIDI to play software instruments.

To verify that Logic Pro is receiving the messages your MIDI controller is sending, show the MIDI activity in the LCD on your control bar as follows:

1. **Control-click an empty area on your control bar and choose Customize Control Bar and Display.**

   A window opens with customization options.

2. **In the LCD column, choose Custom.**
3. **Select the MIDI Activity (In/Out) check box.**
4. **Click the OK button.**

   Your LCD shows MIDI activity. If no MIDI is being received, the display will read No In/No Out.

Changing MIDI channels

MIDI devices send and receive information on 16 different channels. Many Logic Pro software instruments are *multi-timbral*, which means they can play multiple different sounds simultaneously. You can set these sounds to different MIDI channels, which allows you to play them independently using multiple MIDI controllers. Or, if you have a MIDI controller with a split or layer function, you can potentially use a single keyboard to play multiple software instruments simultaneously. You find out how to set MIDI channels for MIDI recording in Chapter 7.

Connecting Your MIDI Devices

MIDI devices can include keyboards, drum pads, alternative controllers such as guitar MIDI systems, and more. Although you don't need a MIDI controller to make music with Logic Pro, it's a lot more fun if you have a way to play software instruments. And trained players can use their skills to input music more quickly and accurately.

Connecting MIDI controllers

Many MIDI controllers will send and receive MIDI through USB, eliminating the need for a third-party interface. Older MIDI controllers may use 5-pin DIN cables and require a third-party hardware interface. Connect your controller to your MIDI interface or your computer's USB port.

Connecting external instruments

If you're connecting a MIDI synthesizer or workstation to Logic Pro, in addition to making MIDI connections, you must be able to monitor the instrument's audio. If your audio interface supports hardware monitoring, you can monitor the instrument through your hardware. If it doesn't, you'll need to add an audio track to your project to monitor the instrument. For more on monitoring audio sources, see Chapter 6.

When you play the keyboard, the instrument sends and receives MIDI simultaneously, so sounds are doubled. Most keyboards have a Local Off function, which you'll want to use to stop the sound from doubling.

Exploring MIDI settings

Global and project settings are similar to audio settings. To open the Global MIDI Settings pane, choose Logic Pro ⇨ Settings ⇨ MIDI. Here's a brief description of four tabs in the MIDI Settings pane:

- **General:** Click the MIDI 2.0 check box if your MIDI devices support the MIDI 2.0 protocol. If you have MIDI communication problems, click the Reset All MIDI Drivers button.
- **Reset Messages:** If your MIDI controllers get stuck, including stuck notes, select the controller you want to reset on this tab.
- **Sync:** If you want to sync Logic Pro with another device or adjust the timing of your MIDI, you can do so on this tab.
- **Inputs:** Use the check boxes to turn the input ports on your hardware devices on or off.

To open the MIDI Project Settings pane, choose File ⇨ Project Settings ⇨ MIDI. Here's a brief description of the four tabs on the MIDI Project Settings pane:

- **General:** On this tab, you can choose to send MIDI settings and fader values after loading your project.
- **Input Filter:** If you're not using certain MIDI functions or don't want these functions to be sent or received, you can select them on this tab, and they will be filtered out.
- **Chase:** Use this tab to set how MIDI events will behave when you start your project after a MIDI event has been triggered. For example, if you play your project in the middle of a pitch bend, selecting the Pitch Bend check box will ensure that the bend is synchronized correctly.
- **Clip Length:** If your project needs to reset specific MIDI messages to zero at the end of MIDI regions, you can do that here.

TIP

You can use the Audio MIDI Setup utility to set up devices connected to your computer. (Audio MIDI Setup is in the Utilities folder of your Applications folder.) You can test MIDI and audio devices, set audio levels, and more.

Now that you know the fundamentals of digital audio and MIDI, you're well on your way to recording your own audio and MIDI. With your hardware set up and your audio and MIDI settings chosen, you're ready to begin recording some great sounds.

IN THIS CHAPTER

» Setting up Logic Pro for recording audio

» Recording audio tracks and multiple takes

» Overdubbing and punch recording audio

» Setting up separate monitor mixes

Chapter 6 Recording Audio

Audio recording was introduced to the original Notator Logic in 1994, about 20 years before the introduction of Logic Pro. "The Sign" by Ace of Base was the number-one song that year. While I can't confirm that the song was referring to the emergence of computer audio recording, I can confirm that it was a breakthrough year for Logic.

Fast forward more than 20 years, and you have a powerful and affordable digital audio workstation that scores of major artists use to create chart-topping hits. In this chapter, you discover how to record audio, build the perfect track from multiple recordings, and much more.

Preparing to Record Audio

In Chapter 5, you connect your audio hardware, set the project sample rate, and select the recording file type. You should confirm that the incoming signal from your audio source (microphone or instrument) is being received by your audio interface by checking the levels of your hardware inputs. (Check your audio interface documentation for details.)

Before you can begin recording, you must create an audio track by choosing Track ⇨ New Audio Track. The new audio track is added to the track list and selected automatically.

TIP

Name your new track something descriptive because the audio files generated from recording will use the track name in the file name. To name your track, double-click the track header or press Shift-Enter and type your track name.

On your new track, select the correct input as follows:

1. **Select the track.**
2. **Open the track inspector by pressing I or choosing View ⇨ Show Inspector.**
3. **Click the input/instrument slot, as shown in Figure 6-1, and choose the correct input.**
4. **Click the input format icon (labeled in Figure 6-1) to toggle between stereo and mono input.**

Testing your recording levels

To test the volume level at which you'll record, enable the track for recording by clicking the record enable icon on the track header (refer to Figure 6-1) or by pressing Control-R. The record enable icon will blink red to let you know that the track is enabled for recording. Play your instrument or speak into your microphone to test the recording level. If your signal is too high or low, adjust the instrument volume or the input level on your audio interface.

WARNING

Don't clip the audio signal! When a signal is too loud and exceeds the limit that digital audio can reproduce, the signal is said to be *clipping*. You can see when your signal is clipping by the peak level display at the top of your track's level meter, as shown in Figure 6-2. When the number above the level meter is positive, the peak level display will become red, indicating that the track is clipping. During the recording phase, the best way to remedy signal clipping is to lower the volume on your instrument or audio interface. (Be sure to check for clipping on your audio interface as well.) Conversely, don't record signals that are too quiet — when you raise their level, you can introduce noise into the mix.

REMEMBER

You must have Enable Complete Features selected in the Logic Pro Advanced Settings pane. Choose Logic Pro ⇨ Settings ⇨ Advanced and select Enable Complete Features.

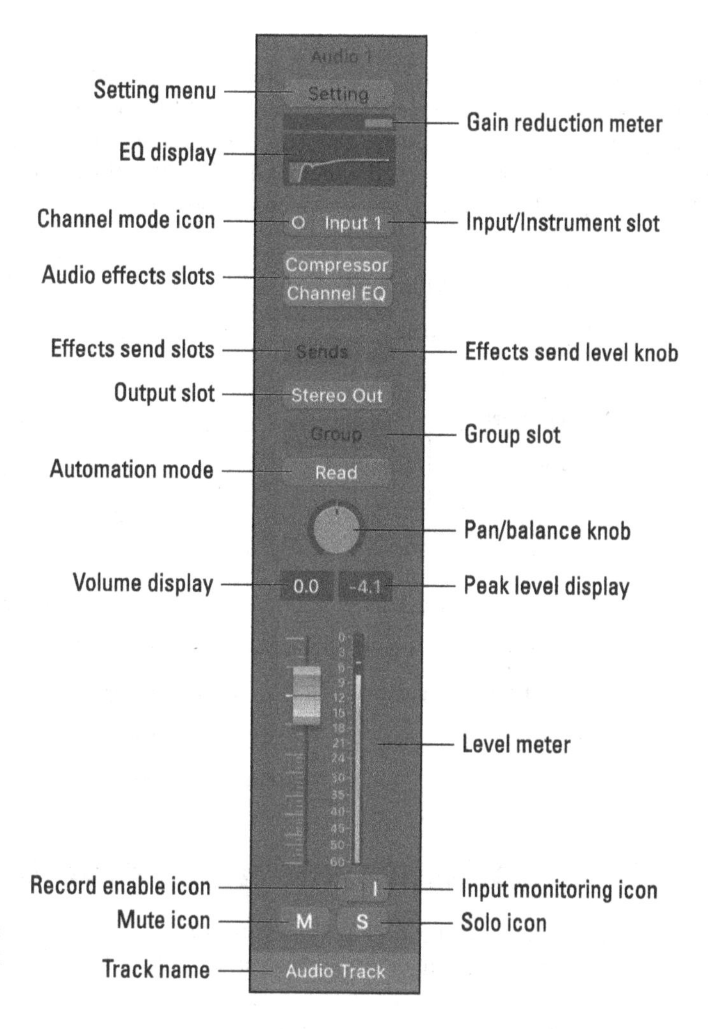

FIGURE 6-1: The Audio Track channel strip.

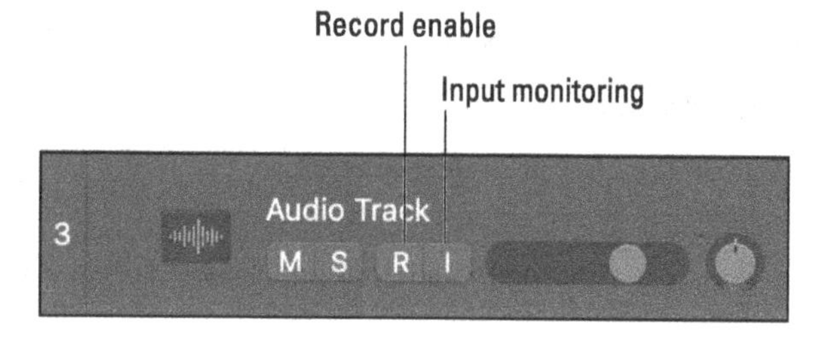

FIGURE 6-2: The track header.

Pre-fader metering is used to test recording levels, which means the level meter shows you the signal level before the fader. You can turn on pre-fader metering by customizing your control bar, as described in Chapter 3, and selecting Pre Fader Metering in the Modes and Functions column. An icon is added to your control bar, as shown in Figure 6-3, to allow you to toggle pre-fader metering. You want your meters to show you the signal pre-fader so that you can be aware of what's being recorded, even if you lower the fader to blend better with the other tracks. If you use post-fader metering, the meter will show you the level of the track after it has been raised or lowered, and you won't know whether it's clipping.

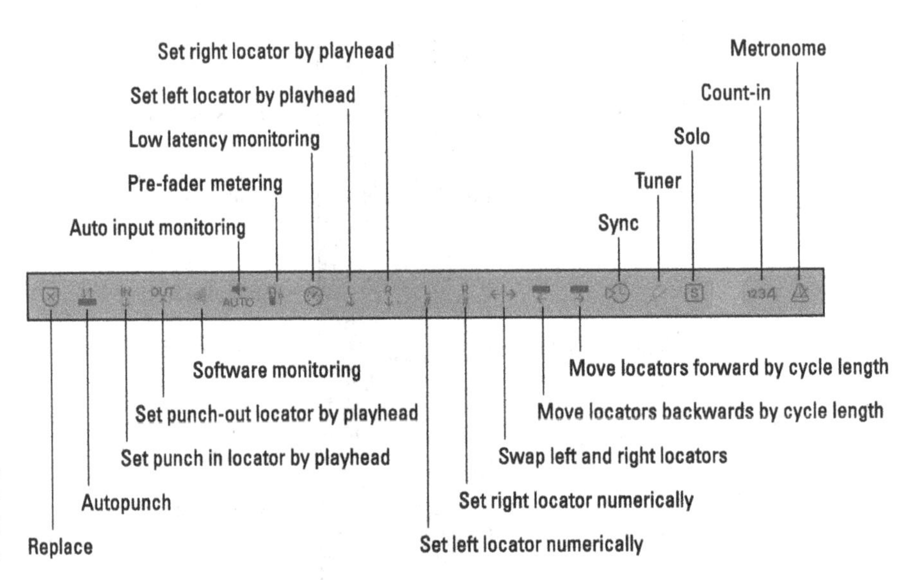

FIGURE 6-3: The control bar modes and functions.

Enabling software and input monitoring

If you're not going to monitor your signal through your hardware, you'll need to turn on software monitoring. Follow these steps:

1. **Choose Logic Pro ➪ Settings ➪ Audio.**

   The Audio Settings window opens.

2. **Click the General tab.**
3. **Select the Software Monitoring check box.**

   You can now use the Logic Pro software to monitor your audio. If you plan on monitoring the signal through your hardware, deselect Software Monitoring.

When a track is record enabled, you'll be able to hear it. If a track isn't record enabled, you won't hear it unless you've turned on input monitoring.

If you're monitoring from hardware and not software, you should turn off input monitoring. Turn it off and on by clicking the input-monitoring icon on the track header (refer to Figure 6-2) or on the channel strip.

Setting up the metronome

By default, Logic Pro will play a metronome as you record. To turn off the metronome while recording, choose File ➪ Project Settings ➪ Metronome and select the Click While Recording check box, as shown in Figure 6-4. If you want to hear the metronome while you play the track, press K to toggle the metronome on and off.

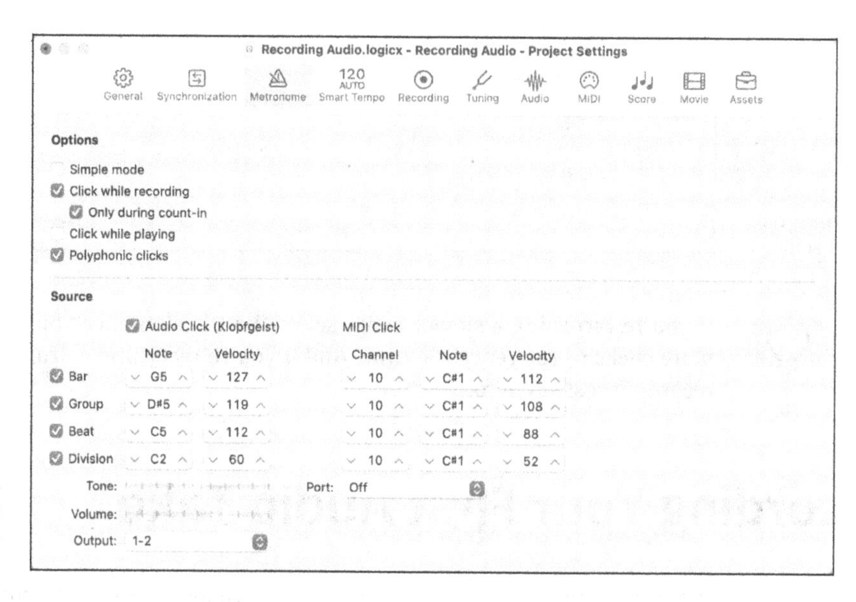

FIGURE 6-4: The metronome Project Settings.

The audio click you hear in the metronome is generated by the Klopfgeist software instrument. You can play it as an instrument (though I don't know why you'd want to). You can also set up an external sound source as your metronome. In Figure 6-4, MIDI channel 10 is set to transmit a MIDI click. MIDI channel 10 is often reserved for drum sounds, so if you have a multi-timbral keyboard workstation, you can set a drum sound to MIDI channel 10 to receive the MIDI click.

If you have a drumbeat that grooves a little differently than the metronome, you may want to hear the click only during the count-in and not while recording. To do so, select the Only During Count-In check box.

TIP

Be careful of metronome blindness, a condition where you don't even realize that the metronome is playing because your brain has tuned it out. It's happened to me and others. The click can be heard over the entire mix, but you forget it's on as you play the track.

If you want to adjust the number of bars that the metronome counts in before recording, choose File ➪ Project Settings ➪ Recording and select the number of bars or beats you want on the drop-down menu, as shown in Figure 6-5.

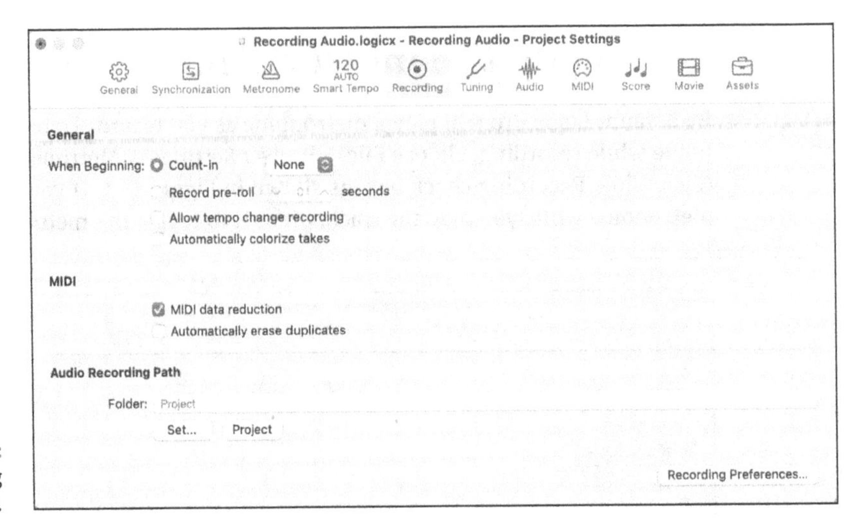

FIGURE 6-5: The recording Project Settings.

TIP

If you're recording a slower song, select the Division check box to give yourself more clicks to reference the time. And if you're recording a drummer, you'll be a recording session hero.

Recording Your First Audio Take

In the recording world, a single recording is called a *take*. Recording a good take is like capturing something special and elusive. Logic Pro helps you capture the moment quickly and easily.

With your track selected and record enabled, you can begin recording as follows:

1. **Place the playhead where you want to begin recording.**
2. **Press R or click the record icon in the control bar transport, as shown in Figure 6-6.**
3. **Wait for the count-in and then start playing.**
4. **When you're finished, click the stop icon on the transport or press the spacebar.**

   A new audio region will fill the area where you began and stopped recording, as shown in Figure 6-7.

To play back what you just recorded, place the playhead at the beginning of your newly recorded region and then click the play icon on the transport or press the spacebar. When you're finished listening to your new recording, press the spacebar again, and the project will stop playing.

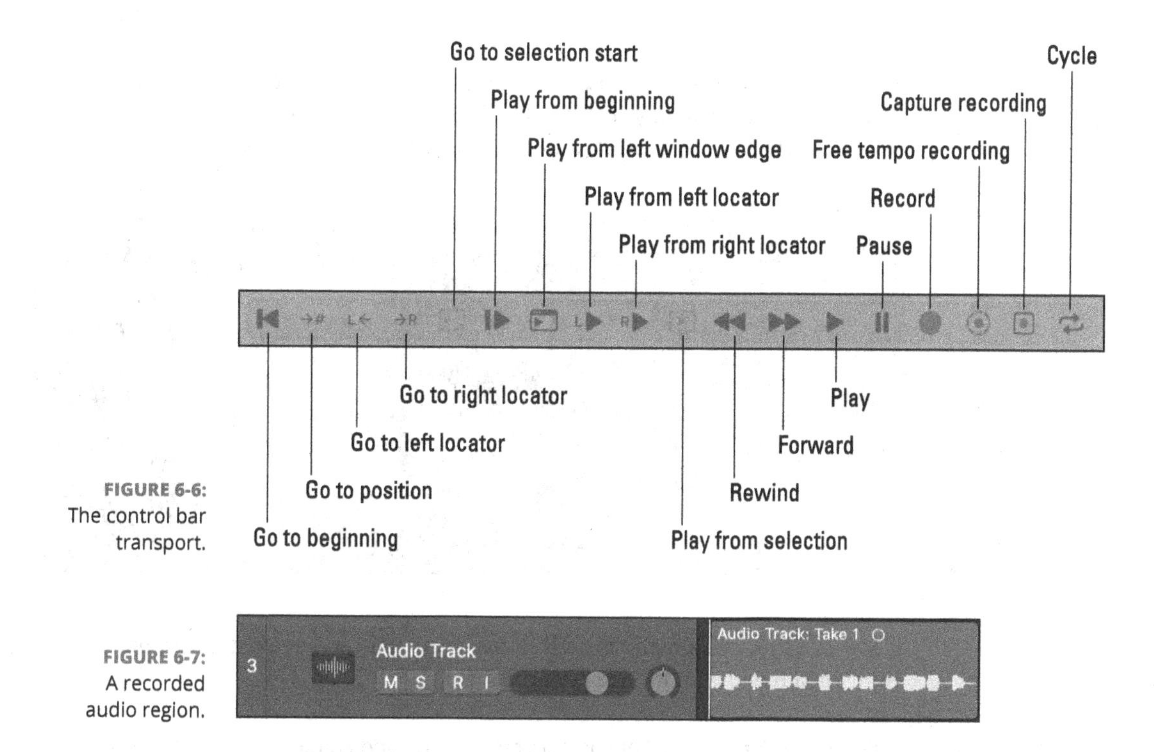

FIGURE 6-6: The control bar transport.

FIGURE 6-7: A recorded audio region.

Recording Multiple Takes in Cycle Mode

You can record additional recordings, or takes, on a track that already contains audio regions. A take folder is created to contain the original and new takes. In cycle mode, new lanes are created each time the cycle passes.

When cycle mode is enabled, playback or recording begins at the left locator and repeats when it reaches the right locator. To set up cycle mode, set the left and right locators by dragging from left to right in the upper half of the ruler in the tracks area. The cycle area will be displayed as a yellow strip in the upper half of the ruler, as shown in Figure 6-8. You can turn cycle mode on and off by pressing C or clicking the cycle icon in the control bar (refer to Figure 6-6).

FIGURE 6-8: Cycle mode.

To record in cycle mode, turn on cycle mode and begin recording as you did previously. After the second take is recorded, a take folder is created in the cycle area, as shown in Figure 6-9, and new lanes are added with each pass through the cycle. This is a great way to get several takes that you can edit into a perfect (or close to perfect) take. You learn how to edit your takes in Chapter 14.

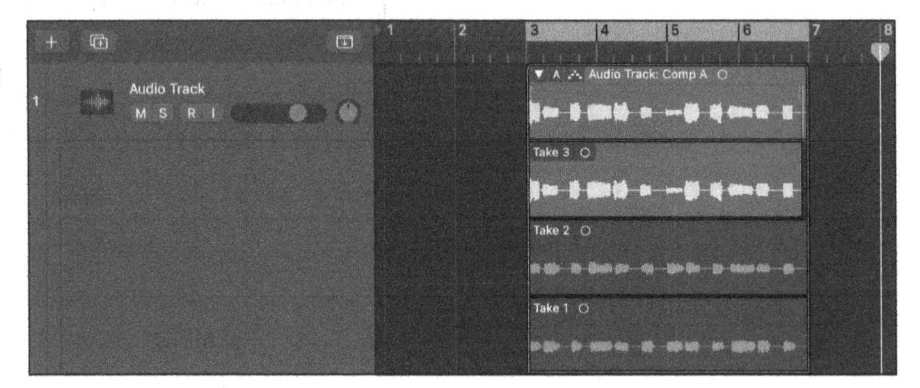

FIGURE 6-9: An audio take folder.

A key to using cycle mode is setting locators. You can set locators in many ways. Here are a few useful locator commands:

- Choose Navigate ⇨ Set Locators by Selection and Enable Cycle (⌘-U) to set the locators and enable cycle mode according to the selected regions.
- Choose Navigate ⇨ Auto Set Locators ⇨ Enable Auto Set Locators to follow your region or marquee tool selections automatically.
- Set the locators manually in the control bar LCD if you have Locators (Left/Right) selected in the Customize Control Bar and Display options (see Chapter 3 for details).

TIP

Open the Key Commands window (Option-K) and type *locators* in the search bar. You'll find dozens of key commands. After you browse the available locator key commands, you'll have an idea of what you can do and how important locators are to a speedy workflow.

Recording Multiple Inputs

You aren't limited to recording one track at a time. You can record multiple tracks by record-enabling several tracks at once and following the same steps as recording a single take. You can also create multiple takes on multiple tracks at once. Say that three times fast.

To record multiple tracks simultaneously, set each track to the correct input, as you did previously. When all track inputs are set correctly, you can begin recording. You can even record multiple takes on multiple inputs at the same time. The number of tracks you can record simultaneously depends on your hardware and computer power.

Punching In and Punching Out

Replacing just a portion of a track is called *punch recording.* You play the track, punch in and record the new part, and then punch out when you're done. It's like punching the work clock for a bar or two. Play, punch in, punch out, and play again.

The first way to set up punch recording is to do it on the fly, as the track is playing, as follows:

1. **Choose Record ➪ Allow Quick Punch-In.**

   This is the default setting.

2. **Choose Record ➪ Record Button Options ➪ Record/Record Toggle.**

   The settings of the record icon and the record key command are updated so that clicking the record icon or using the key command toggles the record state on or off while continuing to play.

3. **Press the spacebar to play your project.**
4. **At the point where you want to begin recording (punch in), press R.**
5. **At the point where you want to stop recording (punch out), press R again.**

   Your project continues playing but no longer records.

While your project is playing, you won't be able to hear the input of the selected track until you begin recording. This setting is the default and is helpful if you need to listen to the recorded track to time the punch in. After you begin recording, you'll be able to hear the input. But if you want to listen to the input during playback as well as during recording, turn off auto input monitoring on the Record menu.

TIP

If you want to listen to the input of tracks that are not record-enabled, click the track's input-monitoring icon (refer to Figure 6-2).

When you need more precision in your punch recording, you can preprogram the punch in and out points. This technique is called *auto-punch recording.* To begin auto-punch recording, follow these steps:

1. **Click the auto-punch icon in the control bar (refer to Figure 6-3).**

   If you don't see the icon, customize the control bar. A second ruler is shown at the top of the tracks area, as shown in Figure 6-10.

2. **Set the auto-punch locators by dragging from left to right in the auto-punch area of the ruler.**

   The auto-punch area will be displayed with a red stripe (refer to Figure 6-10).

3. **Start recording before the punch-in point.**

   Recording begins automatically when the playhead reaches the left auto-punch locator and ends when it reaches the right auto-punch locator.

4. **Stop recording after the punch-in point.**

   A take folder is created, and a new lane is added that includes your auto-punch recording.

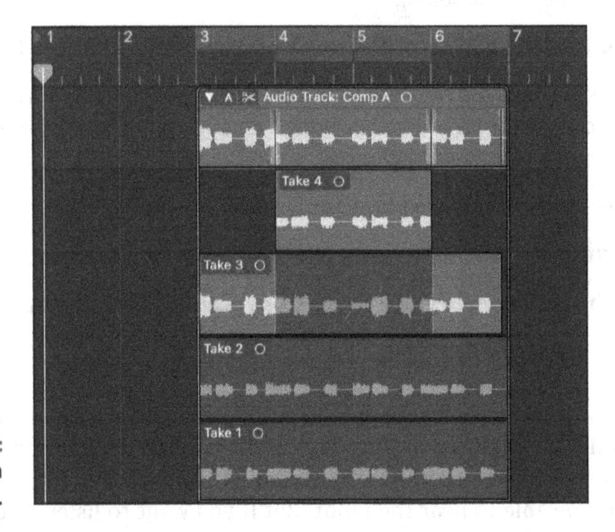

FIGURE 6-10: Auto-punch mode.

Recording with Smart Tempo

If you want to record freely, without a metronome, but have Logic Pro automatically create a tempo track based on the rhythm of your new recording, record with smart tempo. Smart tempo recording is available for only single audio tracks, not multitrack recording, and the timeline of your new recording must not have existing regions.

To record with smart tempo, follow these steps:

1. **In the LCD on the control bar, choose Adapt from the project tempo pop-up menu below the tempo display.**

   The tempo track opens in the global tracks. (Global tracks are covered in detail in Chapter 4.) The tempo and time signature displays turn orange on the LCD.

2. **Begin recording, as you learned previously in this chapter.**
3. **Stop recording when you're finished.**

   Logic Pro analyzes the tempo and adjusts the tempo track to the new recording, as shown in Figure 6-11. A dialog opens asking if you want to open the file tempo editor to adjust the tempo automatically embedded in the audio file. You learn how to use the file tempo editor in Chapter 14.

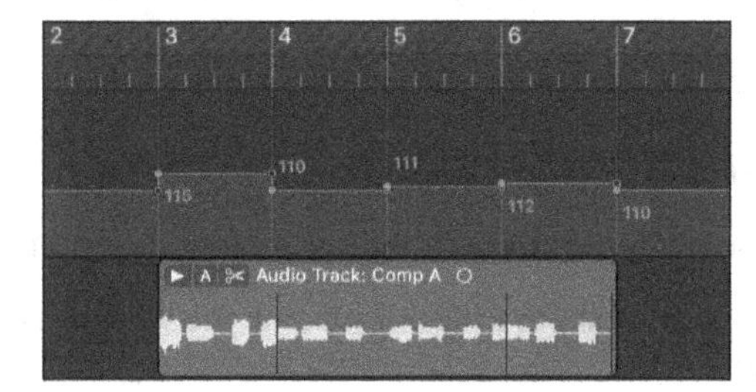

FIGURE 6-11: A region recorded with smart tempo.

Another way to record audio freely, without a metronome or predefined tempo, is to use free tempo recording. After you complete a free tempo recording, you can choose to apply the audio region's tempo to the entire project, apply the project tempo to the recorded audio region, or skip analyzing the tempo or changing the project tempo.

To use free tempo recording, follow these steps:

1. **Click the Free Tempo Recording button in the control bar transport.**

   If the button isn't visible, add it to your transport by right-clicking the control bar and choosing Customize Control Bar and Display. You can also hold down the Record button and choose Free Tempo Recording from the pop-up menu.

   Recording starts immediately and the track is soloed.

2. **Begin recording freely.**

3. **Stop recording when you're finished.**

   A dialog opens asking if you want to apply the region tempo to the project, apply the average region tempo to the project, or apply the project tempo to the region. You can also choose not to analyze the region or change the project tempo.

Setting Up Multiple Monitor Mixes

In the recording world, the *monitor mix* is what the mixing engineer hears from the speakers. It's possible to create multiple monitor mixes if your hardware supports multiple outputs. Small nearfield monitors are the standard choice for getting good sound. Sometimes, a large set of speakers is used to get a full giant sound that impresses the clients and, well, everyone else. A third set of small speakers or even a single mono speaker may be used to imitate cheaper sound systems, though you can use your built-in computer speakers for this job.

To set up additional monitor mixes, follow these steps:

1. **Choose Logic Pro ⇨ Settings ⇨ Audio.**
2. **Click the I/O Assignments tab and then click the Output tab.**
3. **On the Output menu, choose the output pair you want to set up, as shown in Figure 6-12.**

   The outputs you have available depend on your hardware.
4. **Select Mirroring to mirror the stereo output to the selected output.**

   Your mix will now be played through the additional output pairs.

TIP

If you're recording by yourself and using a microphone, you need headphones so that any audio played back won't leak into your recording. But as you add other people to the recording process, you might need to set up different headphone submixes for each person you record. Giving a performer a separate headphone mix, known as a *cue mix*, can improve the precision and emotion of a performance. Vocalists need to hear different things than drummers need to hear. And if you want to get the best performance out of a performer, you need their recording process to be smooth and productive. That's why you need to give performers their own mix the way they want it.

REMEMBER

Your hardware must have multiple outputs to route the submix to the correct output.

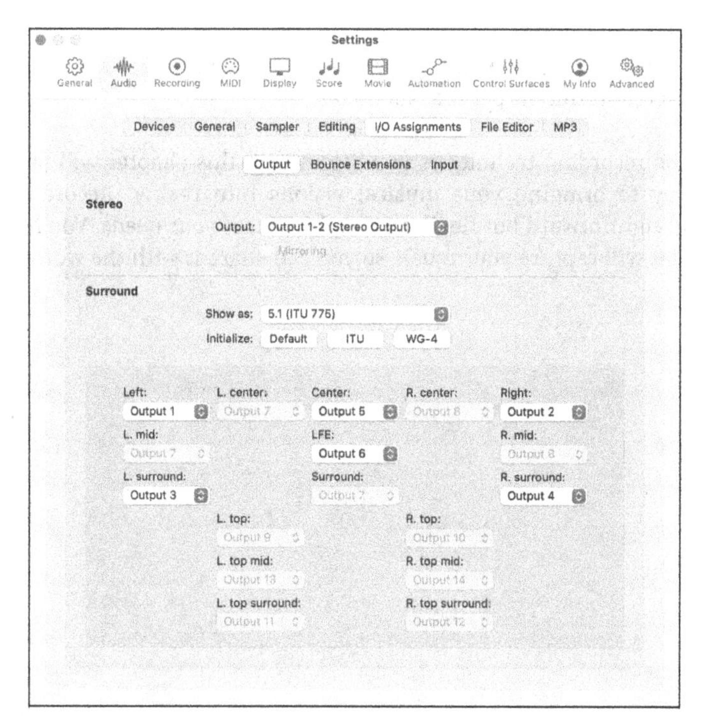

FIGURE 6-12: The I/O Assignments Output tab.

Setting up different submixes involves splitting the audio signal of a track and routing the parallel signal to an auxiliary track via the Send slot. (Refer to Figure 6-1 for the location of the Send slot.) To set up a submix:

1. **Click the Send slot on each track you want to send to the submix.**
2. **On the Bus drop-down menu, choose an unused bus.**

   An auxiliary track is automatically added to the mixer. A bus is used to route audio to auxiliary tracks. To open the mixer and view the auxiliary track, press X.
3. **Click the Send slot again, and choose Pre Fader.**

   Setting the send to pre-fader allows you to adjust the track volume in the monitor mix using the track fader without changing the volume of the signal sent to the auxiliary track. If the send is set to post-fader, any adjustments you make to the track's volume are also sent to the submix.
4. **Adjust the volume of the track in the submix by using the send level knob.**
5. **Click the Output slot on the auxiliary track, and set it to the correct hardware output.**

   You can adjust the overall level of the submix by using the auxiliary track's volume fader in the mixer.

Eight sends are available per channel strip so that you can send the track's signal to eight different parallel submixes.

The recording techniques you discover in this chapter will put you well on your way to bringing your musical visions into reality. Recording in Logic Pro is straightforward but flexible enough to meet your needs. You have a powerful tool that will capture your music so you can share it with the world.

IN THIS CHAPTER

» Setting up Logic Pro to record MIDI

» Recording external instrument and software instrument tracks

» Using musical typing and onscreen keyboards to record MIDI

» Overdubbing and recording multiple MIDI tracks

Chapter 7
Recording MIDI

Logic Pro was created as a powerful MIDI sequencer. MIDI is the ultimate flexible recording medium, enabling you to change what you record until it's perfect. It does have limitations, but it's unlikely that you'll notice them.

In this chapter, you find out how to record MIDI, filter MIDI events, multitrack MIDI recording, use musical typing, and much more.

Preparing to Record MIDI

MIDI is just data. MIDI is similar to sheet music in that it tells the performer precisely what to play and how to play it. Audio is not included in a MIDI region or a MIDI message. What you hear when a MIDI track plays is either the software instrument connected to the track or the audio output from your external instruments. To begin recording MIDI, you'll need to create either an external MIDI track or a software instrument track. Each type of track is explained in this section.

REMEMBER

Enable Complete Features must be selected in the Advanced Settings pane. Choose Logic Pro ⇨ Settings ⇨ Advanced and then select Enable Complete Features.

Recording external MIDI instruments

If you have an external MIDI instrument, such as a keyboard workstation, you must route the audio from the instrument into an audio track in your project. To create an audio track, choose Track ➪ New Audio Track (Option-⌘-A). You should also turn on input monitoring, as explained in Chapter 6, so you can hear the input of your MIDI instrument.

After you've verified that the audio is coming out of your instrument and into the audio track, you create and set up an external MIDI track as follows:

1. **Choose Track ➪ New External MIDI Track (or press Option-⌘-X).**

   A new external MIDI track is added to the track list and automatically selected.

2. **Open the inspector by pressing I or choosing View ➪ Show Inspector.**

   The inspector shows you an external MIDI track channel strip, as shown in Figure 7-1. Note that the controls are different from those available with other types of tracks, and you can't add effects or control the audio in your project. You can, however, control various MIDI functions on the external instrument.

FIGURE 7-1: External MIDI track channel strip.

3. **Display the track inspector by clicking the disclosure triangle above the channel strip, as shown in Figure 7-2.**

   The track inspector area opens, allowing you to adjust the external MIDI track settings.

4. **Select the correct MIDI port and channel in the track inspector.**

   If you have multiple external MIDI instruments connected to a third-party interface, you can choose different MIDI ports to control the instrument independently. Each MIDI port or MIDI instrument can transmit and receive up to 16 different channels. This allows you to control 16 different sounds on each MIDI instrument.

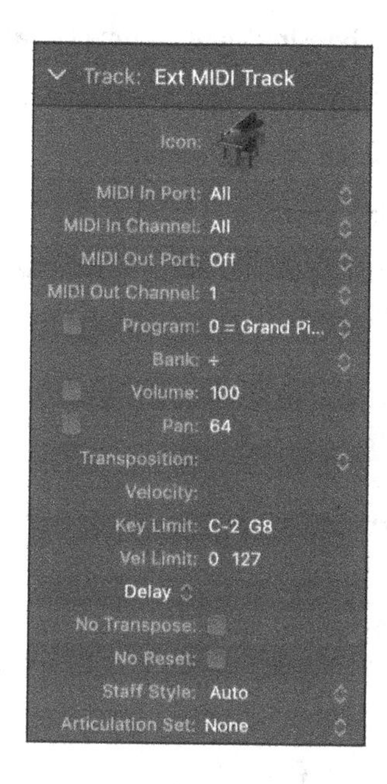

FIGURE 7-2: External MIDI track inspector.

REMEMBER

You need two tracks for external MIDI instruments: an audio track to monitor the instrument and an external MIDI track to record the MIDI. For this reason, I record external MIDI instruments to audio as soon as possible. For starters, leaving tracks as MIDI invites indecision. You can tweak until the end of time, but if you want to be productive, you need to make decisions quickly. Also, because external gear and multiple tracks only complicate your project, recording your external MIDI instruments as audio simplifies the project and guarantees that you'll have the performance captured if a piece of gear changes or breaks later.

TIP

Most keyboard workstations have a local off function. You should turn on Local Off on your external MIDI instrument; otherwise, you'll get doubled audio when both your hands and the MIDI data are playing the sounds on your keyboard.

Recording software instruments

To record a software instrument, you'll need to create a new software instrument track. Follow these steps:

1. **Choose Track ⇨ New Software Instrument Track (or press Option-⌘-S).**

   A new software instrument track is added to the track list and automatically selected.

2. **Open the inspector by pressing I or choosing View ⇨ Show Inspector.**

   The inspector displays a software instrument track channel strip, as shown in Figure 7-3.

3. **Open the library by pressing Y or choosing View ⇨ Show Library.**

   The library menu opens to the left of the inspector.

4. **On the library menu, select the patch you want to use.**

   You can audition patches by selecting a patch and playing your MIDI controller.

5. **Display the track inspector by clicking the disclosure triangle above the channel strip (refer to Figure 7-3).**

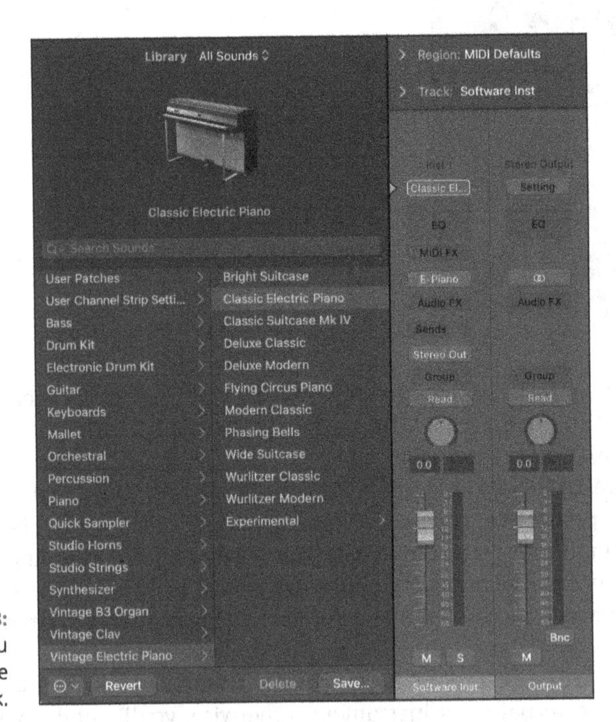

FIGURE 7-3: The library menu and the software instrument track.

The track inspector area opens, allowing you to adjust the software instrument track settings.

6. **Set the correct MIDI channel in the track inspector.**

   The default MIDI channel is set to All, which means your MIDI controller can be set to any MIDI channel and the software instrument track will receive the signal. If you want to set the track to receive from only a single MIDI channel, you need to set it to the correct MIDI channel in the track inspector. Using different MIDI channels is helpful if you have more than one MIDI controller and want them to control specific software instrument tracks.

Recording with musical typing

What if you don't have a keyboard controller handy? No problem. You can use musical typing to play your computer keyboard like a musical keyboard. However, your computer keyboard won't provide the same features as a dedicated MIDI controller, such as velocity sensitivity and aftertouch. To use musical typing, do the following:

1. **Choose Window ➪ Show Musical Typing or press ⌘ -K.**

   The Musical Typing keyboard appears, as shown in Figure 7-4.

2. **Play the notes on the keyboard by pressing the corresponding keys.**

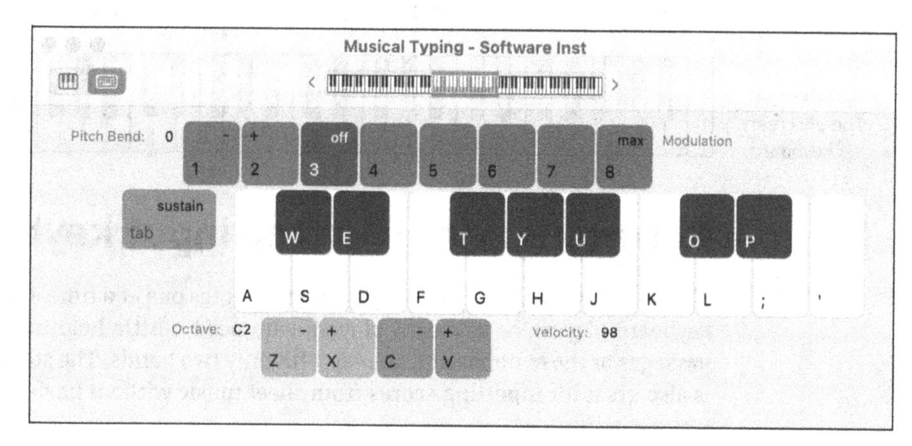

FIGURE 7-4: The musical typing keyboard.

You can do a lot with the musical typing keyboard. It's excellent for spur-of-the-moment recording when you don't have an external MIDI controller connected. Here's a description of the keys and what they do:

- » You have access to almost an octave and a half by using the middle and top rows of alphabet keys.
- » Sustain the notes you play by holding down the Tab key.
- » Pitch bend down and up by using the 1 and 2 keys.
- » Modulate the software instrument by pressing the 4 through 8 keys. Turn off modulation by pressing 3. The parameter that gets modulated will depend on the selected software instrument.
- » Shift the keyboard octave down or up by pressing Z or X, respectively. To change the octave, drag the blue area in the keyboard at the top of the window (as shown in Figure 7-5).
- » Adjust the keyboard's velocity down or up by pressing C or V, respectively.
- » Switch between the musical typing keyboard and the onscreen keyboard by clicking the keyboard icon in the upper-left corner of the interface.

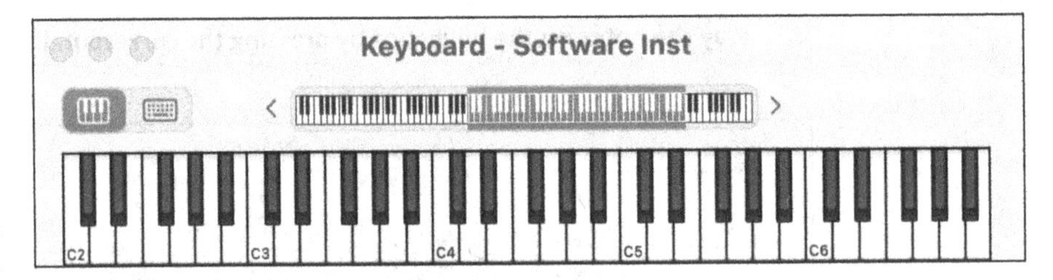

FIGURE 7-5: The onscreen keyboard.

Recording with the step input keyboard

The step input keyboard allows you to enter notes one at a time. It's great for non-keyboard players or keyboard players who need a little help inputting technical passages or those impossible to play with only two hands. The step input keyboard is also great for inputting scores from sheet music without having to learn how to play the part.

A MIDI region must be open in a MIDI editor to input notes with the step input keyboard. To use the step input keyboard, follow these steps:

1. **Double-click the MIDI region you want to edit.**

   The piano roll editor opens at the bottom of the tracks area.

2. **Choose Window ➪ Show Step Input Keyboard (or press Option-⌘-K).**

   The step input keyboard appears, as shown in Figure 7-6.

3. **Place the playhead where you want to begin inputting notes.**

4. **Click the note length and note velocity icons, and then click the key you want to input.**

   A MIDI note event is added to the piano roll editor at the playhead position.

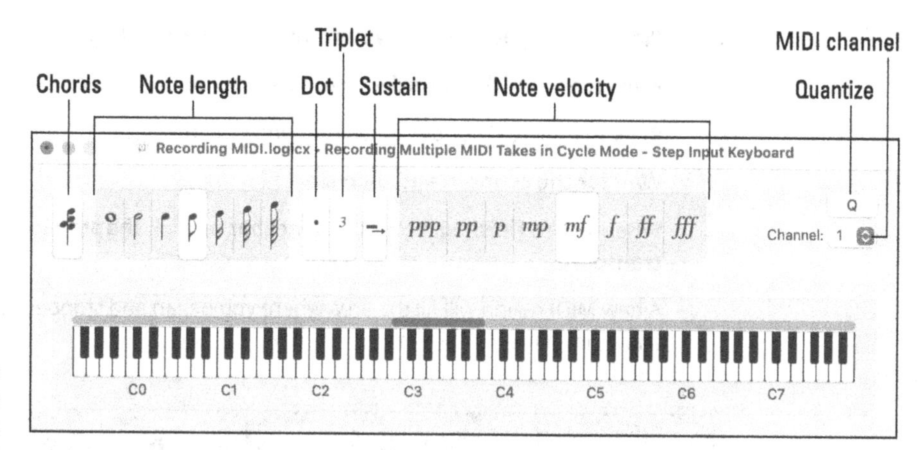

FIGURE 7-6: The step input keyboard.

The step input keyboard has the following advanced features:

- To input notes sequentially on the same beat, select the chord icon on the left side of the step input keyboard.
- Select the dot icon to make the note a length and a half of the currently selected note length.
- Select the triplet icon to make the next three notes you play part of a triplet.
- Select the sustain icon to lengthen the selected note in the editor by the length selected in the Step Input Keyboard.
- Select the quantize icon to snap the next note you input to the nearest division on the grid.
- To define the MIDI channel of the next selected note, choose a value in the MIDI channel drop-down menu.

TIP

The step input keyboard uses traditional notation for velocity values: ppp (16) pp (32), p (48), mp (64), mf (80), f (96), ff (112), and fff (127).

Recording Your First MIDI Take

Recording a MIDI take is similar to recording an audio take. Just do the following:

1. **Select the track, and then record-enable it by clicking the record enable icon on the track header or pressing Control-R.**

   The record enable icon will turn red to let you know that the track is enabled for recording.

2. **Place the playhead where you want to begin recording.**
3. **Press R or click the record icon on the transport in the control bar.**

   Refer to Chapter 3 for the control bar and transport details.

4. **Wait for the count-in and then start playing.**
5. **When you're finished, press the spacebar or click the stop icon on the transport.**

   A new MIDI region will fill the area where you began and stopped recording, as shown in Figure 7-7.

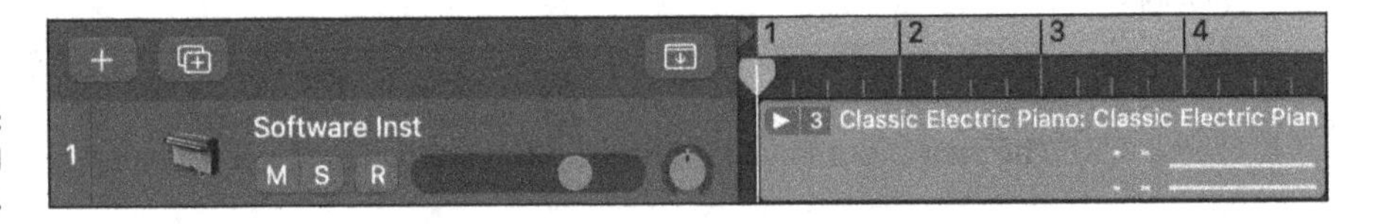

FIGURE 7-7: A recorded MIDI region.

REMEMBER

You can adjust the recording count-in by choosing File ⇨ Project Settings ⇨ Recording ⇨ Count-In.

TIP

As you record a MIDI track, your controller could be sending a lot more data than note on and note off messages. You could be sending aftertouch data, system exclusive data, or other control changes. If you don't need these types of MIDI data in your project, you can filter them. Choose File ⇨ Project Settings ⇨ MIDI, click the Input Filter tab, and then select the MIDI data you want to filter (see Figure 7-8). Recording all MIDI data isn't a big deal, but if you need to edit MIDI in the list editor (described in Chapter 15), having fewer MIDI events makes it easier.

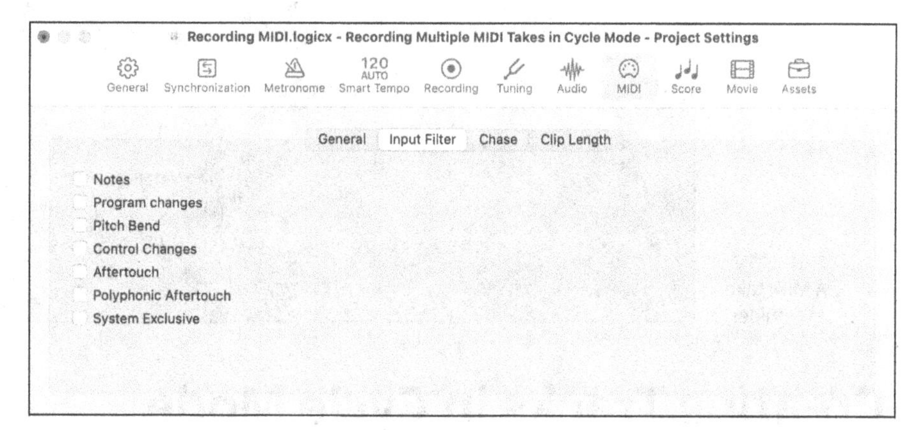

FIGURE 7-8: Project settings for the MIDI input filter.

Recording Multiple MIDI Takes in Cycle Mode

As with audio takes, you can record multiple MIDI takes to build a perfect take. Building a final composite take from multiple takes is called *comping*. To create a take folder when recording in cycle mode, you must first choose Record ⇨ Overlapping MIDI Recordings ⇨ Create Take Folder. The overlapping MIDI recording options will set your project's behavior when you create overlapping MIDI regions.

To record MIDI takes in cycle mode, follow these steps:

1. **In the cycle ruler at the top of the tracks area, drag from left to right to set the cycle locators.**

   Cycle mode is automatically turned on and represented by a yellow strip in the ruler.

2. **Press R to begin recording.**

   After the second take is recorded, a take folder is created, as shown in Figure 7-9, and new lanes are added with each pass through the cycle. All previous takes are muted, so you hear only the current take.

3. **Press the spacebar to end recording.**
4. **Click the disclosure triangle in the upper-left corner of the take folder.**

   The take folder opens and displays all the takes. You can open the take folder also by double-clicking it or by selecting it and pressing Option-F.

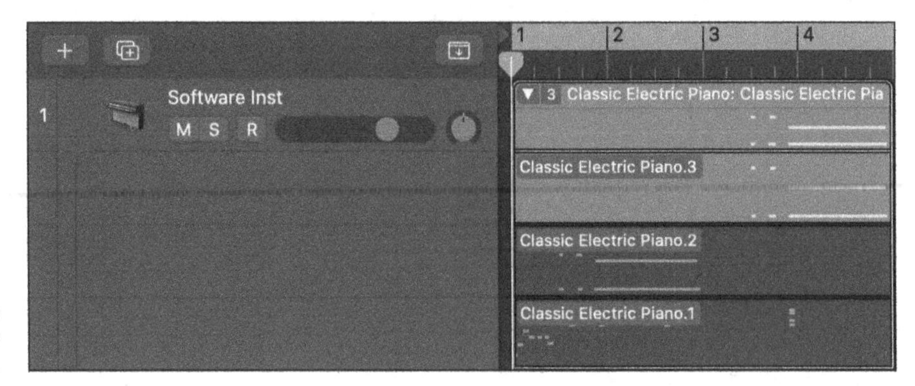

FIGURE 7-9: A MIDI take folder.

Creating Tracks in Cycle Mode

You might want to record a beat or groove and build it one take at a time. Recording this way makes it easier to focus on a single element rather than playing a full part. Logic Pro gives you several options when you record in cycle mode:

- Record ⇨ Overlapping MIDI Recordings ⇨ Cycle On ⇨ Create Tracks creates a new take track with each pass. A take track uses the same channel strip as the original track, so it doesn't use up more resources. For example, if you're recording MIDI drums and want each drum sound on a different track, you can create new tracks in cycle mode until you have a complete beat. You can use the same technique with a multi-instrument (a single software instrument with more than one sound) and create an entire groove from the ground up. Recording this way is like using a looper, with each loop building on the first, except it continues to add layers until you stop recording.
- Record ⇨ Overlapping MIDI Recordings ⇨ Cycle On ⇨ Create Tracks and Mute mutes the previous track after each pass. Use this if you don't want to build on the previous take but do want to create take alternatives. This method is similar to take recording but creates new take tracks instead of creating a take folder. Logic Pro lets you accomplish similar goals in several ways. In many cases, there's no right way to do something — just the way that enables you to get the job done.

Overdubbing MIDI

You may need to add a part to an existing region. Building drum grooves is a perfect example. You might want to join the new recording to the already existing region. In this case, choose Record ⇨ Overlapping MIDI Recordings ⇨ Cycle Off ⇨ Merge. Everything you record on top of a current region will be merged with it.

If you want a similar recording behavior while recording in cycle mode, choose Record ➪ Overlapping MIDI Recordings ➪ Cycle On ➪ Merge. Now all regions will be merged after each pass.

Recording Multiple MIDI Inputs

You can record more than one MIDI track at a time. Record-enable each track that you want to record and then record as usual. Anything you play will be recorded on all record-enabled tracks.

Suppose you want to use two or more controllers to perform different parts on different software instruments. In that case, you'll need to set the tracks and controllers to separate MIDI channels in the track inspector, as described previously in the "Preparing to Record MIDI" section.

TIP

I've saved the best tip for last. Imagine that you're playing your project, and it's sounding good. You're feeling the vibe, so you reach over to your keyboard and improvise a sick riff. "Oh, that was nice! I wish I had recorded that," you think to yourself as you gaze beyond the cold, heartless computer, disappointed. Stop feeling blue. I have great news: You were recording the entire time! You can capture that recording retroactively:

- Press Shift-R to capture the most recent MIDI performance. Depending on your current MIDI recording options, your most recent performance is captured as a region, take, or take track on the selected track.
- You can add the capture recording icon to the control bar. For details on customizing the control bar, see Chapter 3.
- Capture recording works only on MIDI tracks. You can't retroactively capture audio recordings.

As you can see, your creative power is limited only by your imagination. And your imagination is fueled by the possibilities Logic Pro gives you. You have a professional recording studio that can handle any project you can envision.

IN THIS CHAPTER

» Importing audio and MIDI into your project

» Using Apple loops

» Navigating the Logic Pro browsers

» Importing video to your project

Chapter 8
Adding Media to Your Project

What if you don't play an instrument or sing? Are you destined to stare at an empty project? No way. Apple loops and prerecorded media come to the rescue.

Logic Pro includes an enormous amount of media you can add to your project. You can build an entire song with nothing but this media. In this chapter, you learn how to use Apple loops, import audio and video files, and much more.

Adding Apple Loops to Your Project

Apple loops are audio and MIDI files that contain additional metadata, such as the key signature, time signature, and tempo. Logic Pro reads this metadata and adjusts the Apple loop to your project settings. For example, if you take an Apple loop in the key of G at 120 beats per minute and put it into a project in the key of E at 100 beats per minute, the Apple loop will automatically adjust to the project tempo and key. Apple loops are flexible. They do a lot of yoga when you're not using them. It would be a shame for all that stretching to go to waste, so please use Apple loops.

Apple loops can be beats, instrument parts, sound effects, or anything you want to repeat. You can build an entire project with only Apple loops or use them as accents to live instruments. As limber as they are with time and key signatures, they're equally able to fit into your project needs.

Navigating the loop browser

Logic Pro gives you a special loop browser to search and find Apple loops, as shown in Figure 8-1. To open the loop browser, choose View ➪ Show Loop Browser or press O. You can also open the loop browser by clicking the loop browser icon in the control bar. If you don't see the loop browser icon, you can customize the control bar as described in Chapter 3.

FIGURE 8-1: The loop browser.

Here's a description of the loop browser and its functions:

» **View icons:** At the top-left of the loop browser are two view icons that let you switch between button view and column view. The button view, which is the default state, displays clickable keyword buttons to refine your loop search. Shift-click the keyword buttons to select more than one at a time. The column view lets you navigate through loops by category.

- **Loop Packs drop-down menu:** At the top of the loop browser is a drop-down menu where you can select different loop collections installed on your computer. To view only the loops you've created, choose the My Loops category. At the bottom of the menu, you can choose Reindex All Loops to rebuild the catalog. You may want to reindex your loops after you've added loops to your system.
- **Keyword buttons:** While in button view, you can click multiple keyword buttons to filter the search results. The top-left button (the x icon) is the Reset button, which clears all button choices. The top-right button (the heart icon) is the Favorites button, which filters the search results to any loop that has been selected as a favorite.
- **Category columns:** While in column view, you can filter search results by navigating through categories of loops.
- **Scale menu:** Filter your search by scale type, including Major, Minor, Neither, or Both.
- **Signature menu:** You can filter your search by the time signature.
- **Search bar:** Search for loops by name or keyword with the search bar.
- **Results list:** This area displays the loops. It has six columns: loop type, name, beats, favorites, tempo, and key. Loop types are divided into blue audio Apple loops and green MIDI Apple loops.
- **Play In pop-up menu:** The gear icon at the bottom left of the loop browser is the Play In pop-up menu. You can select the auditioned loop to play in the song key (the key of the current project), the original key (the key of the Apple loop), or a specific key.
- **Speaker icon:** You can play or mute the selected Apple loop.
- **Volume slider:** Selecting a loop automatically plays it. The volume slider adjusts the loudness of the loop as you audition it.
- **Count:** The number of loops that fit your search criteria is displayed at the bottom right of the loop browser.

Adding audio loops

Audio loops are audio files and can be added to audio tracks. In the loop browser, you can spot an audio Apple loop by its blue icon containing a waveform. You can edit audio Apple loops just as you can a recorded audio region. To add an audio Apple loop to your project, do one of the following:

- » Drag an audio Apple loop from the loop browser to an empty area of the tracks area or track list, as shown in Figure 8-2. An audio track will be created, and the loop will be added to a region on the track.
- » Drag an audio Apple loop from the loop browser to an existing audio track. The Apple loop will be added to the tracks area at the position where you drop the loop.

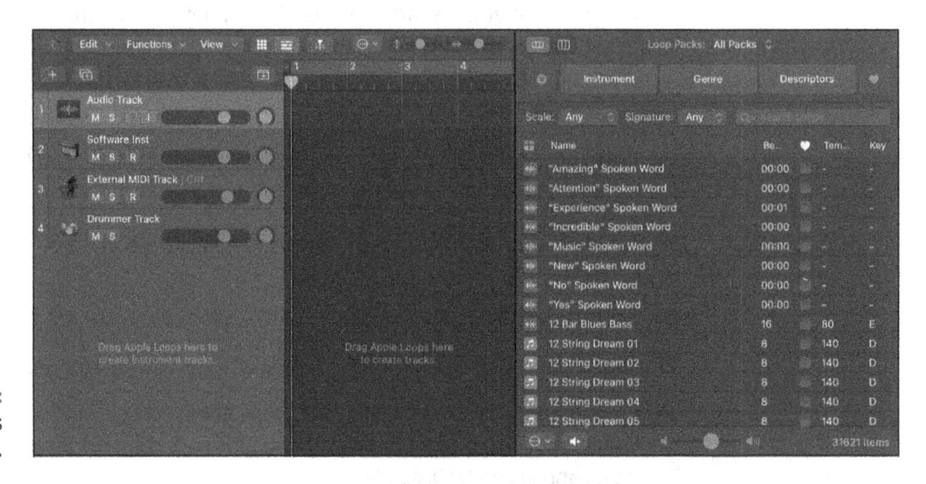

FIGURE 8-2: Drag Apple loops to the tracks area.

REMEMBER

You can't drag an Apple loop to a software instrument track or an external MIDI track from the loop browser. Logic Pro is thoughtful enough to warn you if you try. After the loop is in the tracks area, however, you can drag an Apple loop region to a software instrument track or an external MIDI track, but it will not play. So if you drag an Apple loop into your project and it's not playing, make sure it's on the right type of track.

REMEMBER

If you stretch an audio Apple Loop too far from its original key or tempo, your loop may end up with a pulled muscle. An audio Apple loop doesn't always sound good when it's stretched too far, but it still makes an excellent placeholder until you can replace it.

Adding MIDI loops

MIDI loops can be added to software instrument tracks or external MIDI tracks. You can identify MIDI Apple loops by their green icon in the loop browser. To add MIDI Apple loops to your project, do one of the following:

» Drag a MIDI loop to an external MIDI track. A MIDI region is created on the track with the Apple loop contents.

» Drag a MIDI loop to a software instrument track or an empty area of the tracks area or track list. A software instrument track is created, the corresponding software instrument is added to the track, and the loop is added to a MIDI region on the track.

If you drag a MIDI Apple loop to an audio track, the loop will be converted to audio and added to the tracks area at the position where you drop the loop. You learn how to edit audio in Chapter 14.

Adding drummer loops

Drummer loops can be added to drummer tracks, which you learn about in Chapter 4. You can identify drummer loops by their yellow icon in the loop browser. To add drummer loops to your project, do one of the following:

» Drag a drummer loop from the loop browser to an empty area of the tracks area or track list. A drummer track is created, and the loop is added to a region on the track.

» Drag a drummer loop from the loop browser to an existing drummer track. The drummer loop is added to the tracks area at the position where you dropped the loop.

You can also create your own Apple loops from any audio, MIDI, or drummer region in your project. To create your own Apple loops, select any region and choose File ➪ Export ➪ Region/Cell to Loop Library (Shift-Control-O). The Add Region to Apple Loops Library window appears, as shown in Figure 8-3.

You can create the following:

» **Loop:** A loop will follow the project tempo. Loops are great for regions you want to repeat again and again.

» **One-shot:** A one-shot will not follow the tempo and will play until the region completes. One-shots are great for sound effects or sounds that don't have rhythmic content.

You can also choose the scale, genre, key, and instrument descriptors. Click the Create button, and your Apple loop will be added to the loop browser.

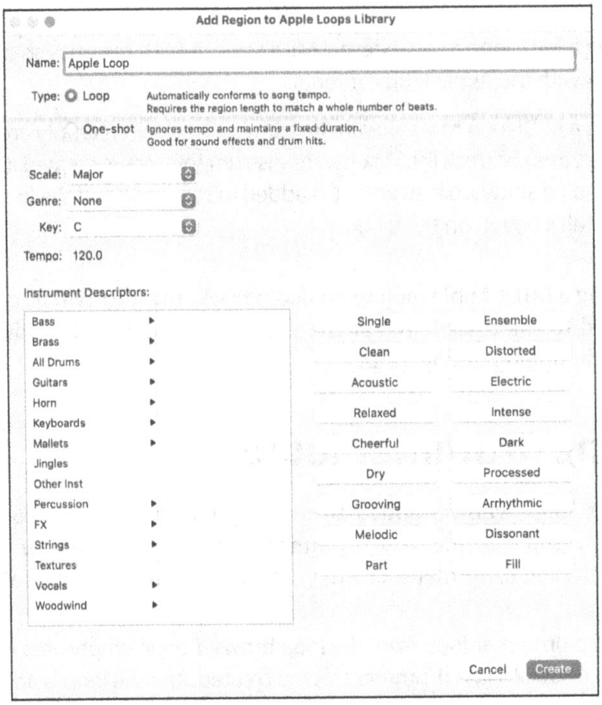

FIGURE 8-3: The Add Region to Apple Loops Library window.

Adding Prerecorded Audio to Your Project

If you have audio files on your hard drive that you want to bring into your project, Logic Pro makes the process a breeze. In most cases, adding audio files to your project is a simple drag-and-drop process from Finder or one of the Logic Pro browsers to an empty audio track.

Using the browsers to find audio files

To open the Logic Pro browsers, choose View ➪ Show Browsers or press F. The browsers open to the right of the tracks area, as shown in Figure 8-4.

Logic Pro gives you two browsers for adding media to your project:

- **Project audio browser:** Add or remove files already in your project.
- **All files browser:** Import files from your hard drive and import settings from other Logic Pro projects. (For details, see Chapter 2.)

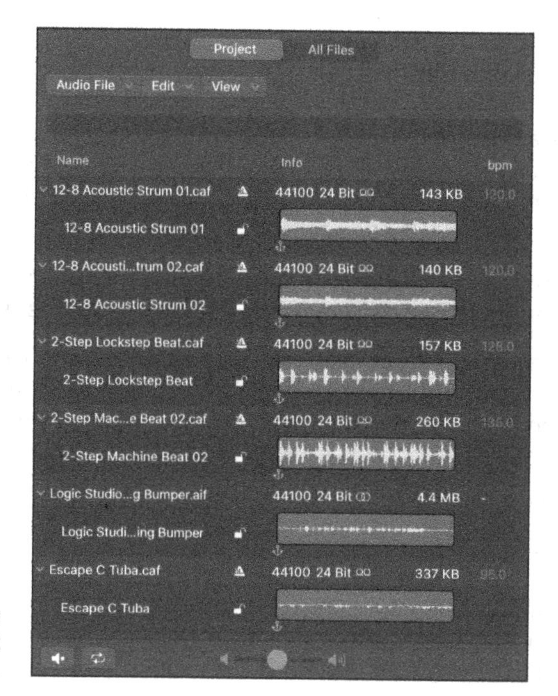

FIGURE 8-4: The project audio browser.

REMEMBER

Enable Complete Features must be enabled to view the project audio browser and all files browser. Choose Logic Pro ⇨ Settings ⇨ Advanced, and select Enable Complete Features.

Adding audio to your project

All the audio you record or import into your project will appear in the audio browser. Click the Project tab in the browser window to see all the audio collected in your project (refer to Figure 8-4).

At the top of the project audio browser are three menus:

- **Audio File:** Add audio files from anywhere on your computer. You can delete, back up, and move files. You can also convert audio files to different formats or convert selected regions to independent audio files on this menu.
- **Edit:** Cut, copy, paste, and delete audio files. You can also select all, used, or unused files on this menu. The project browser has its own undo history, which you can access on this menu.
- **View:** Sort, show, hide, or group your audio files in the browser.

The file list has the following four columns:

- **Name:** Displays the file name with a disclosure triangle next to it. Click the disclosure triangle to display all the regions that use the audio file. Clicking a region will select the region in the tracks area. You can double-click the file and region name to rename them.
- **Icon:** Displays icons that tell you whether the file or region is time-stamped, follows the project tempo, contains tempo information, or is missing.
- **Info:** Displays sample rate, bit depth, input format, and file size. It also displays a waveform with an anchor icon. You can drag the icon to edit the absolute start time of the region. Click the center of the waveform to audition the audio file.
- **bpm:** Displays the tempo of the audio file if the file contains tempo information.

You can audition these files by clicking the play icon at the bottom left of the browser. Click the play icon again to stop. Click the loop icon to repeat the audio file after you click the play icon. A volume slider adjusts the volume of the audio file you're auditioning.

You can drag the audio files and regions to existing audio tracks or an empty area in the tracks area or track list, where a track will be created to contain the region.

Adding audio from the all files browser

The third tab in the browser window is the all files browser. If you want to add audio, MIDI, or movies from your computer, you do it from this browser, as shown in Figure 8-5. The all files browser is like the Finder app.

Here's a description of the elements in the all files browser:

- **Back and forward icons:** Use these icons to navigate backward or forward in your browsing history.
- **Computer, home, and projects icons:** These three icons will take you to your Macintosh HD, user home directory, or Logic Pro projects directory, respectively.
- **List and column view icons:** These icons display your files in list view or column view, respectively.
- **File path selector:** The file path is displayed in this area and is clickable so that you can navigate up the file hierarchy.

- **Search field:** Use the search field to find files in your current folder. The plus icon to the right of the search field allows you to add up to ten search filters.
- **Results list:** Your current search results are displayed in this area. You can sort the three columns — Name, Date, and Size — by clicking the column headers.
- **Action menu:** The gear icon has actions you can apply to the currently selected item in the results list. You can add audio files to the project audio browser, find the selected file in Finder, create new files, and convert ReCycle files to Apple loops. Depending on the file type selected, you can also audition the audio.

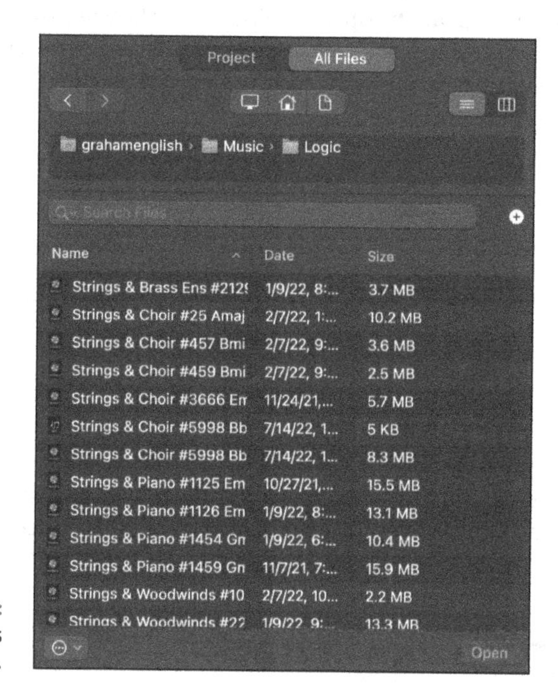

FIGURE 8-5: The all files browser.

After you've located the file you want to add to your project, you can drag it into the project just like the other browsers. You can also import project settings from the all files browser, as described in Chapter 2.

Importing Video to Your Project

Logic Pro isn't just an audio/MIDI sequencer. You can also import video and add your own movie score. Film and TV scoring with Logic Pro is intuitive, and you'll find that adding movies to your project is simple.

Adding a movie to your project

As you previously discovered in the chapter, you can add QuickTime movies to your project using any media browsers. You can also add a movie by choosing File ➪ Movie ➪ Open Movie (Option-⌘-O). Navigate to the movie file in the dialog that appears and click Open. The Open Movie dialog asks you to open the movie; if the movie contains an audio track, you can extract the audio track into your project, as shown in Figure 8-6. Click OK, and the movie is added to your project, and the audio is added to the track list.

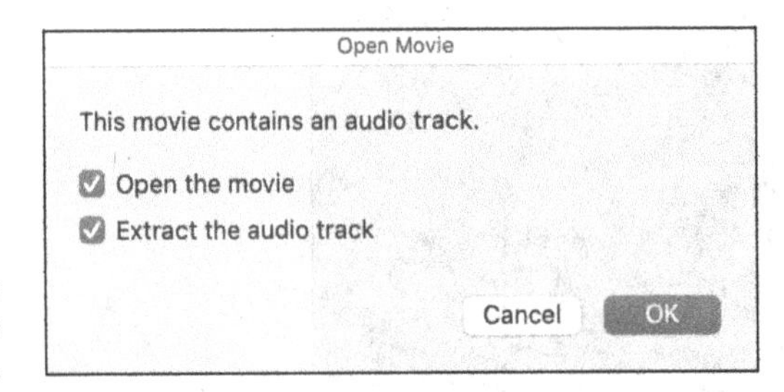

FIGURE 8-6: The Open Movie dialog.

Display the inspector, and the movie appears at the top of the inspector, as shown in Figure 8-7. You can double-click the movie to open the video in a separate window.

Exploring the movie track

After you add a movie to your project, you can display the movie track. Choose Track ➪ Global Tracks ➪ Show Global Tracks, and the global tracks open above the tracks area. (If the movie track isn't visible, choose Track ➪ Global Tracks ➪ Show Movie Track.) The movie track displays thumbnails of the video and is adjustable by dragging the bottom edge of the movie track, as shown in Figure 8-8.

FIGURE 8-7: The movie inspector.

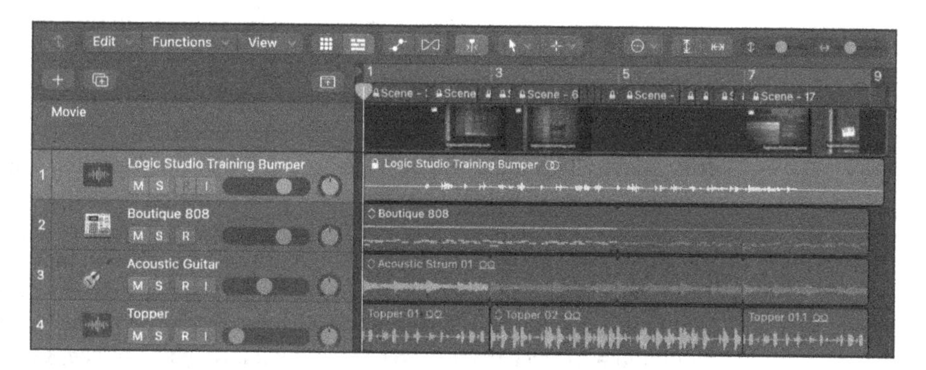

FIGURE 8-8: The global tracks movie track.

Creating movie scene markers

You can create movie scene markers locked to absolute time instead of relative time. This means they won't move if you change the tempo of your project. Movie scene markers are great for timing your music to the scene's rhythm.

To create movie scene markers, choose Navigate ➪ Other ➪ Create Movie Scene Cut Markers and select one of the following options:

- **Auto Range:** Create markers based on the marquee selection, project cycle, selected regions, or all three.
- **Marquee Selection:** Create markers based on the marquee selection.
- **Cycle Area:** Create markers based on the cycle area.
- **Selected Regions:** Create markers based on the selected regions.
- **Entire Movie:** Create markers based on the entire movie.

After you select the option for creating movie scene markers, the movie will be analyzed, and markers will be created, as shown in Figure 8-9. Markers are created based on a fixed threshold for scene cuts and work well for most types of video.

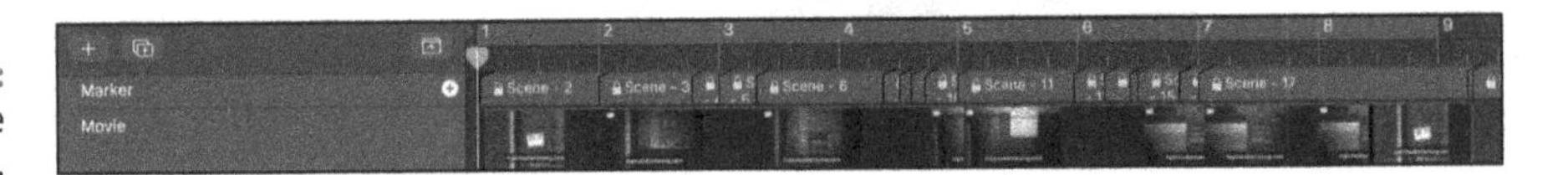

FIGURE 8-9: Movie scene markers.

Importing audio from your movie

If you've already imported a movie but not the audio, or if you want to import audio from a movie without importing the movie, choose File ➪ Movie ➪ Import Audio from Movie. The audio will be added to a track in the track list. Importing audio is useful if you need to build a soundtrack around dialog or sound effects.

As you've learned in this chapter, you don't need to play an instrument to benefit from Logic Pro. You discovered how to import movies so you can compose your own soundtracks. You also learned how to navigate the Logic Pro browsers so you can build a project from prerecorded media. As you can see, Logic Pro can support professional demands, which is why so many musicians rely on it to get great results.

3 Making Music with Virtual Instruments

IN THIS PART . . .

Make beats with Drummer and Ultrabeat.

Understand keyboard and synthesizer fundamentals and edit instrument sounds and effects.

Create virtual models of acoustic instruments and emulate vintage synthesizers.

Play orchestral instruments with expression and define a productive composing workflow.

IN THIS CHAPTER

» Creating beats with your virtual drummer

» Producing acoustic and electronic drum sounds

» Designing grooves with Ultrabeat

» Sequencing drum patterns

Chapter 9

Making Beats with Drum and Percussion Software Instruments

Did your drummer get lost on the way to the studio? Don't you just hate it when that happens? Oh, you have Logic Pro? Never mind.

Listen, I love playing with a live drummer. Some of my best musical partnerships have been with amazing drummers. But I love how Logic Pro gives me a virtual live drummer to inspire me and help turn my rhythmic ideas into reality.

In this chapter, you discover how to use Drummer, Ultrabeat, and Drum Machine Designer. You create beats, build custom drum kits, and use a pattern sequencer. Logic Pro's drum and percussion software instruments are fantastic songwriting and music-producing tools. But whatever you do, don't beat on your computer. It only sounds like a drum — it's not built like one.

Playing with Your Virtual Drummer

Drummer is your virtual session player. Drummer is a combination of track type and software instrument. The track type does the drumming, and the software instrument provides the drum sounds. As described in Chapter 4, a drummer track is a particular track type reserved for use with Drummer.

The drummer track comes with its own drummer editor, which chooses the style of music and the player, and tells the drummer how to play the track. The editor is so simple that anyone can use it.

The Drum Kit Designer or Drum Machine Designer software instrument is automatically added to a drummer track, depending on your drummer selection. (Note that you don't have to use the preloaded software instrument.) Drummer works the same for both software instruments, so I won't describe them separately in this chapter.

Creating a drummer track

To create a drummer track, choose Track ➪ New Drummer Track. A new drummer track is added to the track list, and a default eight-bar region is added to the tracks area, as shown in Figure 9-1.

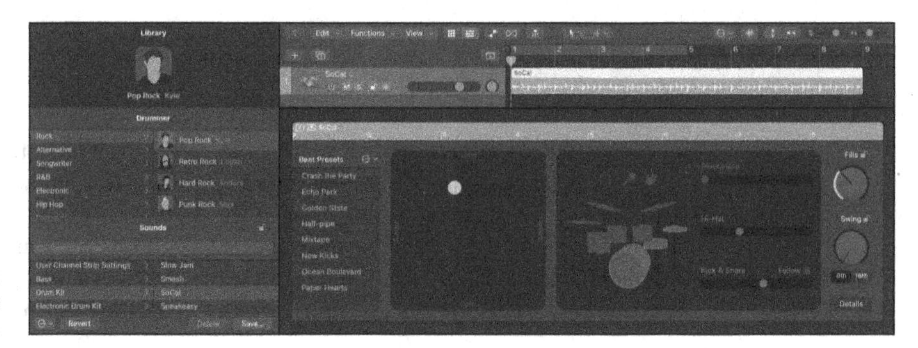

FIGURE 9-1: A drummer track and region.

Creating drummer regions

Although Drummer automatically creates an eight-bar region, you don't have to live with it. By using several regions to build your project rather than one big region, you can more easily change what your drummer plays during different song sections (as you'll soon see).

To create a drummer region, select the pencil tool in the tracks area and click where you would like the region to begin. By default, drummer regions are eight bars long. If you want to change the region's size, drag the region's lower-left or lower-right edge to resize it. You could also split a region (making two regions out of one) by using the scissors tool.

TIP

If you want to force the drummer to hit the crash cymbal at the beginning of a song section, create a new region. You may need to adjust the Fills parameter, as you learn next, but getting Drummer to play a cymbal crash at the beginning of a region is a good reason why you want new regions to start at song sections or groove changes.

Choosing and directing your drummer in the editor

The real power of Drummer's artificial intelligence–like personality is in the drummer editor. To open the editor, double-click a drummer region or select the region and choose View ⇨ Show Editor (E). The editor, shown in Figure 9-2, opens at the bottom of the tracks area.

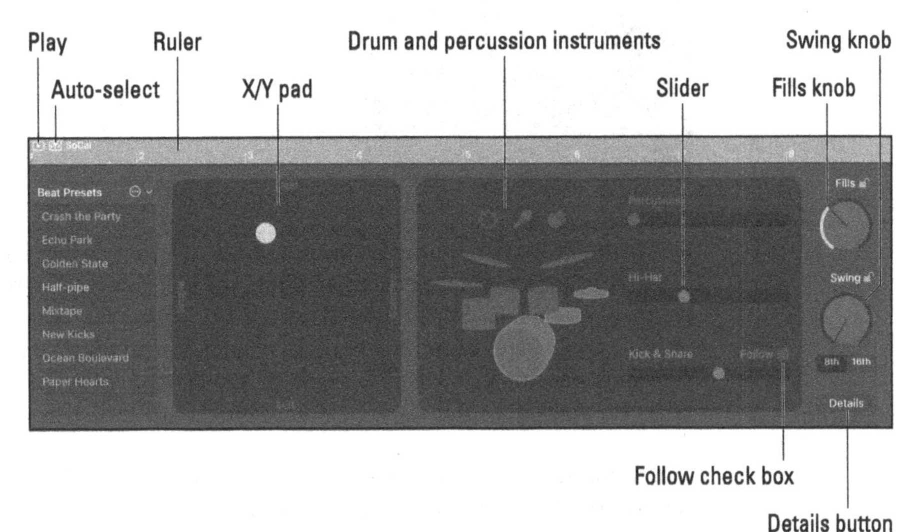

FIGURE 9-2: The drummer editor.

The drummer editor is filled with personality. To get the most out of it, open the library (press Y or choose View ⇨ Show Library) so you can change settings for the entire drummer track. At the top of the area is a headshot and description of the currently selected drummer character, as shown in Figure 9-3. Below the current drummer is the Drummer section, where you can choose the style and drummer character. Each style has several different drummers with names and headshots, and new drummers are occasionally added with software updates. Click the drummer you want for the entire track. The track regions and drummer editor are updated with the style of the selected drummer.

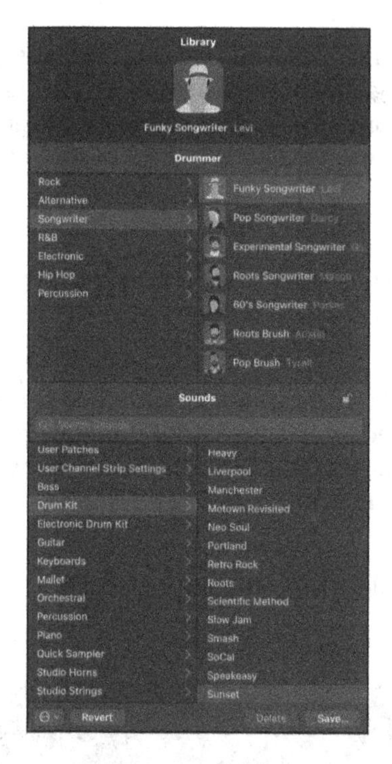

FIGURE 9-3: The drummer library.

Below the drummer character focus area is the Sounds section. Although each drummer character has an associated kit, you can choose different kits independently of the selected drummer character. Click the lock icon next to the Sounds heading to change drummers without changing drum kits. Since Logic Pro first came out, developers have added new drummers and kits, and more are bound to be on the way.

The drummer editor (refer to Figure 9-2) is where you change settings for the selected drummer region. Here's a description of each section:

» **Ruler:** At the top of the region settings are a ruler, a play icon, and an auto-select icon. You can play the region or move the playhead within the region in this ruler. When auto-select is engaged, the drummer editor displays settings for the region at the current playhead position.

- **Beat Presets:** Choosing a drummer character loads a set of beat presets you can click to update the editor controls. At the top of the Presets menu is a drop-down list where you can save, delete, and recall the default preset. You can also refresh the region to make subtle changes to the current region. Finally, you can choose to keep the settings while changing drummers.
- **X/Y pad:** The X/Y pad has a yellow puck that you can move between Loud/Soft and Complex/Simple. The position of the puck makes a big difference in the beat the drummer will play.
- **Drum and percussion instruments:** Click the instruments to select the drum and percussion sounds that will play in the region. Depending on the selected drummer character, this area will show different instruments that are available to play. The sliders to the right of the instruments allow you to choose between groove variations. If you select the Follow check box for the Kick & Snare slider, the slider changes to a drop-down menu, and you can select a track in the project that the kick and snare will follow.
- **Fills:** The Fills knob adjusts the number and length of fills. Click its lock icon to freeze the fills setting when changing presets.
- **Swing:** The Swing knob adjusts the amount of shuffle feel. Click the lock icon to freeze the swing setting when changing presets. You can also click the 8th or 16th buttons to decide whether the swing is based on eighth or sixteenth notes.
- **Details:** Click the Details button to open an additional editing panel, as shown in Figure 9-4. Use this panel to change Feel, Ghost Notes, and Hi-Hat performance. The Feel knob adjusts how Drummer plays relative to the tempo. You can pull back the performance so that it plays behind the beat or push it forward to play in front of the beat. The Ghost Notes knob adjusts how loudly or quietly Drummer plays *ghost notes* (notes that are played at a low volume between the loud notes). The Hi-Hat knob adjusts the amount of closed or open hi-hat that Drummer plays.

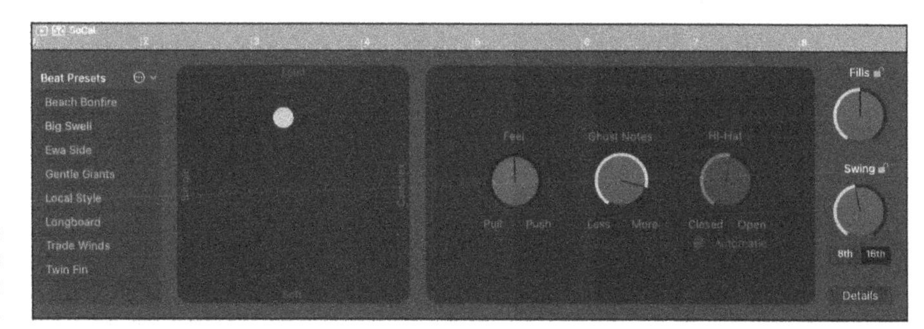

FIGURE 9-4: The Details area of the drummer editor.

Selecting producer kits

Each drummer has a default kit connected to the drummer. If the drummer character is an acoustic drummer (as opposed to an electronic drummer), the kit is automatically loaded into the stereo Drum Kit Designer software instrument. You can upgrade the kit to a special producer kit with more tracks and channel strips for ultimate control. To select a producer kit, do the following:

1. **Choose View ➪ Show Library to open the library.**

   The library opens on the left side of the tracks area.

2. **Choose Drum Kit ➪ Producer Kits, then select a patch.**

   The track is upgraded to a track stack containing several tracks of individual drums. Note that producer kit patches have a + sign in the name to help you tell them apart from non-producer kit patches.

REMEMBER

To have access to all producer kits, choose Logic Pro ➪ Sound Library ➪ Open Sound Library Manager. The Sound Library Manager window opens, as shown in Figure 9-5. Select the Drum Kit check box and click Install. Be mindful of the size. The multi-output drum kits are over 10 gigabytes, but these kits were engineered by golden-eared professionals and are worth every gigabyte.

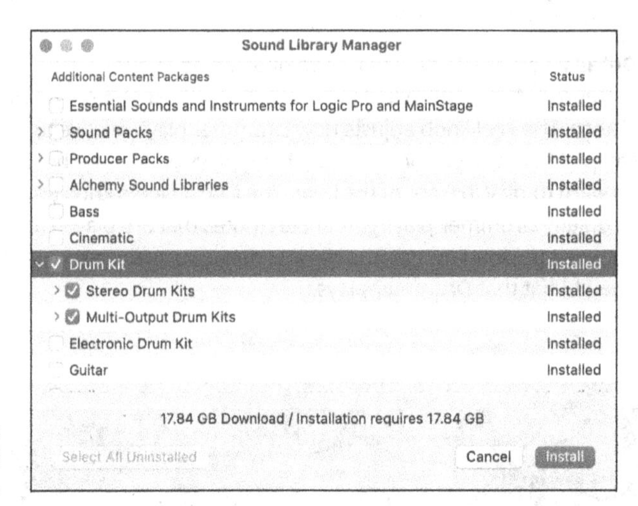

FIGURE 9-5: The Sound Library Manager.

Building custom kits with Drum Kit Designer

The Drum Kit Designer software instrument is automatically added to drummer tracks that use acoustic drummer characters. To open Drum Kit Designer, follow these steps:

1. **Select a drummer track and display the inspector by choosing View ➪ Show Inspector (I).**

   The inspector opens to the left of the tracks area.

2. **Click the instrument slot in the channel strip.**

   The Drum Kit Designer software instrument opens, as shown in Figure 9-6.

FIGURE 9-6: Drum Kit Designer.

Click any drum to play it. As you click a drum, the left side displays the Exchange panel where you can choose different drums, as shown in Figure 9-7. The right side displays the Edit panel, where you can control the selected drum sound.

Each type of drum or cymbal has different parameters. You can tune and dampen every sound and adjust the volume by using the Gain knob. If you've loaded a producer kit (refer to the preceding section, "Selecting producer kits"), you can select whether the sound should be included in the overheads and room microphones or should leak into other drum mics. You can select between two mic setups for the room mics by using the A/B slider.

At the bottom of the software instrument screen is a disclosure triangle that opens the additional settings shown in Figure 9-7. In this area, you can adjust the volume of the drummer's percussion instruments.

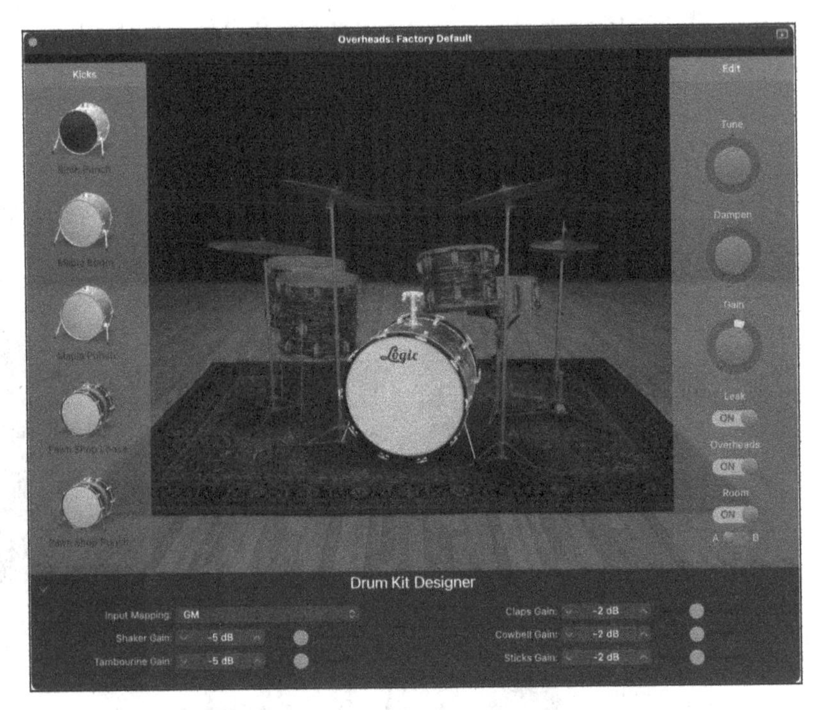

FIGURE 9-7: Drum Kit Designer Exchange and Edit panels.

Converting drummer regions to MIDI regions

If you need absolute control over a drummer region, you can convert it to a MIDI region. Two methods are available for converting a drummer region to MIDI:

- » Drag any drummer region to a MIDI track. The region will automatically convert to a MIDI region.
- » ⌘-click a drummer region and choose Convert ⇨ Convert to MIDI Region. The region will convert to MIDI, and the drummer editor will no longer be available.

TIP

You can convert a MIDI region on a drummer track back to a drummer region by ⌘-clicking the region and choosing Convert ⇨ Convert to Drummer Region. However, any edits made to the MIDI region will not be saved after the conversion.

Creating Beats with Ultrabeat

Ultrabeat is a 25-voice drum synth and pattern sequencer that operates similarly to a traditional hardware drum machine. Drum voices 1–24 are assigned to the first 24 MIDI keys (C1 to B2), and the 25th drum voice is assigned to the 25th MIDI key and above (beginning at C3), so it can be played chromatically, making it ideal for bass sounds, pads, or leads.

Ultrabeat is capable of doing a lot more than drumbeats and bass lines. Each of the 25 voices is a complete synth and gives you flexible control over each voice. In this section, you discover how to design drum sounds and patterns and incorporate Ultrabeat into your projects.

Exploring the Ultrabeat interface

To open Ultrabeat, follow these steps:

1. **Choose Track ⇨ Create New Software Instrument Track.**

   A new software instrument track is added to the track list.

2. **Choose View ⇨ Show Inspector (I) to display the inspector.**

   The inspector opens on the left side of the tracks area.

3. **Click the right side of the instrument slot and then select Ultrabeat (Drum Machine).**

   The Ultrabeat interface opens in a new window, as shown in Figure 9-8.

The Ultrabeat interface is divided into three sections:

- **Assignment:** The left side of Ultrabeat is where you select, mix, and assign drum sounds.
- **Synthesizer:** The largest area of Ultrabeat is where you design the currently selected drum sound.
- **Step Sequencer:** The bottom section of Ultrabeat is where you create drum patterns for the currently selected drum and control the Ultrabeat sequencer.

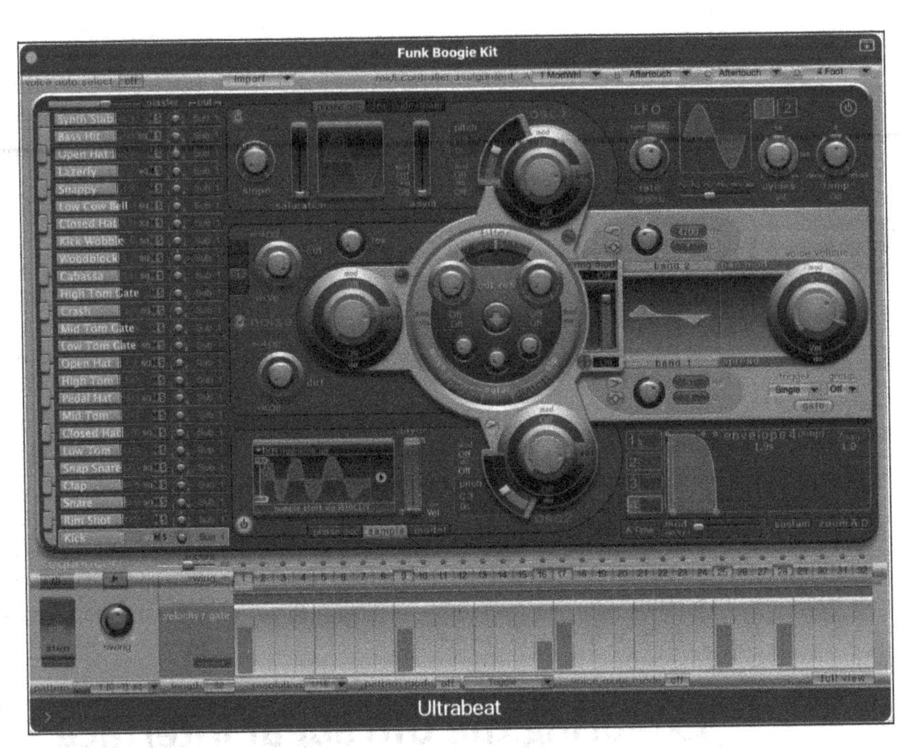

FIGURE 9-8: The Ultrabeat interface.

At the top of the Ultrabeat interface is a menu with several functions:

- **Voice auto select:** Turning on voice auto select allows you to select drum voices with your MIDI controller. This feature is useful when going back and forth between two sounds. You can use one hand to select the drum voice with your MIDI controller while editing the synth section with your other hand.
- **Import:** Click the Import button to import sounds and sequences from other Ultrabeat drum kits and Sampler instruments. For more on the Sampler software instrument, see Chapter 11.
- **MIDI controller assignment:** You can assign MIDI controllers to the four controller slots, which allows you to modulate the synthesizer section.

Choosing sounds in the assignment section

You use the assignment section, shown in Figure 9-9, to select and mix drum sounds. Here's a description of the assignment section parameters:

- **Master volume slider:** This slider at the top of the assignment section controls the volume of the entire drum kit.

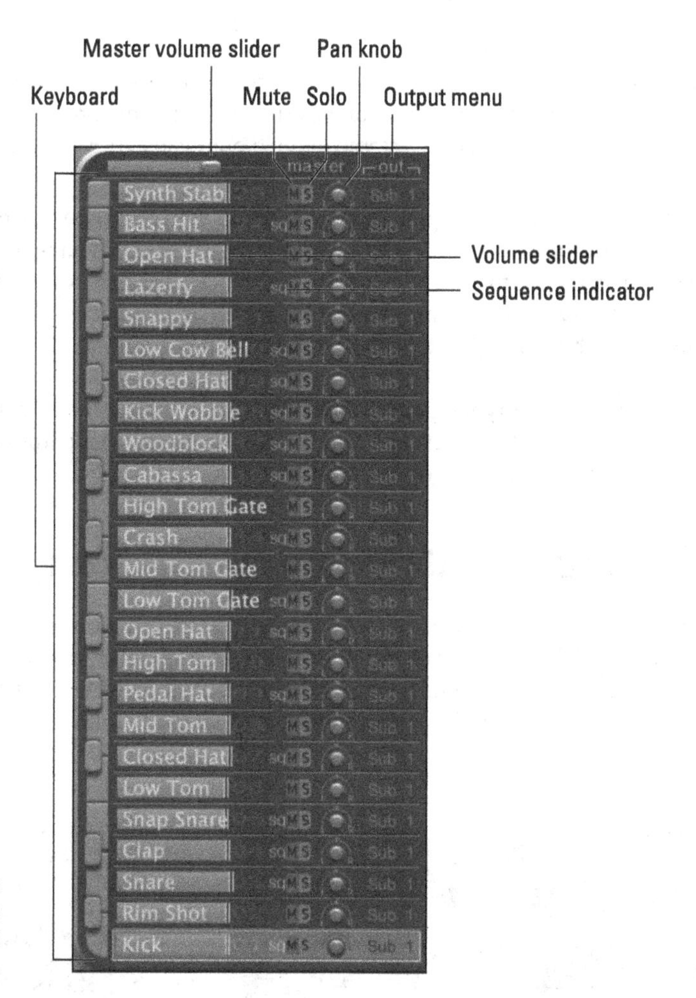

FIGURE 9-9: The Ultrabeat assignment section.

- **Keyboard:** The vertical keyboard is playable; each key lights when a drum voice is played.
- **Volume slider:** You can drag the blue slider on each drum voice left and right to lower and raise, respectively, the sound's volume. Control-click the volume slider to display a menu where you can copy and paste voices and sequences and initialize the drum sound to one of several default sounds.
- **Sequence:** If the drum sound is associated with a sequence, you see a sequence indicator next to the drum sound number.
- **Mute and solo icons:** Click the mute (M) icon to silence the drum sound. Click the solo (S) icon to silence all other drum sounds.
- **Pan knob:** Use the pan knob to place the drum sound in the stereo spectrum.

» **Output menu:** The Output menu is to the right of the pan knob. Click the menu and choose the output of the drum voice. The options change depending on whether you load a stereo or multi-output instance of Ultrabeat.

Drum voices can be dragged and dropped. If you drag one voice onto another voice, the drum voices and synth settings are swapped. If you hold ⌘ while dragging and dropping the drum voice, the sequences are also swapped. Option-drag to copy (not swap) the sound but not the synth or sequencer settings. Hold down Option-⌘ while dragging to copy the sound and sequences.

TIP

Don't have a magnifying glass around? You can increase the size of the interface for easier viewing by selecting a higher number on the View drop-down menu at the top of the software instrument interface.

Shaping sounds in the synthesizer section

You shape the selected drum sound in the synthesizer section, as shown in Figure 9-10. The synthesizer signal flow moves from left to right. Don't let all the buttons and knobs intimidate you. At the most basic level, you generate sounds at the far left, filter sounds in the middle, and shape the envelope and volume of sounds at the far right.

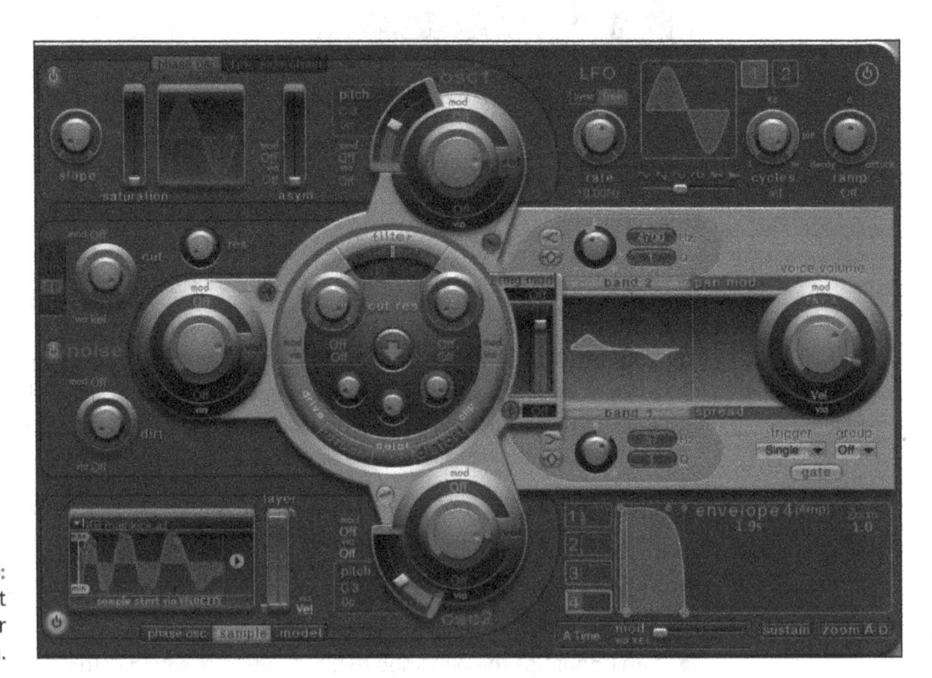

FIGURE 9-10: The Ultrabeat synthesizer section.

Here's a brief description of the Ultrabeat sound generators:

- **Oscillator 1:** The top sound generator is called oscillator 1. You turn it on and off with the power button in the upper left. You can choose between a phase oscillator, FM synthesis, or external side-chain input. As you choose each synthesis mode, the oscillator 1 parameters change to give you further tone-shaping capabilities.
- **Oscillator 2:** The bottom sound generator is called oscillator 2. You turn it on or off with the power button in the lower left. You can choose between a phase oscillator, a sampler, or component modeling. Choosing each synthesis type updates the oscillator 2 parameters.
- **Noise generator:** The middle section is a noise generator you can turn on or off with the power button at the left. There are no additional parameters aside from what you are given.

The scope of this book couldn't possibly touch on every parameter in each sound generator. However, I don't want to leave you hanging, so here are a few things you can try with each tone generator:

- Oscillator 1 is excellent for synth sounds. Set it to Phase Osc, turn the Slope knob and Saturation slider all the way down, and turn the Asym (asymmetry) slider all the way up. Depending on the envelope parameter settings, you should have a useful synth bass sound. If you ⌘-click the drum voice name in the assignment section and choose Init ➪ Bass, Ultrabeat will produce a bass synth that you can begin tweaking.
- Oscillator 2 works great as a sampler. Set it to Sample, click the Sample Name display, and choose Load on the drop-down menu. Navigate to an audio file on your hard drive and the sample will load, allowing you to play it from your MIDI controller. Alternately, you can ⌘-click the drum voice name in the assignment section and choose Init ➪ Sample; Ultrabeat will set the oscillator to a pre-loaded sample that you can shape or replace.
- The noise generator is excellent for creating snare sounds and is also useful when combined with the other oscillators. To hear a noise generator in action, ⌘-click the drum voice name in the assignment section and choose Init ➪ Snare. Turn off oscillator 1 and notice how much of a snare sound is generated by noise.

The rounded filter section in the center of Ultrabeat, shown in Figure 9-11, gives you more tonal-shaping capabilities. The top half of the filter section is dedicated to the filter, while the bottom half controls a distortion circuit.

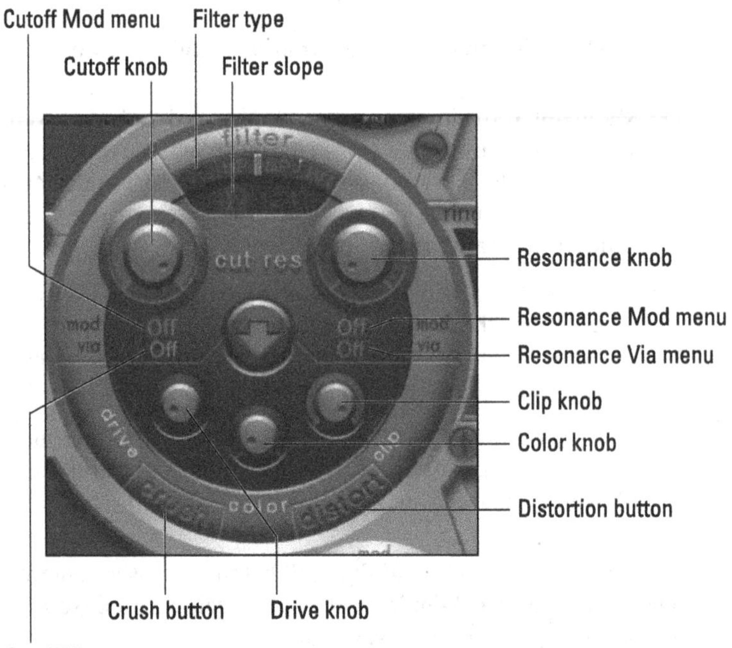

FIGURE 9-11: The Ultrabeat filter section.

Here's a brief description of the parameters you can use to adjust your sound:

- **Filter on/off:** Click the filter name at the top of the filter section to turn the entire filter section on or off.
- **Filter type:** Click the filter type buttons to change the filter type to LP (low pass), HP (high pass), BP (band pass), or BR (band rejection).
- **Filter slope:** Click the filter slope buttons to change the filter slope to 12 dB or 24 dB. The higher the filter slope, the more extreme the filtering.
- **Cutoff:** Rotate the Cutoff knob to adjust the cutoff frequency of the different filter types. Rotate the knob to the left to lower the frequency and to the right to raise the frequency.
- **Resonance:** Rotate the Resonance knob to adjust how the cutoff frequency affects the filters.
- **Mod and Via pop-up menus:** Click the Mod menu to choose a modulation source for the cutoff and resonance parameters. Click the Via menu to select a MIDI controller assignment or velocity to control the modulation source.
- **Crush button:** Click the Crush button to turn on the bit crusher. The bit crusher lowers the digital resolution of the sound, creating a digital distortion effect. If the Crush button is engaged, the Distortion button is disengaged.

» **Distortion button:** Click the Distortion button to distort your sound by using an analog-modeled overdrive effect. If the Distortion button is engaged, the Crush button is disengaged.

» **Drive knob:** Rotate the Drive knob to adjust the amount of distortion.

» **Color knob:** Rotate the Color knob to change the tone of the distortion. Rotate it to the right to make the sound brighter, and rotate it to the left to make the sound darker.

» **Clip/Level knob:** Rotate the Level knob to set the level of distortion while in distortion mode. If crush mode is engaged, this knob becomes a Clip knob and sets the threshold before bit crushing begins. *Bit crushing* is a distortion effect that reduces the digital audio's resolution.

The right section of the Ultrabeat interface, shown in Figure 9-12, is dedicated to amplifying, modulating, and shaping the envelope of the drum sound. Here's a brief description of the three sections on the right side of the Ultrabeat synthesizer:

» **LFO:** The top area is where you change how sound parameters are modulated by the LFOs (low frequency oscillators), envelope generators, velocity, and MIDI controllers.

FIGURE 9-12: The Ultrabeat output, LFO, and envelope section.

- **Output:** The middle area is where you change the volume, two-band EQ, and stereo panning of the drum sound.
- **Envelope:** The bottom area is where you adjust the ADSR (attack, decay, sustain, and release) envelope for the drum sound.

Here's how you can shape your sound with the output section:

- Use the two-band EQ to brighten or darken the sound. The controls for the two bands are located above and below the EQ curve display. Each EQ has two types available: a shelving EQ and a peak EQ. Choose the type of EQ by clicking one of the buttons to the left of the Level knob. Use the shelving mode (the top button) to remove frequencies above or below the level set in the Hz (frequency) field. Use the peak mode (the bottom button) to raise or lower the frequencies around the level set in the Hz (frequency) field. Set the Q field to change the width of the EQ boost or cut. Use the Level knob to the left of the Hz and Q fields to set the amount of EQ boost or cut. Finally, click either EQ band names to turn the EQ on or off.
- Use Pan Mod (modulation) or Spread to adjust the drum sound in the stereo field. (Only one of them can be selected at a time.) Clicking the name of the Pan Mod or Spread parameter turns it on or off or toggles between the two options while displaying the parameters.
- Use the envelope section to graphically shape the drum sound's attack, decay, sustain, and release (ADSR). Each sound has four envelopes numbered 1 through 4, with 4 hardwired to the sound's volume. The other three envelopes can be modulated by various Ultrabeat sound parameters. Click any envelope number and then drag the graphic handles to shape the envelope.

As you can see, Ultrabeat is a beast of a drum synth. It's safe to say that most Ultrabeat users get a lot out of it without ever touching most parameters. The easiest way to find sounds you like is to load different presets from the top of the software instrument interface. Ultrabeat comes with dozens of kits that sound amazing without any tweaking. But if you feel like tweaking drum sounds, Ultrabeat will keep you thoroughly entertained.

Sequencing patterns in the step sequencer

The Ultrabeat step sequencer, shown in Figure 9-13, is used to edit patterns for the currently selected drum sound and to control entire Ultrabeat sequences. Most Ultrabeat presets will load several patterns into the sequencer to get you going.

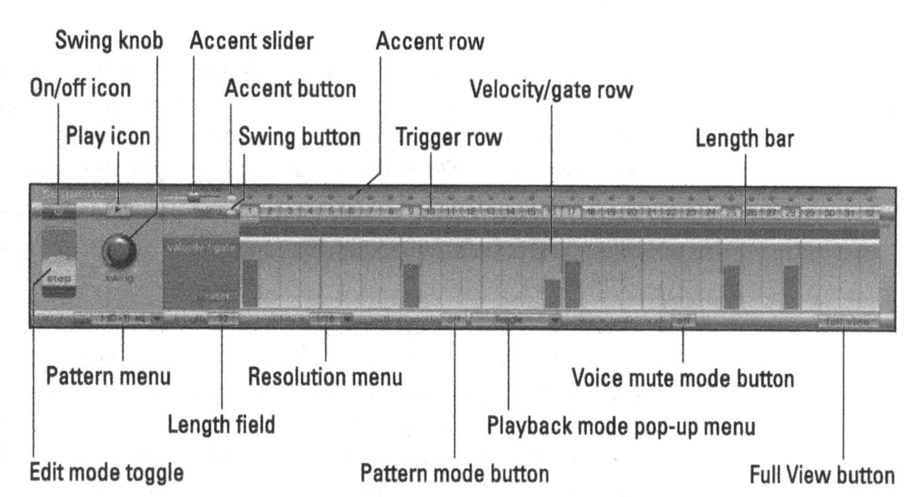

FIGURE 9-13: The Ultrabeat step sequencer.

REMEMBER

You don't have to use the integrated step sequencer in Ultrabeat. Like other software instruments, you can play Ultrabeat's drum sounds from MIDI regions. But if you enjoy step sequencing, you'll like using Ultrabeat's sequencer. As you see in this section, exporting beats into the tracks area is easy.

The left side of the step sequencer contains the following global parameters:

- **On/off icon:** Click the on/off icon to turn the pattern sequencer on or off.
- **Edit toggle:** Click the Edit toggle to switch between voice mode and step mode. While in the default voice mode, any edit you make to a drum's sound is global. While in step mode, you can automate the sound parameters with each step of the pattern sequencer.
- **Play icon:** Click the play icon to start and stop the pattern sequencer.
- **Swing knob:** The Swing knob adjusts the amount of shuffle feel for any drum sound with swing enabled.

Below the global parameters is a small section for the following pattern parameters:

- **Pattern menu:** Click the menu to choose one of the 24 patterns. If a pattern slot has a pattern available, it will be marked with the letters *sq*. You can copy, paste, and clear patterns by Control-clicking the menu.
- **Length field:** Double-click the length field to enter the length of the pattern.
- **Length bar:** The length bar spans the entire length of the top of the pattern sequencer. Drag the right edge of the bar to change the length of the pattern.

- **Resolution menu:** The resolution menu determines the note length value of the pattern. A setting of 1/8 means that each step of the grid represents an eighth note, and 1/16 represents a sixteenth note. Other available resolutions are 1/12 (triplet), 1/24 (sextuplet), and 1/32 (thirty-second note).
- **Pattern mode button:** Turning on pattern mode allows you to start and select patterns with MIDI notes. Notes that will trigger patterns are referenced in the Pattern menu.
- **Playback mode pop-up menu:** The playback mode determines how the pattern is played when an incoming MIDI note is received. One-Shot Trig mode starts the pattern and plays it once before stopping. Sustain mode plays the pattern until the note is released. Toggle mode plays the pattern until the next note is played and immediately switches to the new pattern. (If the same note is triggered, the pattern stops.) Toggle on Step 1 is similar to Toggle mode except the pattern changes the next time beat 1 is reached instead of immediately.
- **Voice mute mode button:** When voice mute mode is on, playing MIDI note C1 and above will mute the corresponding sound in Ultrabeat. Playing the note a second time will unmute the sound.
- **Full View button:** Pressing the Full View button gives you an overview of the entire pattern in the main synth display.
- **Accent button:** Click the Accent button to turn on the accent function. When accents are on, clicking the accent buttons above the pattern steps globally accent that beat in the pattern.
- **Accent slider:** Drag the Accent slider to the right to raise the level of the accents.
- **Swing button:** The Swing button turns the swing function on or off for individual sounds.

The largest area of the pattern sequencer is dedicated to the step grid, which plays the drum patterns. The step grid has the following parameters:

- **Trigger row:** The trigger row contains numbers that identify the beat. Click the number buttons to play the currently selected sound on that beat. Click the button again to turn the sound off. Control-click the trigger row to display a menu of available functions for automatically creating beats and copy, paste, and clear functions.
- **Velocity/gate row:** The velocity/gate row sets the length and velocity of the note. Drag the height to set the note velocity and the width to set the length. Control-click the velocity/gate row to display a menu of velocity functions, including a randomize feature.
- **Accent row:** The accent row is above the trigger row. Click the beats you want to accent.

TIP

Click the Full View button on the bottom right of the pattern sequencer to turn the entire Ultrabeat interface into a pattern sequencer, as shown in Figure 9-14. Click any grid area to turn a drum sound on or off on that beat. Visualizing the entire pattern with all the drum sounds makes programming beats a breeze.

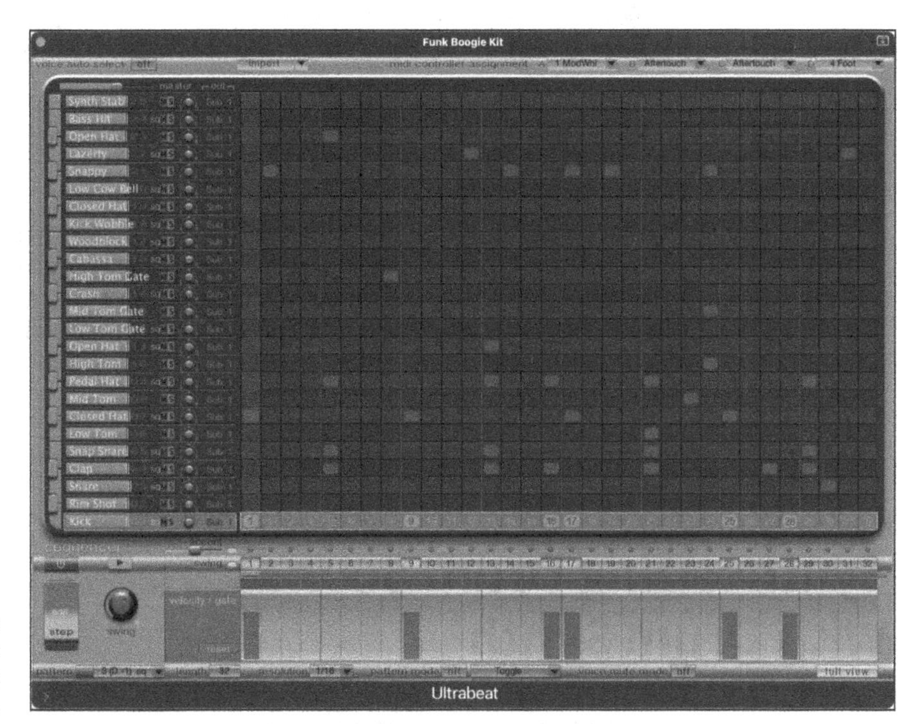

FIGURE 9-14: The Ultrabeat step grid in full-view mode.

Controlling patterns with MIDI

You can trigger the Ultrabeat patterns by using MIDI. Along the bottom row of parameters in the step sequencer is the Pattern Mode on/off button and playback mode menu. To trigger the patterns:

1. **Turn on the Ultrabeat pattern sequencer.**
2. **Turn on pattern mode.**
3. **Play the key associated with the pattern.**

   The pattern will begin playing. To find the key associated with the pattern, click the pattern menu to see the list of patterns and corresponding MIDI keys. You can decide how the patterns behave when triggered from the playback mode pop-up menu, described in the previous section.

TIP

Click the Voice Mute Mode button to mute single drum sounds with MIDI keys. Use the two octaves above the pattern trigger keys (C1–B2) to mute and unmute drum sounds. Muting drum voices is useful for remixing and adding interest to your drum arrangements.

Exporting patterns into the tracks area

The Ultrabeat pattern sequencer is a fun and excellent tool for building beats. But sometimes, you might want more control over the MIDI patterns. Exporting your Ultrabeat pattern into the tracks area is simple.

Drag the pattern icon from the bottom of the global parameters section into the tracks area. A MIDI region the length of the pattern will be added to the tracks area. If you're going to trigger the Ultrabeat sounds from a MIDI region instead of the pattern sequencer, make sure you turn off the pattern sequencer in Ultrabeat. For a video that details how to make beats and patterns with Ultrabeat (and Drummer), visit `https://logicstudiotraining.com/chapter9`.

TIP

Software instrument tracks with Ultrabeat get their own set of smart controls.

Synthesizing Drum Sounds with Drum Synth

Drum Synth is a versatile software instrument that can be used to create electronic drum sounds. Drum Synth gives you access to a powerful drum synthesis engine with the controls you need to create sounds quickly.

Exploring the Drum Synth interface

Drum Synth, shown in Figure 9-15, has four global parameters:

» **Group type pop-up menu:** Click the group type pop-up menu to choose between the four sound groups: Kicks, Snares and Claps, Percussion, Hats and Cymbals. Depending on the chosen group type, parameters unique to the group will be shown on the interface.

- **Sound type pop-up menu:** Click the sound type pop-up menu on the upper left of the interface to choose the preset sound types. Updating the sound type also updates the parameters on the interface.
- **Instrument icon:** Click the instrument icon to play the sound.
- **Key tracking button:** When key tracking is on, you can play the sound chromatically. Only the root note will play when set key tracking is off, regardless of the note played.
- **Mode pop-up menu:** Three playback modes are available. Mono allows only one note to be played at a time. Poly allows you to play multiple notes at a time. Gate requires a held key for the sound to be heard.

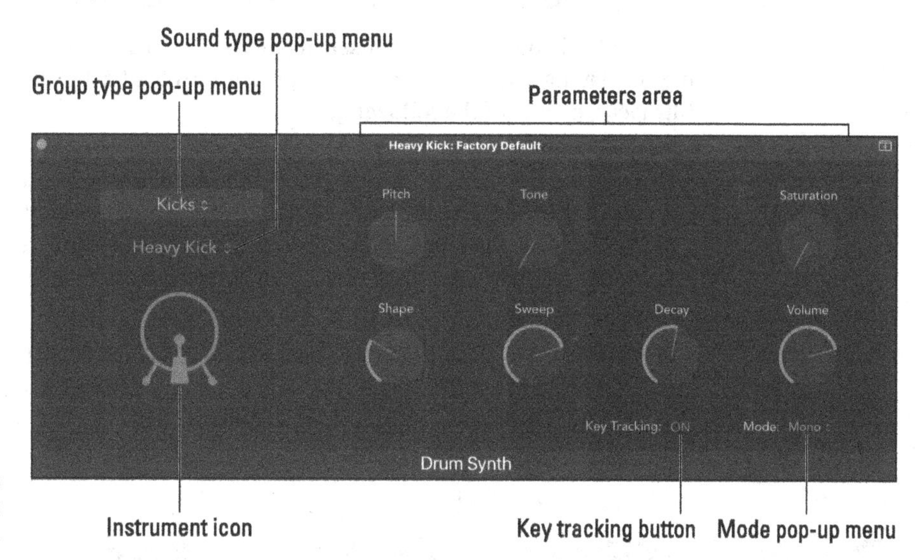

FIGURE 9-15: The Drum Synth interface.

Editing Drum Synth sounds

Drum Synth provides a variety of ways to edit the sound. The first step is to select a group type from the Group Type pop-up menu. The available sound types will be updated in the Sound Type pop-up menu. Next, choose a sound type from the list. You can now edit the parameters on the interface to shape the sound. For example, the parameters for the Snares and Claps group type are shown in Figure 9-16.

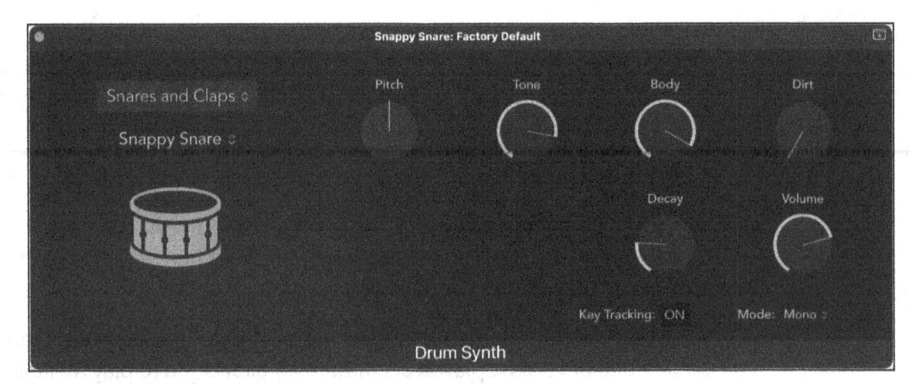

FIGURE 9-16: The Snares and Claps group type.

Loading Drum Synth settings into Ultrabeat

You can load your Drum Synth settings into Ultrabeat for further sound processing or playing with the Ultrabeat sequencer. To load a Drum Synth setting into Ultrabeat, do one of the following:

» From the preset menu at the top of the Ultrabeat interface, choose Load and then navigate to the Drum Synth presets located on your hard drive at ~/Music/Audio Music Apps/Plug-In Settings/Drum Synth. The Drum Synth setting will load into the Ultrabeat assignment section, where you can further edit the sound.

» Click the instrument slot in the channel strip of a Drum Synth track and choose Ultrabeat from the pop-up menu. Ultrabeat will automatically load the Drum Synth setting into the assignment section.

Designing Electronic Drums Kits with Drum Machine Designer

Drum Machine Designer is a software instrument interface for building electronic drum kits and customizing drum sounds. Drum Machine Designer loads automatically with many of drummer's library presets. Drum Machine Designer includes the following components:

» **Drum Machine Designer interface:** Contains a drum grid for choosing drums and smart controls for shaping sounds, which you explore in the next section.

- **Ultrabeat software instrument:** The default sound engine used by Drum Machine Designer is Ultrabeat, but any software instrument in your library can generate sounds.
- **Track and channel strip group:** Adding Drum Machine Designer to your project automatically creates tracks in the tracks area so you can record each drum sound separately. Channel strips are added to the mixer so you can shape the sound of each drum. You learn how to use the mixer in Chapter 16.

You can load Drum Machine Designer in the following ways:

- Create a drummer track (choose Track ➪ New Drummer Track) and select any drummer character from the Hip-Hop or Electronic menus in the library, as shown in Figure 9-17.
- Create a software instrument track (press Option-⌘-S or choose Track ➪ New Software Instrument Track) and select any patch from the Electronic Drum Kit ➪ Drum Machine Designer menu in the library. (Press Y to open the library.)
- Create a software instrument track and select Drum Machine Designer from the instrument slot in the channel strip inspector. (Press I to open the inspector.)

FIGURE 9-17: Drum Machine Designer on a drummer track.

Exploring the Drum Machine Designer interface

The top half of the Drum Machine Designer interface contains a drum grid, as shown in Figure 9-18. Each cell contains a drum sound. The lower half of the interface contains smart controls that dynamically update depending on the selected cell. If you click the header at the top of the plug-in interface, the smart controls area will update to show controls that affect the entire drum kit.

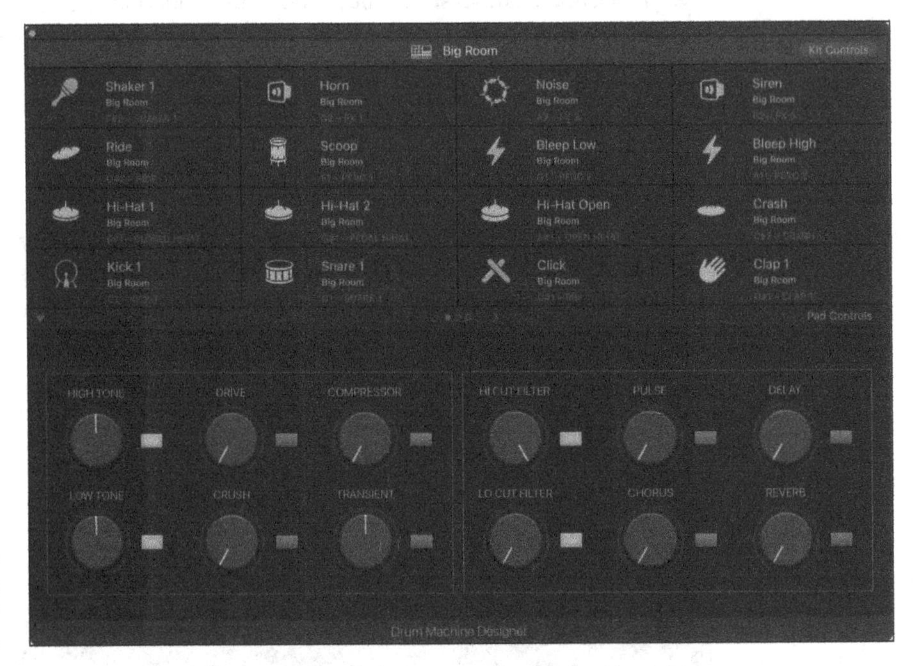

FIGURE 9-18: The Drum Machine Designer interface.

Playing drums and kit pieces

Click any of the cells in the drum grid to play the drum sounds. You can also play the drum sounds with your MIDI controller. Each drum cell has a corresponding MIDI note visible at the bottom-right corner of the cell to help you determine which notes to play on your controller. If Drum Machine Designer is loaded on a drummer track, you can play the drum sounds with the drummer editor, as you learned previously in this chapter.

Editing drum sounds

You can change drum kits and individual drum sounds by selecting patches in the library (Press Y to display the library). Click the plug-in header and select a patch in the library to change the entire drum kit. Click a cell and select a patch to change a single drum sound.

Adjust individual drum sounds by clicking a cell and using the smart controls. When you hover your cursor over a drum cell, mute and solo icons are visible. Click the mute icon to silence the drum sound, and click the solo icon to mute all other sounds. You can also drag audio files onto the cells and save them as patches for use in other kits.

TIP

The drum grid has multiple pages of drum cells. You can reorder your drum cells by dragging them to new locations. Reordering your drum cells won't change the corresponding MIDI note or the sound, but it can help you organize your kit and get the sounds you need on the same page.

Wow. Drums are a big deal. And Logic Pro puts many resources into making beats and banging drums. But more importantly, the intuitive interfaces and raw power of Logic Pro make it easy for you to give your drums the attention they deserve. You can control the smallest detail or have Drummer and Ultrabeat play for you.

Go play!

IN THIS CHAPTER

- » Loading and playing the vintage instruments
- » Editing instrument sounds and effects
- » Producing classic and creative tones
- » Understanding keyboard and synthesizer fundamentals

Chapter 10
Playing Virtual Vintage Instruments

The vintage instruments included in Logic Pro might make you want to put drink rings and cigarette burns on your laptop because these classic keyboards can evoke the sounds of yesteryear. But they also can freshen today's music — you'd be hard-pressed to throw a dart at the Billboard charts and not find one or more of these instruments somewhere in the track.

The sounds of these vintage keyboards are heard in almost all musical genres. Music producers depend on them, and musicians adore playing them. And what's not to love? They're simple to use, and their tonal capabilities are limitless. Perhaps best of all, you can get great sounds with these software instruments without the hassle of vintage gear upkeep and the two or three strong bodies required to transport them.

In this chapter, I show you how to load, edit, and play the vintage software instruments in Logic Pro. You find out how to use them creatively so they can meet your demands. You also pick up some tips on making them sound authentic and classic. If a client ever craves a vintage keyboard, you'll be able to deliver the right sound.

Taking Stock of Vintage Instruments

Logic Pro gives you five virtual vintage software instruments:

- **Vintage B3 Organ:** This software instrument is modeled on a Hammond B3 organ and Leslie speaker. The Hammond B3 is an electric organ that generates sound from tone wheels and spinning discs that determine the pitch and transmit the signal to a magnetized rod and coil. The Leslie speaker contains two speakers: The top passes the sound through a rotating horn, and the bottom passes through a rotating drum.
- **Vintage Clav:** This instrument is modeled on a Hohner Clavinet D6 keyboard. The Clavinet (Clav for short) has guitar-like strings that are struck by a hammer. Magnetic pickups transmit the signal through a filter section to shape your sound.
- **Vintage Electric Piano:** This one simulates three famous electric pianos — Fender Rhodes, Wurlitzer, and Hohner Electra-Piano. These electric pianos have hammers that strike metal tines or reeds. All three electric pianos are similar in sound and have metallic bell-like qualities.
- **Vintage Mellotron:** The original Mellotron uses magnetic tapes and a tape head to play back recordings of acoustic instruments, such as voices, strings, and other orchestral instruments.
- **Retro Synth:** It's four classic synthesizers in one — analog, sync, wavetable, and FM. Each type of synthesis gives you a completely different sound.

Loading and playing vintage instruments

To use the vintage instruments in Logic Pro, you must first create a software instrument track: On the main menu, choose Track ➪ New Software Instrument Track. The default software instrument is the vintage electric piano. You can add a vintage instrument to the selected software instrument track in two ways:

- On the main menu, choose View ➪ Library and select a preset (called a *patch* in Logic Pro) from the list of vintage instruments.

TIP

Choosing Retro Synth patches from the library can be tricky because all synthesizer patches in Logic Pro are grouped into a single Synthesizer menu. It's faster to choose Retro Synth by using the next method. Later in the chapter, I show you how to save your own patches so you can quickly call up Retro Synth from the library.

» Choose View ⇨ Show Inspector and select the vintage instrument in the software instrument slot. (See the section on the inspector in Chapter 3 for details.)

To play your software instrument, connect your MIDI controller (as described in Chapter 1), select the track, and start playing your MIDI controller. If you don't have a MIDI controller connected, you can choose View ⇨ Show Musical Typing to input notes with your computer keyboard or View ⇨ Show Keyboard to input notes with your mouse or trackpad.

Choosing vintage instrument presets

Every software instrument has a preset menu at the top of its interface. These presets enable you to load and save your instrument settings.

REMEMBER

A preset saves only the software instrument settings. Other settings on the track's channel strip, such as volume and effects, are not saved. If you want to save the entire channel strip, including the software instrument settings, save your settings as a patch in the library (see Chapter 4).

TIP

When editing a factory preset, it's best to make your changes to a copy. I like to keep factory presets in their original state and make my edits on copies. Also, the user presets you create appear on the main menu, not on a submenu, saving you a click down the road.

TIP

Save your default preset for instant recall when loading instruments. On the preset drop-down menu at the top of the software instrument interface, choose Save as Default. I have a starting position that I use on my genuine Hammond B3 that I've copied over to the vintage B3 and saved as my default. Now, whenever I call up the vintage B3, it's exactly the way I like it.

Spinning Your Tone Wheels with the Vintage B3

The Hammond B3 is an important part of music history, and with the vintage B3, you can put its sonic stamp on your own music. Since the Hammond B3 organ is found in country, gospel, reggae, rock, jazz, pop, and more, it's safe to say that you'll want to use the vintage B3.

FOR THE LOVE OF HAMMOND!

Every instrument has a story. The tale of my first Hammond organ is a journey through a blizzard in whiteout conditions. On the way home, with hundreds of pounds of organ in the back of my Chevy Bronco and less than 10 feet of visibility, I almost drove into the wrong lane of an on-ramp on the opposite side of the highway.

Looking back, I don't think I'd do anything differently. You couldn't keep me from taking that organ home where it belonged. It was a Hammond L-100 with a Leslie 122 speaker, both painted white to mimic the ones owned by The Beatles. The owner chopped it in half so it could be transported easily. But there was nothing easy about carrying those giant chunks of organ up three flights of stairs into my tiny studio apartment.

Tubes and tone wheels are worth the pain. Today, I have a Hammond B3 and Leslie 122 in my living room, and each day I say "Good morning!" to them while I'm having my first cup of coffee. I'm biased, but I don't think it's possible to have too much organ in your tracks. A Hammond B3 is always *just right*.

A Hammond B3 is almost always paired with a Leslie speaker. The Leslie's rotating design throws the organ sound throughout a room to make it sound bigger and omnipresent, like a pipe organ. The vintage B3 has a built-in Rotor Cabinet, which emulates the Leslie speaker (built by Don Leslie).

The vintage B3 software instrument and Rotor Cabinet simulation emulate the real thing to the point that you can edit the age of a single component to make your organ sound fresh from the factory or 50 years old.

In this section, I show you how to use the vintage B3 drawbars and other organ parameters to shape your sound. You set up your default organ settings just like a professional B3 player. You also learn how to play the vintage B3 with multiple keyboard controllers for an authentic Hammond playing experience.

Understanding drawbars

Before I show you how to tweak the sound you can get from the vintage B3 organ, you should create a new instrument track and select a preset for the vintage B3, as follows:

1. **On the main menu, choose Track ⇨ New Software Instrument Track.**

   A new track appears in the main window track list.

2. **Choose View ➪ Show Library.**

   The Library window appears.

3. **Click the Vintage B3 Organ menu and select a preset.**

   For example, select Classic Rock Organ to create a sound similar to what organists used in 60s and 70s rock music.

4. **Access controls for the vintage B3 in one of two ways:**

   - Choose View ➪ Show Smart Controls to access a set of drawbars, Rotor Cabinet controls, and effects controls.
   - Choose View ➪ Inspector and click the center of the instrument slot to see the full software instrument interface.

REMEMBER

Your drawbars are your main tools for tone shaping. You can always get to them quickly by using your track's smart controls or by clicking the Main button on the top control bar of the vintage B3 organ interface.

Take a look at your three sets of drawbars — Upper, Pedals, and Lower — as shown in Figure 10-1. Pull the drawbar out (by pulling down) to make the drawbar louder, and push the drawbar in (by pushing up) to make the drawbar quieter or silent. Drawbars are like mixer faders but in reverse.

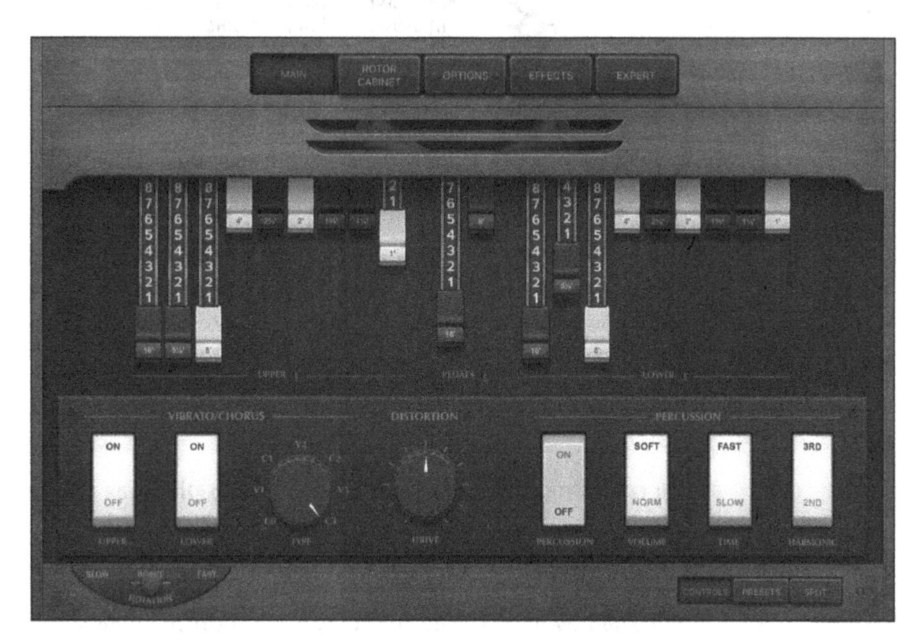

FIGURE 10-1: Main window of the vintage B3 Organ.

The upper and lower nine drawbars correspond to the upper and lower keyboards of a Hammond B3 organ. These two keyboards are called *manuals*. B3 players usually play bass lines and rhythm parts on the lower manual and solos and melodies on the upper manual, though anything goes into the hands of a skilled musician. The pedal drawbar corresponds to the B3 foot pedals. Each set of drawbars controls the volume of different aspects of the pitch you are playing.

Sine waves and the harmonic series

The Hammond B3 tone wheel was designed to create a pure sine wave, the simplest sound in existence. Playing a key on the B3 spins a tone wheel that generates a pure pitch. Every pitch corresponds to a frequency, which is the number of cycles the sound completes in one second. All sounds except a pure sine wave contain different frequencies based on the mathematical harmonic series.

The third drawbar of the Hammond B3 is close to a pure sine wave. The other drawbars are frequencies of the harmonic series, as shown in Figure 10-2, and control the complexity of the sound so that you can create and mold sounds similar to a flute, an oboe, a trumpet, or even a string section.

FIGURE 10-2: The drawbars in relation to the harmonic series.

You can think of the drawbars in two sets of divisions that outline how you use them to shape the sound of the organ. The first set is divided by function and moves from left to right:

» **Subdrawbars:** The first two drawbars control the volume of frequencies below the fundamental pitch you are playing.

Technically, the second drawbar controls the volume of a frequency above the pitch you're playing, but it's mathematically related to the first drawbar, so they're joined at the hip. These two subdrawbars are used to add bass to or take bass from the overall sound.

- **Foundation drawbar:** The next four drawbars control the volume of the fundamental frequency and frequencies above it.

  The third drawbar controls the volume of the fundamental pitch. Pull it out all the way by itself, and you hear a single frequency, a simple sine wave. The third drawbar is almost always used to some degree.

  The fourth, fifth, and sixth drawbars control the second, third, and fourth harmonics. The fourth drawbar gets pulled out a lot in rock. The fifth through sixth drawbars are commonly used in gospel or whenever you want more upper harmonics in your sound.

- **Brilliance:** The seventh, eighth, and ninth drawbars control the fifth, sixth, and eighth harmonics (the seventh harmonic is skipped).

  The ninth drawbar is often pulled out in jazz solos and gives your tone a whistle. The other two drawbars are pulled out when you want extreme highs, as in a full organ sound. Another popular setting called the *squabble* uses the last four drawbars pulled all the way out along with the first drawbar.

The second set is divided by color, with each color describing the function in the harmonic series:

- **White (consonance):** The white drawbars correspond to consonant harmonics, which are perfect octaves of the fundamental pitch. Pulling all the white drawbar out gives you a tall sound of four stacked octaves.
- **Black (dissonance):** The black drawbars correspond to the dissonant harmonics, which include fifths and a major third. Don't let the name *dissonance* fool you. Adding these harmonics makes the sound big and rich.
- **Brown (sub):** The brown drawbars correspond to the subharmonics. These subfrequencies, especially the octave below the fundamental, are used frequently because they allow you to get the quintessential B3 bass sound played with the left hand on the lower manual.

  Even though the second brown drawbar is considered a subtone, it's a fifth higher than the fundamental pitch you're playing. Pull this drawbar out all the way for great solo sounds. It's often pushed in about halfway for chordal accompaniment and bass lines.

Common drawbar setups

Drawbar notation is simple to understand. There are nine drawbars, and each has eight increments of volume. A drawbar setting that includes only the third drawbar (the fundamental pitch) would look like this: 008000000. If you also pulled out the first drawbar, the setting would be 808000000.

Following are some classic drawbar settings.

- **Lower manual: 848000000.** Use this drawbar setting for playing bass lines with your left hand and accompaniment with your right. You can experiment with the level of the second drawbar until it sounds good to you. You can also pull out the ninth drawbar to 1 or 2 to add just a touch of brightness. This drawbar setting is versatile.
- **Upper manual: 888000000.** This setting is a subtle variation of the lower manual setting, but it makes a big difference. Pulling the first three drawbars all the way out gives the upper manual just enough volume to cut through the lower manual and a full band. Pull out the last drawbar between 2 and 4 to add a high whistle tone and cut through the mix even more. If you need more volume and presence to cut through the mix, pull out the fourth drawbar to 8.
- **Pedals: 80.** The pedals have only two drawbars, and the second drawbar isn't used much. Most organists push in both pedal drawbars all the way when a bass player is present. However, when there's no bass player, pull out the first bass drawbar to between 6 and 8 and blend it with the rest of the organ and band. Because you play the pedals with your feet, it's uncommon to hear intricate and fluid lines from them unless you're one of the few living virtuosos. For authenticity, keep your pedal bass lines simple.

Getting the vintage vibe with vibrato/chorus and percussion

In addition to the drawbar, the B3 has two blocks of effects that you can apply to your sound: vibrato/chorus and percussion.

Vibrato/chorus

Vibrato introduces pulsating pitch variation to varying degrees, with V1 the slowest and V3 the fastest. *Chorus* is created by mixing the vibrato with the unaffected tone. Chorus has three settings, C1 through C3, which correspond to the three vibrato settings.

You can control the vintage B3 vibrato/chorus in two ways:

- **Software instrument interface:** Click the Main button at the top of the vintage B3 interface and then click the Controls button at the bottom. A panel opens with the vibrato/chorus controls on the left and percussion controls on the right. In the center of the panel is a Drive knob that works only if you have distortion turned on in the Effects panel of the vintage B3.

  Use the toggle buttons to turn on or off the vibrato/chorus effect on the upper or lower manual. Choose the vibrato/chorus setting (V1 through C3) with the rotary knobs.
- **Smart controls:** To display smart controls in the tracks area of the Logic Pro main window, click the Smart Controls button at the top of the main window or choose View ⇨ Show Smart Controls. Or, if you want smart controls in a separate window, choose Window ⇨ Open Smart Controls. The smart controls have a toggle button that turns vibrato/chorus on or off for the upper manual. For a smart controls refresher, refer to Chapter 3.

Chorus is used more than vibrato, but the warbly vibrato makes some classic movie-theater organ sounds. The C3 chorus setting probably has the most followers because it fattens up your organ tone.

TIP

Using the chorus effect with the Leslie Rotor Cabinet gives you a sound with a lot of motion. I usually start with the Leslie set to spin slow or set to brake so it doesn't spin at all. I play with the chorus set to C3 for a fat sound with lots of motion. Then when I kick the Leslie on fast, I get a powerful shift with dramatic effects. It's an excellent technique for building musical climaxes.

Percussion

The percussion effect is a unique bell tone on top of the organ sound. *Percussion* is monophonic, meaning it plays only one note at a time. You must let go of the first key if you want the percussion to sound on the second key. This effect works well with a short, staccato playing style but can also sound great at the beginning of phrases that are connected and legato.

You need to know a couple of things about the percussion effect as it relates to a genuine Hammond B3. On a real B3, percussion works only on the upper manual. The software replicates this characteristic, but you can remove this limitation by clicking the Options button on the vintage B3 interface and changing the Perc On Preset parameter to All. Also, on a real B3, turning on the percussion turns off the ninth drawbar. The software doesn't re-create this behavior, so if you're a Hammond B3 player looking for authenticity, you'll have to turn off the ninth drawbar when you turn on percussion.

TIP

If a jazz musician has percussion turned on, it's usually the third harmonic set to fast and soft. You can create a cool piano effect by pushing in all the drawbars and playing the second harmonic set to slow and normal. Try it out.

Playing with presets

The genuine Hammond B3 has a lower octave of reverse-colored keys on both the upper and lower manual that select hardwired drawbar presets. This function is duplicated on the vintage B3. Click the Presets button on the bottom of the vintage B3 interface, and you'll see the drawbar presets laid out like a keyboard octave. You can use your keyboard controller to select these presets. The lowest note, C, clears all the drawbar settings, and no sound is heard. Keys C# through B choose presets that instantly change your drawbar. You can also select these presets manually by clicking them.

Edit the drawbar presets by selecting a preset and adjusting the drawbar. Note that these presets only store and recall the drawbar positions, not vintage B3 effects or other parameters. The original B3 has presets that emulate strings, flute, clarinet, and other acoustic instruments. You can get an amazing assortment of sounds with the vintage B3 just by playing with the presets. If you want a vintage B3 patch with the original Hammond B3 presets, visit `https://logicstudiotraining.com/b3patch` and I'll hook you up.

Spinning the Leslie

The Leslie speaker is a huge part of the Hammond B3 sound and experience. The Rotor Cabinet in the vintage B3 instrument is a fantastic emulation of the iconic Leslie speaker.

You can control Rotor Cabinet parameters in the smart controls and the vintage B3 software instrument interface:

- **Smart controls:** To access the smart controls in the tracks area, choose View ➪ Show Smart Controls. You can choose the Rotor Cabinet's rotation speed with the lever in the lower-left corner of the smart controls.
- **Software instrument interface:** Click the Rotor Cabinet button on the top control bar of the vintage B3 interface to control the Leslie speaker options.

A Leslie speaker has a rotating horn at the top and a rotating drum at the bottom. At the bottom left of the interface, click the Rotation toggle switch to change the rotation speed to Slow, Brake (off), or Fast. At the far right of the interface, switch the mic position to Rear to see the horn and drum animated in real time as you adjust the speed settings. You'll also hear a significant variation in the different mic positions. If you want to mellow the sound, set the mic position to Front, as shown in Figure 10-3.

FIGURE 10-3: The vintage B3 Rotor Cabinet options panel.

You can also change the cabinet type to other Leslie models. The Proline cabinet is a black Tolex-covered cabinet, similar to a guitar amp, with a more open sound. With the Real Cabinet type selected, you can click the microphones to see a few mono and stereo microphone choices. It's common to record the drum mono and horn stereo. Because the drum carries low frequencies, a stereo spread isn't as dramatic and can take up too much bass energy in a mix.

Choose the stereo width and balance of your mics on the left side of the Rotor Cabinet control panel. Edit the motor acceleration speed to customize the transition from slow to fast. The Motor Control setting enables you to spin the horn and drum independently.

Going deep into expert options

Three more panels of controls (Options, Effects, and Expert) give you ways to customize the B3 that would be difficult to achieve with the Hammond B3. The

Options and Expert panels let you customize how the B3 sounds and behaves, and the Effects panel adds perfectly paired effects to your sound as follows:

- **Options:** These settings control the tuning, volume, and expression level of the vintage B3.

  A key characteristic of the B3 is the volume pedal. You'll find yourself constantly adjusting the volume during a performance, and the Expression knob lets you adjust the sensitivity of your connected expression pedal.

  Another important part of the B3 sound is the key click. On the vintage B3, you can adjust several aspects of its sound in the Options and Expert panels. The Options panel also gives you finer control of percussion and vibrato/chorus.

  The Morph function allows you to smoothly switch between two presets, much like an organist would achieve with a hand pushing and pulling drawbar. You can adjust the Morph characteristics on the Options panel, and you can control it from the Presets area of the Main panel.

- **Effects:** This panel has four perfectly paired effects for the B3 — reverb, equalizer, wah wah, and distortion.

  A Reverb control gives you a choice of seven types of reverb. The Equalizer can tame or boost your lows, mids, and highs. The wah wah is one of those effects you don't usually get to try on a real B3 unless it's been retrofitted to interface with external effects devices. The same goes for distortion. While you can get the Leslie speaker to overdrive, adding five types of distortion to a B3 is only possible with a retrofit. Not so for the vintage B3.

- **Expert:** These settings turn your B3 into a factory-fresh classic or a used-and-abused road warrior. You can adjust the age of the filters, the amount of drawbar leak, and the key click color. You can further adjust pitch, sustain, and choose your hardware controller.

  The organ model options let you adjust the balance between pedals, lower manual, and upper manual. The shape slider changes your organ from a B3 to a Farfisa, Solina, or Yamaha organ. You can also adjust the bass filter and add another octave to the low end of the upper and lower manuals. If you need to lower your CPU usage, you can slide Maximum Wheels to the left, reducing the number of emulated tone wheels.

Controlling the manuals and foot pedals with MIDI controllers

If you have more than one keyboard controller or a dual-manual controller, you can get the full B3 experience by setting each controller to a different MIDI channel and playing the upper and lower manuals together. Click the Split button on the lower right of the Main panel to set the MIDI channels. Don't forget the pedals!

Funking Up the Vintage Clav

The original Hohner Clavinet keyboard has a solid link to some funky keyboard players, such as Stevie Wonder and Herbie Hancock. It has a percussive guitar-like sound that works great in syncopated funk. It also sounds great in rock, which Led Zeppelin proved in songs such as "Custard Pie" and "Trampled Under Foot." From Bob Marley to Pink Floyd, the Clavinet (Clav for short) is a dominant sound in popular music. But searching for a vintage instrument in mint condition can be difficult and expensive. Good thing you'll never have to search hard for the funky tone of a Clavinet as long as you have Logic Pro.

To play the vintage Clav, create a new software instrument track (as described previously in the "Loading and playing vintage instruments" section) and choose a patch on the vintage Clav menu in the library.

Choosing your Clav type

You can choose 14 Clav types on the pop-up menu on the left of the top control bar. Some types are based on actual instruments, but others are very different from the original instruments.

Classic I and II are models of the original Hohner Clavinet D6. Vintage I and II are modeled after road-worn Clavs. Other emulations include a harpsichord, dulcimer, sitar, and a bell-like model. These variations will give you more than just the classic Clavinet tone.

Picking your pickups

A real Clavinet is a simple instrument with a simple control set. It has upper and lower sets of magnetic pickups that transmit the sound of hammers hitting guitar strings. You can play one pickup at a time or both.

On the Main panel of the vintage Clav interface, shown in Figure 10-4, you see the Pickup Position window in the upper left. You can select each pickup and drag it along the strings to create new sounds. I wish I could do that with a real Clavinet.

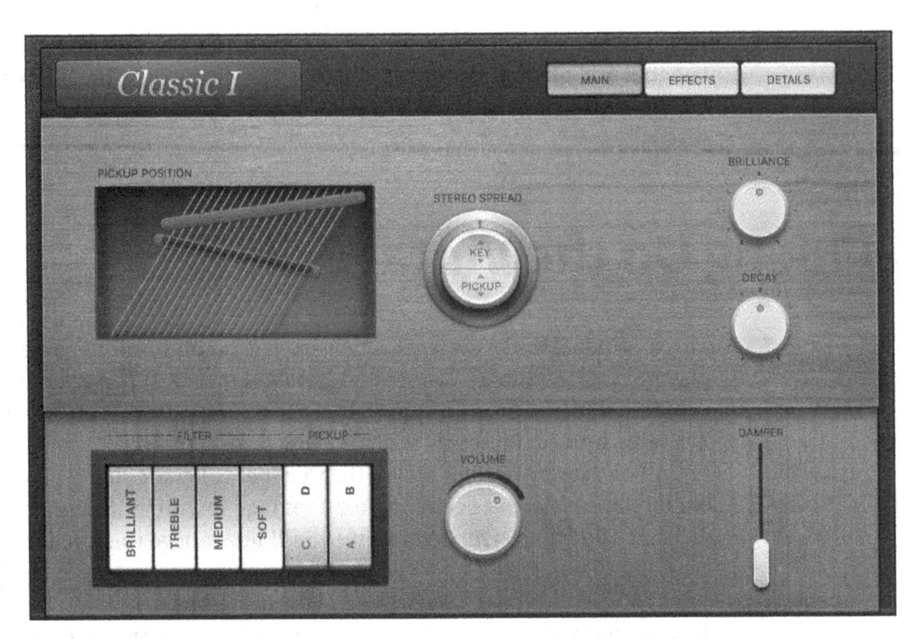

FIGURE 10-4: The vintage Clav main window.

Below the pickups are the pickup selector toggles. When you click them, the pickups change configuration as follows:

- **Position A/C:** Lower pickup only. This setting, often called the *rhythm pickup,* activates the pickup below the strings. It's one of the most common pickup settings because of its strong fundamental tone.
- **Position B/C:** Upper pickup only. This setting, often called the *treble pickup,* activates the pickup above the strings. It's used less often than position A/C due to a weaker fundamental tone, but it can be helpful when you need more treble to cut through a mix.
- **Position A/D:** Both pickups, out of phase. This setting activates both upper and lower pickups out of phase, which cancels specific frequencies, resulting in a thin and biting tone. This setting is rarely used but it's perfect for cutting through a busy mix and when you want the Clavinet to occupy a smaller sonic space in the mix.
- **Position B/D:** Both pickups, in phase. This setting activates both upper and lower pickups at the same time. It's the biggest tone of all the settings, with lots of presence, and probably the second most commonly used setting.

Filtering your pickups

Your filter buttons are to the left of the pickup buttons. Your pickup selection greatly affects how the filters affect your sound. On a real Clavinet, you must select a filter (the bottom half of the button must be down), or the Clav won't make a sound. But on the vintage Clav, when no filter is selected (the top half of each button is down), it makes a pure Clavinet sound with no filtering. That means if you want to emulate a real Clavinet, at least one of the filters should be engaged.

Dampening and spreading your tone

The Clavinet also has a Damper slider that removes the sustain of the strings and creates a shorter plucked sound, similar to a guitar being palm muted. You can engage the damper gradually by using the Damper slider at the bottom right of the interface.

Getting classic sounds with effects

A Clavinet would not be complete without a bevy of effects ready and waiting, especially the wah-wah effect. Click the Effects button on the control bar to reveal the following four vintage Clav effects:

- **Wah:** The wah-wah pedal, made famous by guitarists such as Jimi Hendrix, is a perfect partner to the Clav. Choose from eight types of wah-wah effect. Control the range and envelope with the rotary knobs. Add an expression pedal to your MIDI controller to control its funky filter sweep with your feet.
- **Compressor:** A compressor squashes your sound and reduces the dynamic range. It makes loud sounds softer and soft sounds louder and is a helpful tool for taming peaks and raising the overall sound level. Turn the Ratio knob to the right to add more compression.
- **Distortion:** Distort your pure Clav tone into a subtle crunch or an in-your-face growl. Adding distortion is excellent for rock rhythms and saturated lead tones. Adjust the Gain and Tone knobs to your taste.
- **Modulation:** You can modulate your vintage Clav with a phaser, flanger, or chorus. A *phaser* is a narrow-band EQ cut swept back and forth through the frequency spectrum. A *flanger* creates washy phase cancellations throughout the frequency spectrum. *Chorus* uses multiple delays for thickness and richness. Adjust the rate and intensity with the knobs. Clicking the note symbol in this section syncs the timing of the effect to the tempo of your song.

Extending your Clav

Click the Details button on the control bar, and an extended parameter panel gives you even more control of your Clav. The Shape slider adjusts the sound of the hammers, and the Click parameters adjust how the hammers sound attacking and releasing the string.

Playing with the Key Off parameter is worthwhile because a big part of the Clavinet sound is the click made by releasing the key. Your MIDI controller must transmit release velocity values for the Key Off parameter to work. The String and Pitch parameters make a Clavinet owner like me grateful. You can adjust your strings without having to change them and without the painstaking tuning before every session. The Misc panel provides options for using your MIDI keyboard to control the vintage Clav, including the velocity curve and number of voices. If you're pushing your computer resources, pull the Voices number down to free up some CPU.

TIP

You can adjust the stereo spread of the keys or the pickups using the Stereo Spread button. Select the key or pickup half of the button and drag it up or down to adjust the stereo spread.

Getting the Tone of Tines with the Vintage Electric Piano

The sound of the Rhodes and Wurlitzer electric pianos are iconic and flexible. No matter what genre of music you enjoy, you'll be able to find a place in your track for some vintage electric piano. To use the vintage electric piano, create a software instrument track as described previously in the "Loading and playing vintage instruments" section. The default instrument is the vintage electric piano.

Exploring tines, reeds, and tone bars

In a Rhodes electric piano, every key strikes a long, thin, metal tuning-fork–like bar called a *tine*. The Wurlitzer is similar to a Rhodes but uses steel instead of metal tines. A third electric piano available, the Electra-Piano, is modeled after the Hohner Electra-Piano and sounds similar to the Rhodes and Wurlitzer.

You can choose between 19 electric piano models on the Model pop-up menu at the top of the vintage electric piano interface, shown in Figure 10-5.

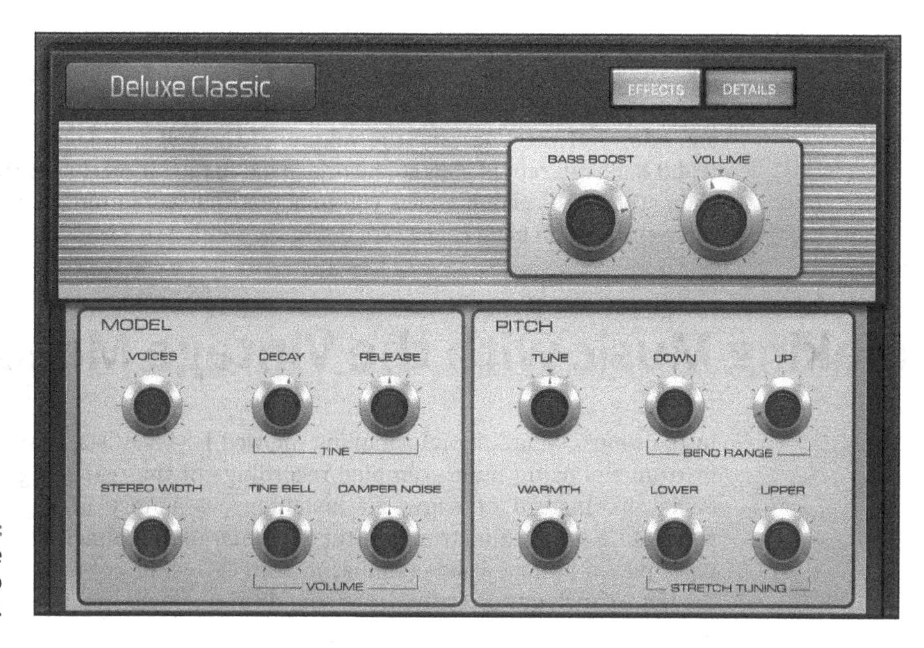

FIGURE 10-5: Vintage electric piano main window.

Producing great sounds with timeless effects

Every electric piano model gives you access to the following classic and great sounding effects:

- **EQ:** A simple EQ can adjust your bass and treble frequencies.
- **Drive:** Two types of drive distort and crunch your electric piano.
- **Chorus, phaser, and tremolo:** The chorus, phaser, and tremolo modulate your sound into a shimmery and pulsating piano. Chorus uses multiple delays to give you a thick and rich sound. Phasing is a narrow-band EQ cut swept back and forth through the frequency spectrum. Tremolo varies the volume and is a standard effect built-in to a Wurlitzer. Click the note symbol in the Phaser and Tremolo sections to sync the timing of the effect to the tempo of your song.

Playing with additional parameters

The vintage electric piano is the most straightforward keyboard of the bunch, but it still gives you modifications that would be hard to achieve with a real electric piano. Click the Details button in the control bar to access the advanced parameters.

Like the other vintage instruments, you can adjust the number of voices if you find yourself running out of computer processing power. You can adjust the decay and release of the tines and the volume of the bell and damper noise. A Stereo Width knob can rein in your sound or expand to the far reaches of the stereo spectrum. The Pitch section allows you to adjust the overall tuning, warmth, stretch tuning, and pitch bend range. Try doing that with a real Rhodes!

Making Music with the Vintage Mellotron

In the 1960s, before digital sampling allowed keyboardists to simulate acoustic instruments using multi-sampled recordings of the real thing, there was the Mellotron. Instead of using an instantaneous library of digital samples to re-create a complex instrument, the *Mellotron* used a bunch of electromagnetic tapes played by tape heads (similar to reel-to-reel or cassette tapes). And even though technology has evolved, the sounds of the Mellotron remain timeless.

The Vintage Mellotron re-creates many of the endearing quirks of the original, such as pitch fluctuations and imperfect recordings. You get the vibe of a unique instrument along with its instantly recognized flutes, strings, and choirs. Many limits are removed, such as the 8-second limit of the original Mellotron tapes, allowing you to loop the sounds indefinitely. The Vintage Mellotron captures all of the original's lo-fi character but puts it in a software instrument environment, so you have flexible control and pristine sound.

To add the Vintage Mellotron to your project, do either of the following:

- » Select a software instrument track (see Chapter 4) and choose any patch in the Vintage Mellotron menu of the library. (Press Y to open the library.)
- » Click the instrument slot in the channel strip inspector of a software instrument track and choose Vintage Mellotron. (Press I to open the inspector.)

The interface is divided into upper and lower sections, with an extended parameters section you display by clicking the disclosure triangle, as shown in Figure 10-6.

The upper section of the interface is where you choose sounds. Click the Sound A and Sound B pop-up menus to choose your instrument, and use the Blend A/B knob to set the balance between the two sounds. The lower section shapes the sound further. Use the Tape Speed and Tone knobs to change the instrument's sound quality. Use the Volume knob to raise or lower the overall volume of the Vintage Mellotron. The extended parameters area allows you to change the speed of the attack and release of the sound. You can also adjust the pitch bend range and choose sounds to be played with a fixed velocity.

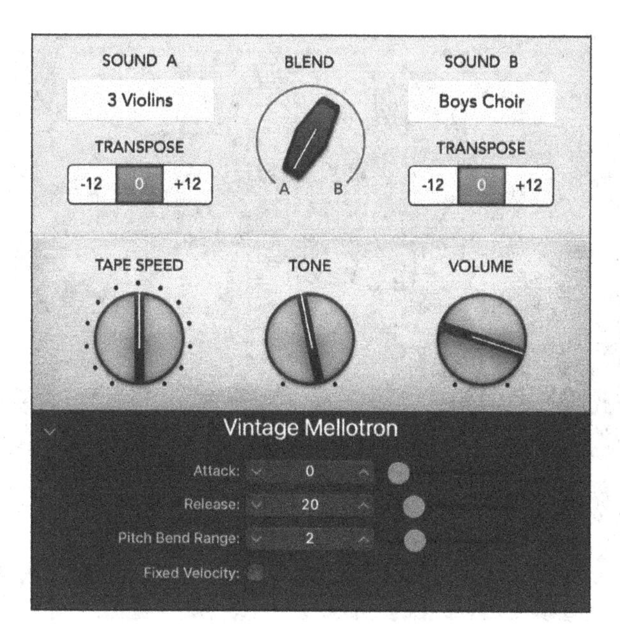

FIGURE 10-6: The Vintage Mellotron main window.

Fusing Four Synths with Retro Synth

Retro Synth is four synths in one. It's a 16-voice synthesizer that can easily switch between four of the most popular types of synthesis. From deep bass to screaming leads, Retro Synth has you covered. The best part is how easy it is to dive in and get good sounds quickly. The second best part is that all these synths don't collect dust in a wing of your rock star mansion.

Choosing your flavor of synthesis: Analog, sync, wavetable, and FM

The upper-left section of the Retro Synth interface, shown in Figure 10-7, allows you to choose between the following synth engines:

- **Analog:** Analog synthesis is found in classic synthesizers. It's great for leads, basses, and pads.
- **Sync:** Sync is a more aggressive type of synthesis. It's less suitable for pads and just right for leads and basses.
- **Wavetable:** Wavetable synthesis is used for real instrument sounds in addition to synthesized sounds. It's perfect for synth/acoustic hybrids.
- **FM:** Frequency modulation (FM) synthesis creates classic digital sounds. It's excellent for bells, electric pianos, and bass sounds.

FIGURE 10-7: The Retro Synth main window.

Controlling your synth parameters

After choosing a synthesis type, you can further shape your sound in the Oscillator and Filter sections:

- **Oscillator:** The Retro Synth oscillators generate the waveforms that form your basic synth sound. The controls in the Oscillator section change depending on which type of synthesis you have selected. Synths are fun when you simply play with the knobs and sliders and see what happens. The Oscillator section is where you do your primary tone shaping.
- **Filter:** The Filter section shapes the sound coming from your oscillators. Choose the type of filter in the pop-up title bar at the top of the section. You can choose from low-pass, high-pass, band-pass, band reject, or peak filters. Each filter allows specific frequencies to pass through and can be further modified with the other filter controls.
- **Amp:** The Amp section sets the global synth volume. You can also mix a sine wave with your synth sound to make it bigger.
- **Glide/Autobend:** Choose either Glide or Autobend in the pop-up title bar at the top of the section. Glide controls how the pitch of a note bends into the pitch of the following note. Autobend controls how a note bends when you first play it.

- **Global:** Click the Settings button on the bottom right of the Retro Synth interface to get to the global settings. Here you can adjust pitch settings, the stereo spread and choose how many voices can be played at once. You can also adjust how your MIDI controller interacts with the Retro Synth.

Modifying synth effects

On the top right of the Retro Synth interface is a simple Effects section where you can choose between a chorus and flanger:

- **Chorus:** Chorus gives you a thick and rich sound by the use of multiple delays. You can adjust the mix of the chorus as well as the chorus rate.
- **Flanger:** Flanging combines the original signal with itself, creating washy phase cancellations throughout the frequency spectrum. Like with the chorus, you can adjust the mix and rate.

Modulating the synth

Modulation alters your sound to make it interesting and exciting. You can use modulation to create vibrato, modify the filters, affect the volume, and many other less-than-realistic effects.

You can modulate your oscillator waveforms with the following parameters:

- **LFO:** Your low frequency oscillator (LFO) is a waveform that will modulate your sound. Choose different waveforms and rates to create unique modulations. You can control the LFO from your MIDI controller's modulation wheel or with *aftertouch* (pressure applied to a key on a keyboard while the key is being held down) on the source pop-up menu on the bottom right of the LFO section or both.
- **Vibrato:** Add vibrato to your synth sound by using the same controls as the LFO.
- **Filter envelope:** The filter envelope adjusts the attack, decay, sustain, and release (ADSR) of your filter. Drag the envelope handles in the display to adjust your filter envelope.
- **Amp envelope:** Similar to the filter envelope, the amp envelope adjusts the ADSR of the overall Retro Synth volume. Faster attack times create instantaneous sounds, while slower attack times make the sound appear gradually.

» **Controller:** Click the Settings button on the bottom right of the Retro Synth interface to adjust global parameters such as tuning and the number of voices. The right side of the Settings section includes parameters for your MIDI controller modulation wheel, aftertouch, and velocity.

One of the best ways to learn how to program a synthesizer is to open a preset that you like and see how the sound is created. Compare two different sounds to see which parameters are affecting the sound. And don't be afraid to fiddle with the knobs. It might look vintage, but it won't break!

IN THIS CHAPTER

» Playing the Logic Pro synths

» Understanding synthesis fundamentals

» Getting classic synth sounds

» Modeling acoustic instruments

Chapter 11

Sound Design with Synths and Samplers

Synthesizers are crucial to popular music, and Logic Pro gives you enough synth power to consider selling every keyboard you own. If you don't own any keyboards, you might have just saved thousands of dollars by buying Logic Pro. You now own several instruments modeled after classic synths and some innovative sound design tools that will enable you to create sounds you've never even imagined. And some of the fun things you'll discover about playing with synthesizers are all the happy accidents you encounter along the way.

In this chapter, you discover how to emulate classic synths, create sampler instruments, and model acoustic sounds and sounds that would be hard to create in the analog world. You'll also learn basic synthesis fundamentals to help you navigate these synths and design your own sounds.

Exploring the Logic Pro Synths

With Logic Pro, you get a whole bunch of instruments that could easily replace every synth and keyboard you own. They're powerful and flexible — and they sound amazing. Also, they can seem daunting to program when you look at all the controls and parameters you can adjust. In this section, you get a tour of the instrument interfaces and parameters. For a more detailed video demonstration of what these synths can do, visit `https://logicstudiotraining.com/chapter11`, where I show you how synthesizers work and how to use these synths in your music.

REMEMBER

To play the Logic Pro synths, you must create a software instrument track and select the instrument from the channel strip instrument slot as follows:

1. **Choose Track ➪ New Software Instrument Track (or press Option-⌘-S).**

   A new software instrument track is added to the track list.

2. **Choose View ➪ Show Inspector (or press I).**

   The inspector opens to the left of the track list.

3. **Click the right side of the instrument slot and choose the instrument you want.**

   The software instrument interface opens.

Before you begin the tour of Logic Pro's fabulous synth collection, it's important to understand some basic synthesis terms:

- **Oscillator:** A synthesizer *oscillator* produces a continuous signal that forms the basis for your sound. Oscillators can produce several waveform shapes with different tonal qualities. Oscillators are the most important part of the synthesizer because they create the sound that the other synth parameters will shape.
- **Modulation:** A static synth sound gains interest when it's varied in some way. *Modulation* is the process of varying synthesizer parameters. Vibrato is a common example of modulation.
- **Filter:** Synth sounds are shaped through the use of filters. *Filters* remove parts of the frequency spectrum, allowing you to contour the sound.
- **Envelope:** A synth *envelope* shapes the beginning, middle, and end of your sound. The most common envelope adjusts the attack, decay, sustain, and release (ADSR). For example, a piano has a fast attack, fast decay, medium sustain, and fast release.
- **LFO:** A *low frequency oscillator (LFO)* is a signal, usually below the audible frequency spectrum, that modulates a signal. LFOs are used to alter the original signal in some way. A common use of an LFO is to create vibrato.

The EFM1 FM synth

The EFM1 FM synth, shown in Figure 11-1, sounds like the 80s classic Yamaha DX7, one of the most popular digital synthesizers of all time. The EFM1 uses FM (frequency modulation) synthesis to get digital sounds such as electric pianos, bells, organs, basses, and other cool and complex sounds. The EFM1 can play 16 simultaneous voices and, unlike the DX7, is easy to program.

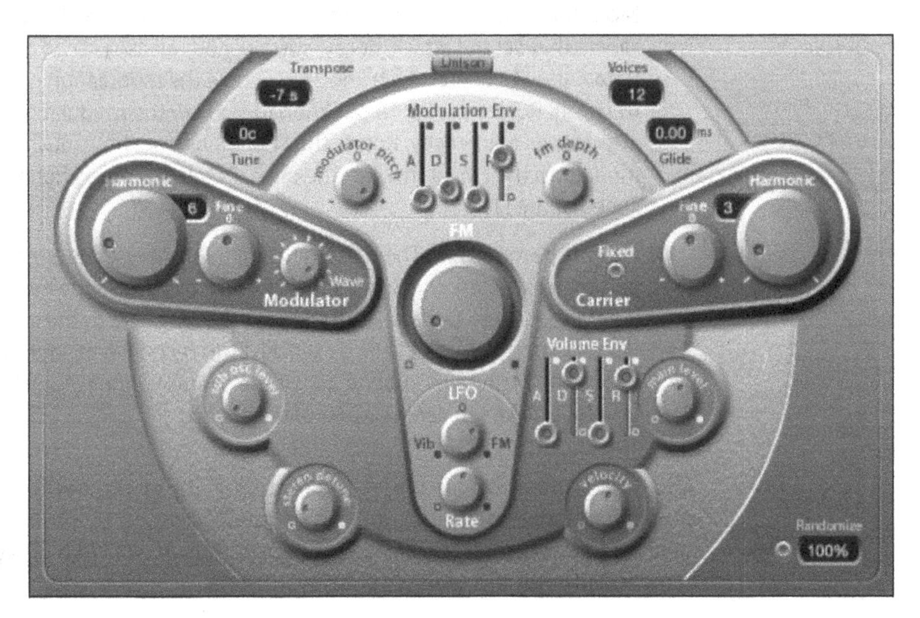

FIGURE 11-1: The EFM1 FM synth.

To design sounds with FM synthesis, you set the modulator and carrier parameters and then alter the FM intensity. The tuning ratios between the modulator and carrier set the harmonic overtones. The FM intensity sets the level of the overtones. Here's a description of the EFM1 parameters:

- **Modulator parameters:** The modulator parameters are on the left side of the EFM1 interface. Rotate the Harmonic knob to set the modulator signal's tuning ratio. Rotate the Fine tune knob to adjust the harmonics. Turn the Wave knob all the way to the left to set the modulator waveform to the traditional FM sine wave or anywhere to the right for additional waveforms. Rotate the large center FM knob to adjust the FM intensity.
- **Carrier parameters:** The carrier parameters are on the right side of the EFM1 interface. Rotate the Harmonic knob to set the carrier signal's tuning ratio. Rotate the Fine tune knob to adjust the harmonics. Click the Fixed Carrier button to avoid modulating the carrier by the keyboard, pitch bend, or LFO.

» **Global parameters:** In the top section of the EFM1, you can set global parameters. Click the Transpose field and Tune field to change the pitch of the EFM1. Click the Voices field to choose how many notes can be played simultaneously. Click the Glide field to set the time it takes to slide from one pitch to another, also known as *portamento.* Click the Unison button to layer voices and make the sound richer, reducing by half the number of voices that can be played simultaneously.

» **Modulation parameters:** In the center of the EFM1 are Modulation Envelope sliders that set the attack, decay, sustain, and release (ADSR) of the sound. Rotate the Modulator Pitch knob to set how the modulation envelope affects the pitch. Rotate the FM Depth knob to set how the modulation envelope affects the FM intensity. Rotate the LFO knob to set how much the LFO modulates the FM intensity or the pitch. Rotate the Rate knob to set the speed of the LFO.

» **Output parameters:** The bottom half of the EFM1 is dedicated to output parameters. Rotate the Sub Osc Level knob to increase the bass response. Rotate the Stereo Detune knob to add a chorus effect to the sound. Rotate the Velocity knob to set the velocity sensitivity in response to your MIDI controller. Rotate the Main Level knob to adjust the overall volume. Adjust the Volume Envelope sliders to set the ADSR of the sound.

TIP

Click the Randomize button at the bottom right of the EFM1 interface to create random sounds. Adjust the amount of randomization by clicking the Randomize field and setting the percent of randomization. If you like crazy digital sounds, 100 percent randomization is your best friend.

REMEMBER

You don't have to be a programming genius to get great sounds from Logic Pro's synthesizers. Every synth comes with a menu of presets at the top of the interface. Load a sound you like, twist some knobs and have some fun. Below the preset menu are other useful buttons such as Copy, Paste, Undo, and Redo. The Compare button allows you to compare your edited settings with the saved settings so you can edit as much as you want but always get back to your starting point.

The ES1 subtractive synth

The ES1 synthesizer, shown in Figure 11-2, creates sounds by using *subtractive synthesis,* in which you start with an oscillator and a suboscillator and then subtract parts of the sound to shape it. The ES1 is modeled after classic analog synths and is excellent at creating basses, leads, pads, and even percussion sounds.

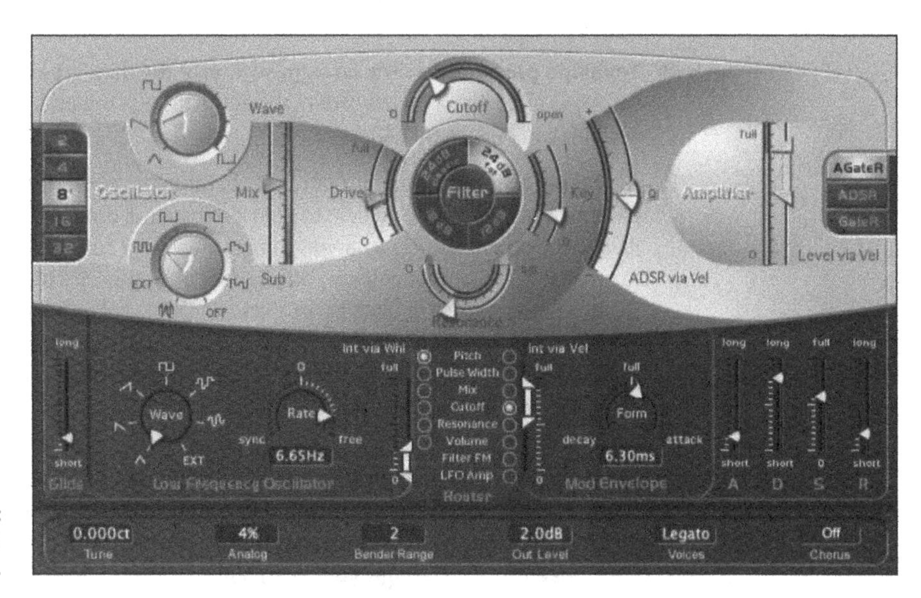

FIGURE 11-2: The ES1 subtractive synth.

A description of the ES1 parameters follows:

- **Oscillator parameters:** The left half of the ES1 interface gives you the oscillator parameters that define your basic sound. Click the buttons on the left to choose the octave. Rotate the Wave knob to set the oscillator waveform. Rotate the Sub knob to set the suboscillator waveform. Adjust the Mix slider to set the mix between the two oscillators.
- **Filter parameters:** The center section of the ES1 filters the two oscillator waveforms. Adjust the Cutoff slider to set the cutoff frequency of the low-pass filter. Adjust the Resonance slider to set the quality of the frequencies around the cutoff frequency. Click one of the four Slope buttons to choose how extreme the low-pass filter affects the signal. Adjust the Drive slider to affect the resonance setting and to overdrive the filter. Adjust the Key slider to set how the pitch adjusts the filter. Adjust the ADSR via Velocity slider to set how the filter is affected by note velocity.
- **Amplifier parameters:** The right sections of the ES1 adjust the volume level and performance. Adjust the Level via Velocity slider to set how the volume is affected by note velocity. Click the Amplifier Envelope buttons to set how the ADSR envelope affects the volume.
- **Modulation parameters:** The largest section of the dark-green area of the ES1 adjusts how the sound is modulated. The Glide parameter sets the speed of the portamento. Rotate the Wave and Rate knobs to set how the Low Frequency Oscillator (LFO) stimulates the sound over time. The Modulation Envelope sets how the modulation fades in or out. The Router parameters set the targets of the LFO and Modulation Envelope.

» **Envelope parameters:** The far right section of the dark-green area adjusts the ADSR envelope. Use the sliders to set the time of the attack, decay, sustain, and release (ADSR).

» **Global parameters:** The bottom row of parameters controls the ES1 global parameters. Click the Tune field to adjust the overall tuning. Click the Analog field to introduce random changes to the tuning and cutoff frequency, similar to an analog circuit that changes due to heat and age. Click the Bender Range to adjust the amount of pitch bend. Click the Out Level to adjust the overall volume. Click the Voices field to set the number of voices the ES1 can play simultaneously. Click the Chorus field to choose the type of built-in chorus effect that will thicken the sound.

The ES1 is an excellent instrument for getting the feel of analog synthesis. Many of the synths that follow have similar parameters. Getting the hang of setting oscillator waveforms, filters, envelopes, and modulators will help you take command of the synths and design your own sounds.

TIP

Some of these software instruments haven't had their interfaces updated since Apple introduced hardware Retina displays, which are capable of extremely smooth and crisp graphics. The consequence is fuzzy graphics with controls and text that can be difficult to read. At the top right of the software instrument is a View pop-up menu that can change the window size. If you're having trouble seeing something, make the window bigger.

The ES2 hybrid synth

The ES2, shown in Figure 11-3, is like a combined EFM1 and ES1 synth plus wavetable synthesis. A *wavetable* is made up of many different waveforms that evolve from one to another or blend at once, creating complex digital sounds. Although the ES2 can produce sounds similar to the EFM1 and ES1, it shines at creating pads, sonic textures, and synthetic sounds that evolve over time.

I skip many of the parameters that are similar to the ES1 and EFM1 and focus on the unique features of the ES2:

» **Oscillator parameters:** The three numbered oscillators on the upper-left side of the ES2 interface choose the basic sound. The triangle-shaped area to the right of the three oscillators blends them together.

» **Filter parameters:** The round section in the center of the ES2 adjusts the filters that shape your synth sound.

» **Amplifier parameters:** The top-right section contains the ES2 volume level. You can add a sine wave to the output section using the Sine Level knob.

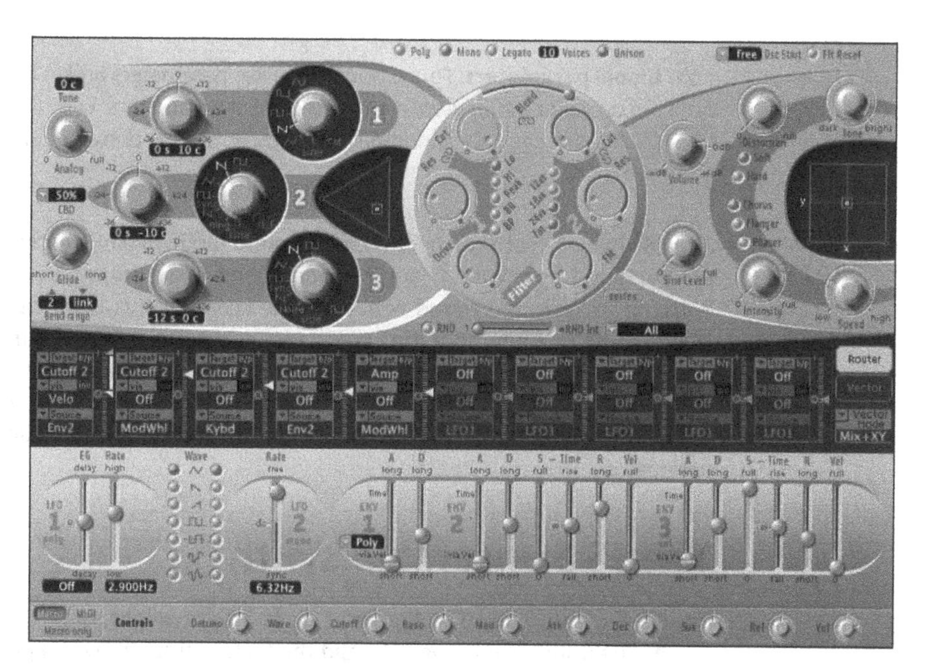

FIGURE 11-3: The ES2 hybrid synth.

- **Effects parameters:** To the right of the ES2 parameters are several built-in effects, including distortion and chorus, as well as a flanger and a phaser.
- **Planar pad:** The X/Y pad to the right of the amplifier parameters can control two parameters simultaneously. The planar pad parameters are chosen in the modulation router or vector envelope.
- **Modulation router and vector envelope parameters:** The dark-blue strip of the ES2 controls the modulation sources and targets and the vector envelope generator. You can toggle between the modulation router and vector envelope using the Router and Vector buttons on the right.
- **Modulation parameters:** Below the modulation router and vector envelope parameters are the modulation parameters. Adjust the two LFOs and three envelopes to modulate the ES2 modulation targets. You set the modulation sources and targets in the modulation router.
- **Macro controls and controller assignment parameters:** The bottom strip of buttons and knobs are where you set the macro controls and MIDI controller assignments. Click the Macro or MIDI button to toggle between the two types of controls. Click the Macro Only button to hide all ES2 parameters except the preprogrammed macro controls, which are helpful when you want to adjust the ES2 sounds globally. The MIDI controller assignments allow you to map controls on your MIDI controller to the parameters of the ES2.

» **Global parameters:** Found above the filter parameters and to the left of the oscillator parameters are the ES2 global parameters. You can tune the instrument, set the number of voices, adjust the portamento speed, and more.

TIP

The ES2 hybrid synth can be used in surround mode to pan your sound throughout the surround spectrum if you're monitoring your project in surround sound. Working in surround sound is beyond the scope of this book, but I thought you would want to know that the Logic Pro designers have seemingly thought of everything. To switch between stereo and surround, click the instrument slot and choose ES2 (Synthesizer 2) ⇨ 5.1. To get to the surround parameters, click the disclosure triangle at the bottom of the ES2 interface to display the advanced parameters.

The ES E ensemble synth

The ES E synth, shown in Figure 11-4, is a lightweight, eight-voice subtractive synth. The E stands for *ensemble,* and the ES E is great for warm pads such as analog brass and strings. Best of all, it's much easier to program than the ES1 or ES2.

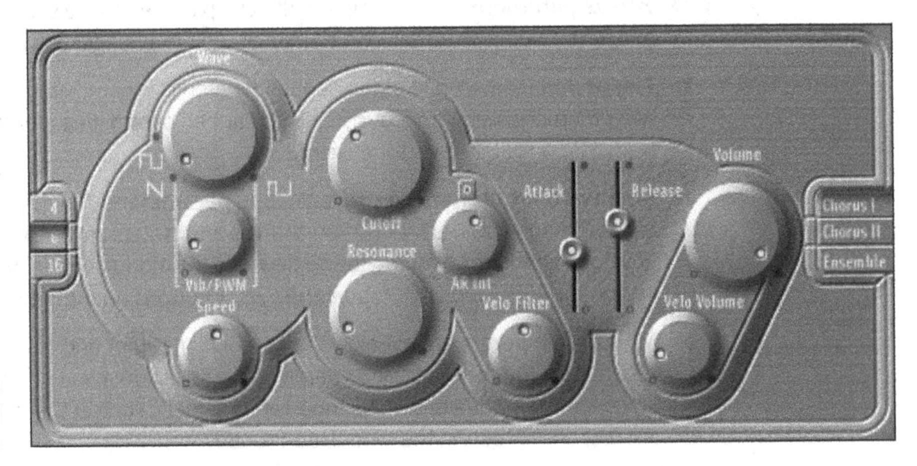

FIGURE 11-4: The ES E ensemble synth.

Here's a description of the ES E parameters:

» **Oscillator parameters:** The left side of the ES E interface adjusts the oscillator parameters. Click the buttons on the far left to choose the octave of your sound. Rotate the Wave knob all the way to the left to generate a *sawtooth wave,* which is bright with strong odd and even harmonics and excels at generating rich pads. The rest of the wave range generates *pulse waves,* which are hollow sounding with strong odd harmonics and can create excellent reedy sounds such as woodwinds.

- **LFO parameters:** The knobs below the wave parameter adjust the LFO settings. The LFO modulates the oscillator waveform. Rotate the Vibrato/PWM (pulse wave modulation) knob to set the modulation intensity. Rotate the Speed knob to set the LFO speed.
- **Filter parameters:** The low-pass filter parameters are to the right of the oscillator and LFO parameters. A low-pass filter allows low frequencies to pass through while reducing the higher frequencies. Rotate the Cutoff knob to set the cutoff frequency, and rotate the Resonance knob to raise or lower the frequencies around the cutoff frequency. Rotate the Attack/Release Intensity knob to adjust how the envelope generator affects the filter. Rotate the Velocity Filter knob to adjust how velocity affects the filter.
- **Envelope parameters:** To the right of the filter parameters are the envelope parameters. Adjust the Attack and Release sliders to set the level of your sound over time. A low attack setting will result in a more immediate sound, and a higher setting will result in a slow fade up to the final volume. A high release setting will cause the sound to slowly fade when you release the key, and a lower setting will cause the sound to fade quickly.
- **Output parameters:** To the right of the envelope parameters are the output parameters. Rotate the Volume knob to adjust the overall ESE volume. Rotate the Velocity Volume knob to adjust the velocity sensitivity.
- **Effects parameters:** To the right of the envelope parameters, you can choose a built-in effect. Choose between Chorus I, Chorus II, and Ensemble to thicken your sound.

The ES M mono synth

The ES M, shown in Figure 11-5, is another lightweight subtractive synth. The *M* stands for *mono*, meaning the ES M can play only one note at a time. Monophonic synths such as the ES M are perfect for bass and lead sounds. Like the ES E, the ES M is simple to program and features a stripped-down set of controls. Both the ES E and ES M are great instruments for learning the basics of synthesis.

A description of the ES M parameters follows:

- **Oscillator parameters:** The left side of the ES M adjusts the oscillator parameters. Click the numbered buttons on the far left to choose the octave. Rotate the Mix knob all the way to the left to select a sawtooth wave and all the way to the right to select a rectangular wave. Rotate the Mix knob between the two positions to mix the sawtooth and rectangular waves. Rectangular waves, like pulse waves, are reedy, nasal, and great for synth bass sounds. Rotate the Glide knob to adjust the speed of the portamento.

- **Filter parameters:** To the right of the oscillator parameters are the filter parameters. Rotate the Cutoff knob to adjust the cutoff frequency of the low-pass filter. Rotate the Resonance knob to boost or cut the frequencies around the cutoff frequency. Rotate the Filter Intensity knob to adjust how the envelope generator modulates the cutoff frequency. Rotate the Filter Decay knob to adjust the filter envelope decay time. Rotate the Filter Velocity knob to adjust how velocity affects the filter.
- **Volume parameters:** To the lower right of the filter parameters are the output parameters. Rotate the Volume knob to adjust the overall volume. Rotate the Volume Decay knob to adjust how the sound decays over time. Rotate the Volume Velocity knob to adjust how volume responds to velocity. Rotate the Overdrive knob to add distortion to your sound.

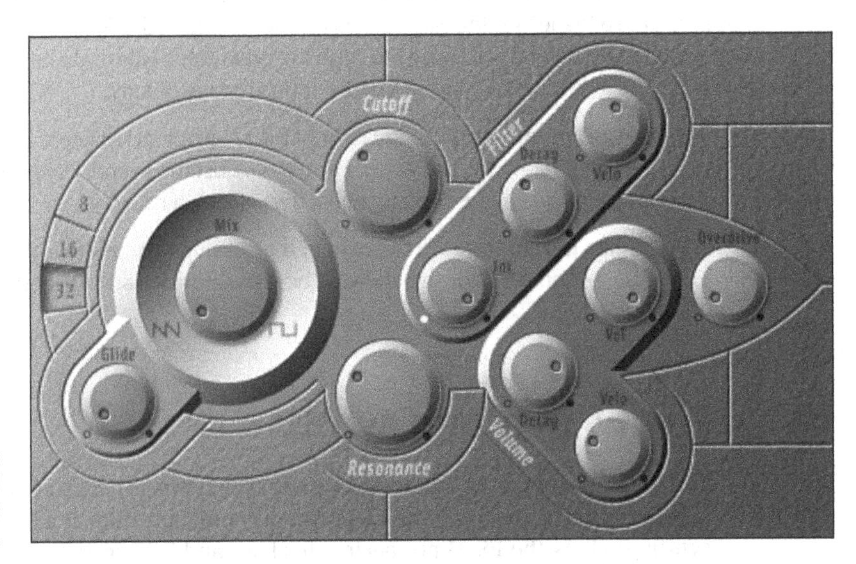

FIGURE 11-5: The ES M monophonic synth.

TIP

Click the disclosure triangle at the bottom of the interface to view the extended parameters. You can adjust the pitch bend amount and fine-tuning in this area.

The ES P poly synth

The ES P, shown in Figure 11-6, is another lightweight subtractive synth. The *P* stands for *polyphonic*; you can play eight voices at once. The ES P is modeled after classic 80s synths and does a great job of creating analog pads, bass, and brass sounds.

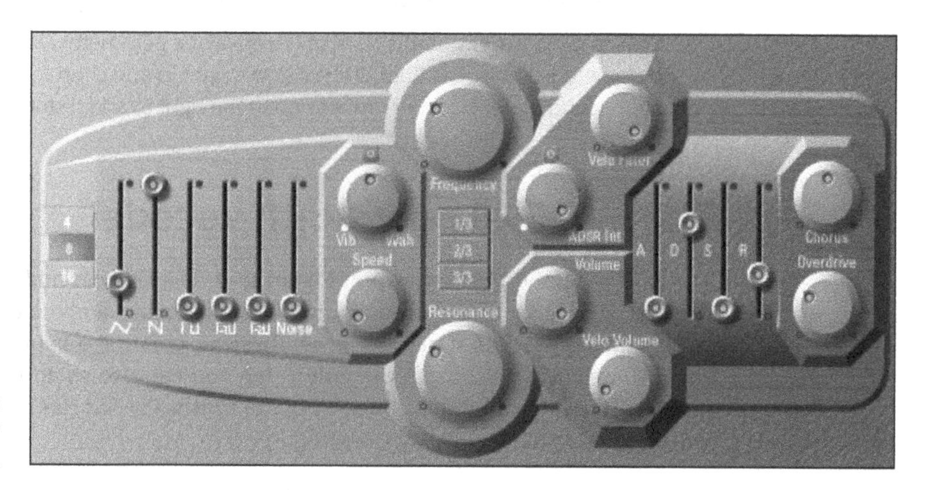

FIGURE 11-6: The ES P polyphonic synth.

Here's a description of the ES P parameters:

- **Oscillator parameters:** The left side of the ES P adjusts the oscillator parameters. Click the numbered buttons to choose the octave. The Oscillator sliders are used to mix the six oscillators. From left to right, you can set the level of a triangle wave, sawtooth wave, rectangle wave, suboscillator -1 (one octave below), suboscillator -2 (two octaves below), and noise generator.
- **LFO parameters:** To the right of the oscillator parameters are the LFO parameters. Rotate the Vibrato/Wah knob to adjust the amount of vibrato or wah-wah effect. Rotate the Speed knob to adjust the speed of the vibrato or wah.
- **Filter parameters:** To the right of the LFO parameters are the filter parameters. Rotate the Frequency knob to set the cutoff frequency of the low-pass filter. Rotate the Resonance knob to boost or cut the frequencies around the cutoff frequency. Click the 1/3, 2/3, or 3/3 buttons to adjust how the pitch affects the cutoff frequency modulation. Rotate the ADSR Intensity knob to adjust how the envelope generator affects the cutoff frequency modulation. Rotate the Velocity Filter knob to set how velocity affects the filter.
- **Volume parameters:** To the lower right of the filter parameters are the volume parameters. Rotate the Volume knob to adjust the overall volume. Rotate the Velocity Volume knob to adjust how the velocity affects the volume. Lower levels mimic classic synthesizers without velocity-sensitive keyboards; higher levels make notes louder if the key is struck harder.

» **Envelope parameters:** To the right of the volume parameters are the envelope parameters. Adjust the attack, decay, sustain, and release parameters (ADSR) to adjust the ES P envelope.

» **Effects parameters:** To the right of the envelope parameters are the effects parameters. Rotate the Chorus knob to the right to add chorus and thicken your sound. Rotate the Overdrive knob to the right to add distortion.

The EVOC 20 poly synth vocoder

The EVOC 20 poly synth, shown in Figure 11-7, is a vocoder and a 20-voice synthesizer. A *vocoder* (voice encoder) takes an incoming audio signal, typically a voice, and applies this signal to the synthesizer, creating a hybrid vocal synthesizer. However, a voice isn't the only thing you can use as an input. You could input a drum loop or an instrument into the synthesizer or run the synth without any input as a stand-alone synthesizer.

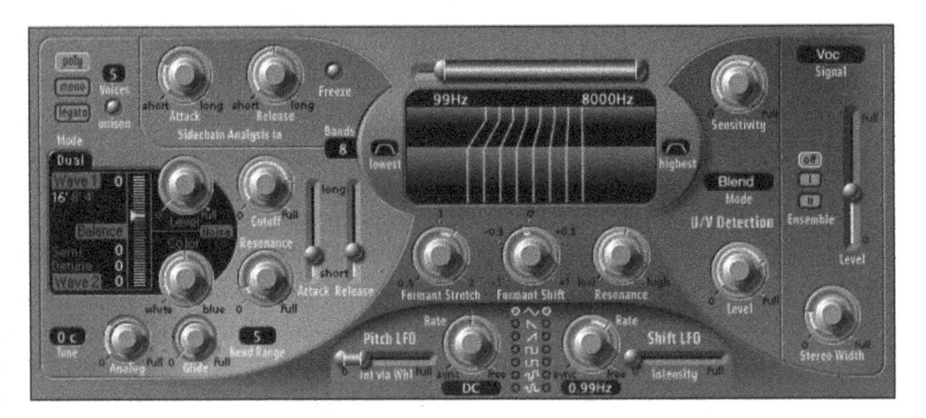

FIGURE 11-7: The EVOC 20 vocoder synth.

To use the EVOC 20 PS as a classic vocoder, do the following:

1. **On the Side Chain menu in the EVOC 20 PS plug-in header, choose the input source.**

   The source can be a live input, an audio track, or a bus. The classic vocoder effect uses a live input or prerecorded vocal track.

2. **Mute the input source so you hear only the output from the EVOC 20 PS.**
3. **Play your MIDI controller simultaneously with the input source.**

   The EVOC 20 PS synthesizes your input source.

Here's a brief description of a couple of important EVOC 20 PS parameters:

- **Side-chain analysis parameters:** The upper-left area of the EVOC 20 PS adjusts the side-chain parameters. Rotate the Attack knob to set how fast or slow the synth reacts to the beginning of the input signal. Rotate the Release knob to adjust how fast or slow the synth reacts to the end of the input signal. Click the Freeze button to hold the current input signal indefinitely.
- **U/V detection parameters:** The right side of the EVOC 20 PS adjusts the U/V (unvoiced/voiced) detection parameters. The human voice comprises voiced sounds such as vowels and unvoiced sounds such as plosives, fricatives, and nasals. Rotate the Sensitivity knob to adjust how sensitive the EVOC 20 PS is to voiced and unvoiced input signals. Click the Mode field to choose how unvoiced sounds are synthesized. Rotate the Level knob to adjust the volume of the unvoiced content.

TIP

You'll get great results if your input source is a constant volume with lots of high-frequency content. Be sure that your input source's volume doesn't vary too much. You can also EQ the input source to boost the high-frequency content. You learn how to EQ signals in Chapter 17.

Modeling Sounds Using Sculpture

Sculpture, shown in Figure 11-8, is a component-modeling synthesizer that creates virtual models of the components of acoustic instruments, such as the necks and bodies of string instruments and the strings themselves. Sculpture can also model the location of these components, so you can put them indoors, outdoors, or in synthetic locations. Sculpture can also synthesize sounds like the other Logic Pro synths described in this chapter.

In this section, you learn how to model instruments and synthesize sounds with Sculpture.

Understanding sound modeling

Sound modeling uses math to simulate the stimulation of physical objects. To create a complete sound model, the math has to consider the object being stimulated and the object doing the stimulating. The object being stimulated can have many components, such as a guitar string, a guitar neck, and a guitar body.

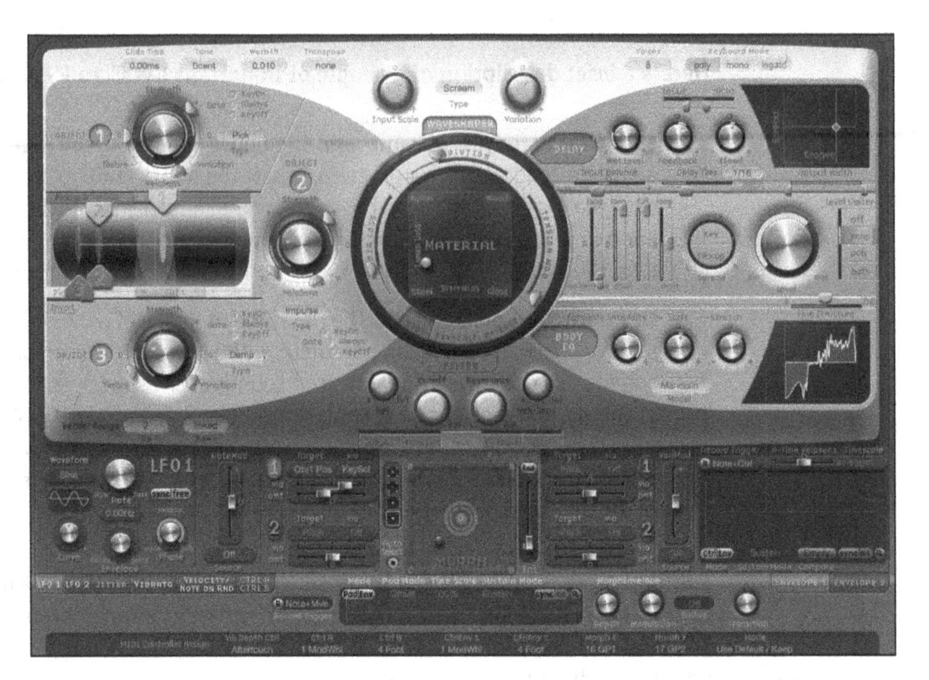

FIGURE 11-8: The Sculpture-modeling synth.

Sculpture gives you the following parameters for detailed sound modeling.

- **String:** In Sculpture, the stimulated component is called a *string*. The string parameters are adjusted in the round Material pad in the center of the interface. You can move the puck in the Material pad to change the string parameters to nylon, wood, steel, or glass. The left/right axis adjusts the stiffness of the material. The up/down access adjusts *inner loss* (dampening caused by the string material). You also have sliders to adjust the resolution, tension modulation, and *media loss* (string dampening due to the surrounding environment). Click the buttons at the bottom of the Material pad to show or hide the Keyscale and Release parameters. Keyscale changes how the string reacts depending on the pitch, and Release affects how the string behaves when released.
- **Objects:** The components that stimulate the string are called *objects* in Sculpture. Objects are adjusted on the left side of the interface, and three objects are available. Click the numbered object buttons to turn an object on or off. One object must be selected for the string to sound when played. Click the object Type buttons to choose different materials and methods of stimulating the string. Each object includes a knob to adjust the object's strength and sliders to adjust the timbre, velocity sensitivity, and variation. Click the Gate mode buttons to set the object to stimulate the string at Key On, Always, or Key Off.

» **Pickups:** Sculpture uses two pickups to take the sound generated by the objects and string and input that signal into the rest of Sculpture's signal processing. Pickups are analogous to a guitar's pickups. Adjust the pickup properties in the cylindrical area in the Object section. Drag the pickup A and B sliders at the top and bottom of the cylinder to adjust the pickup position. Drag the numbered object sliders to adjust the position of the objects in relation to the pickups.

» **Body:** Change the body type Sculpture uses to shape the sound in the lower-right area of the interface. Click the Model button to choose the body type. Click the Body EQ button to turn the entire body parameter on or off. Adjust the intensity, shift, and stretch of the body using the knobs.

Exploring the Sculpture interface

In addition to component-modeling parameters, Sculpture can filter and modulate your sound. Plus, you can add built-in effects. Here's a brief description of some Sculpture parameters:

» **Global parameters:** The buttons and input fields across the top of the Sculpture interface are global parameters. Click the Glide Time field to adjust the portamento speed. Click the Tune field to fine-tune the instrument. Click the Warmth field to detune the voices and thicken the tone slightly. Click the Transpose field to transpose the instrument in octaves. To the right side of the global parameters are the Voices field and Keyboard Mode buttons. The Voices field sets the number of voices that can be played simultaneously. Click the Keyboard Mode buttons to select Polyphonic, Monophonic, or Legato. In monophonic mode, the attack is always retriggered. In legato mode, notes aren't retriggered unless you release the key before playing the next key.

» **Waveshaper parameters:** Directly above the string material parameters are the waveshaper parameters. The *waveshaper* adds nonlinear curves to your sound and can add rich saturation or raw distortion. Click the Type pop-up menu to choose the waveshaper curve. Rotate the Input Scale knob to raise or lower the input signal. Rotate the Variation knob to alter the shape.

» **Filter parameters:** Directly below the string material parameters are the filter parameters. Click the filter type buttons to select the filter mode. Rotate the Cutoff and Frequency knob to adjust the filter. Rotate the Key Tracking knob to adjust how the filters respond to the key that's played. Acoustic sounds generally have mellower low notes and brighter high notes, and key tracking allows the filter to respond to the pitch. Rotate the Velocity Sensitivity knob to adjust how velocity affects the filter. A harder velocity usually makes an acoustic sound brighter, and a higher value will give you more velocity sensitivity.

- **Modulation parameters:** At the bottom-left section of the interface are the modulation parameters. In addition to common LFO, vibrato, and envelope modulation sources, Sculpture provides unique ways to modulate your sound and create constantly evolving textures to make your sound more expressive. Click the modulation type buttons at the bottom to open the parameters for that modulation type. The two Jitter modulators provide continuous and random alterations of your sound. The Velocity/Note On Random parameters can give your sound acoustic realism by emulating the subtle changes a musician would make while playing.
- **Output parameters:** To the far right of the string material parameters are the output parameters. The Level knob adjusts the overall volume. Click the Level Limiter mode buttons to process the volume level depending on the mode. Mono mode limits all the voices; Poly mode limits each voice separately.
- **Morph parameters:** Directly below the filter parameters are the morph parameters. Morphable parameters can be stored in a point in the Morph pad, allowing you to smoothly change the sound over time. All morphable parameters are indicated by orange values rather than blue or turquoise values.
- **Delay parameters:** The upper-right section of the Sculpture interface is dedicated to the delay parameters. The delay effect copies the sound and repeats it based on your chosen parameters. Click the Delay button to turn on the delay effect. Drag the Delay Time slider to choose when the sound repeats. Rotate the Wet Level knob to adjust the volume of the delay. Rotate the Feedback knob to adjust how many times the delay will repeat.
- **MIDI controller parameters:** At the very bottom of the Sculpture interface are the MIDI controller parameters. Click the MIDI controller name pop-up menu to choose the MIDI controller, or select Learn to listen for the next MIDI controller you touch.

Sampling with Sampler and Quick Sampler

The Sampler software instrument, shown in Figure 11-9, plays audio files known as *samples*. Samplers are useful for re-creating acoustic instruments because you're playing back recorded audio files. But you can also sample synthetic sounds or mangle acoustic samples until they're no longer recognizable to create unique sounds.

FIGURE 11-9: The Sampler software instrument.

In this section, you learn how to use Sampler, import third-party sample libraries, create your own sampler instruments, and much more.

Importing sample libraries

Sampler has an enormous library of sampler instruments, but you can also import third-party sample libraries. Sampler (formerly known as EXS24) has a large user base, so finding sample providers online isn't difficult. In addition to the native Sampler instrument format, you can import SoundFont2, DLS, and Gigasampler files.

To import a third-party sample library, follow these steps:

1. **Copy the sample files to a subfolder of your choice in any of the following locations:**

   ```
   Macintosh HD/Users/<USERNAME>/Music/Audio Music Apps/
     Sampler Instruments/
   Macintosh HD/Users/<USERNAME>/Library/Application Support/
     Logic/Sampler Instruments/
   Macintosh HD/Library/Application Support/Logic/Sampler
     Instruments/
   ```

2. **At the top of the Sampler interface, click the preset menu and choose your preset.**

Your imported sampler instrument appears on the menu so that you can select and open it.

Converting regions to sampler instruments

You can create your own sampler instrument from any audio region in your project. For example, you can take a drum loop or an instrument riff and slice it into several pieces that you can play with your MIDI controller. You can also trigger the entire region as a loop or a one-shot sample, which means the region plays just once instead of repeating. To convert audio regions to sampler instruments, do the following:

1. **Select an audio region in your project.**
2. **Control-click the region and choose Convert ⇨ Convert to New Sampler Track (or press Control-E).**

   The Convert Regions to New Sampler Track dialog opens.
3. **Select Create Zones from Regions or Create Zones from Transient Markers.**

   Select Regions to create a loop or one-shot sample. Select Transient Markers to slice the region into several different samples that you can trigger with your MIDI controller.
4. **Name the instrument and select the note trigger range.**

   The note trigger range defines the MIDI notes that will trigger the samples.
5. **Click OK.**

   A new software instrument track is added to the track list loaded with Sampler. A MIDI region containing the trigger notes is also added to the tracks area.

TIP

If you convert a drumbeat to a sampler instrument, you can play the beat chromatically with your MIDI controller. A cool trick is rearranging parts of the beat to create new beats or drum fills. If you converted an instrument part to a sampler instrument, you can rearrange the part and come up with new instrument parts. You could also sample your boss saying, "No one leave until the work is done," and rearrange it to say, "The work is done. Leave."

Controlling sample parameters

At the top of the Sampler interface are navigation bar shortcut buttons for the Synth, Mod Matrix, Modulators, Mapping, and Zone panes. Clicking a button scrolls the interface to the corresponding pane. Clicking the yellow LED inside the button shows or hides the pane. Double-clicking a button shrinks or expands the pane. Option-clicking a button expands the pane while hiding the others.

Like the other Logic Pro synthesizers, Sampler sounds can be shaped by filters and modulation. Here's a description of some Sampler parameters:

- **Synth pane:**
  - **Synth pane header:** The Details button shows or hides additional synthesis parameters such as pitch bend, polyphony limit, and glide, which is the time it takes to glide from one pitch to the next. The Velocity Random and Sample Select Random fields are useful for making your samples sound more natural and less rigid.
  - **Pitch:** Adjusts the pitch in semitones or cents (1/100th of a semitone).
  - **Filter:** The dual-filter section can filter the signal in series (one after the other) or parallel (both filters at once). Each filter has an on/off button and knobs to adjust the cutoff, resonance, and drive. The filter type pop-up menu contains several types of filters, many of which are modeled after classic analog synthesizers.
  - **Amp:** Adjusts the volume and pan.
- **Mod Matrix pane:**
  - **Mod Matrix header:** The Filter By pop-up menu and on/off button allow you to show only filtered results. You can add and remove up to 20 modulation routings with the – and + buttons.
  - **Modulation routing row:** Each row contains an on/off button and pop-up menus for the modulation source, target, amount, and via (an additional source of modulation intensity, such as a MIDI controller's velocity or mod wheel).
- **Modulators pane:**
  - **Modulators header:** You can use the +ENV and +LFO buttons to add up to five envelopes and four LFOs, respectively as modulation sources or targets.
  - **Envelope:** The first envelope is always linked to the amplitude parameter, while the remaining four envelopes can target any available parameter. Each envelope contains an envelope type pop-up menu that contains several types, including the standard ADSR envelope. Drag the envelope nodes to adjust the shape. Use the Vel slider to change how velocity affects the envelope.
  - **LFO:** Choose the LFO waveform from the waveform pop-up menu. To the right of it is a note icon that synchronizes the LFO to the tempo of your

project when pressed. Use the Rate field to set the LFO speed and the Fade field to set the fade in or fade out time. The buttons to the right of the Fade field allow you to edit the fade in or fade out times. The Mono button makes the LFO act identically for all voices, while the Poly button makes the LFO act independently for each voice. To the right of the Mono/Poly buttons are the Unipolar and Bipolar buttons, which allow you to choose between running the LFO in one pole (positive) or both poles (positive and negative). Finally, press the Key Trigger button to restart the LFO when you press a key instead of allowing it run freely.

- **Mapping and Zone panes:**
  - The mapping pane sets and controls group and zone parameters. The zone pane allows you to edit individual samples. Learn more about these parameters in the next section, "Editing sampled instruments."

Editing sampled instruments

There may come a time when you want to create your own sampler instruments or edit a copy of the current instrument. A sampler instrument is made up of zones and groups. A *zone* is the location of a single sample, whereas a *group* can contain many zones. You can edit all the zones in the group simultaneously.

To visualize how zones and groups fit together, consider a piano sampler instrument. Each piano key is sampled at multiple velocity levels. These sampled notes (zones) are grouped according to their velocity level for organization and so the group can be edited as a whole. You can have a group with loud samples and a group with quiet samples, for example, and edit each group and all the zones in it.

To edit sampler instruments, use the Mapping and Zone panes, shown in Figure 11-10. At the top of the Mapping pane is a menu bar with Edit, Group, Zone, and View pop-up menus:

- The Edit menu allows you to select, cut, copy, paste, and delete groups.
- The Group menu is where you create, duplicate, and merge round robins. A *round robin is* when each successive note triggers a different sample, which helps samples of acoustic instruments sound more natural.
- The Zone menu contains various commands and functions to create new zones, load audio files, and edit zones.
- The View menu is where you can show and hide parameters.

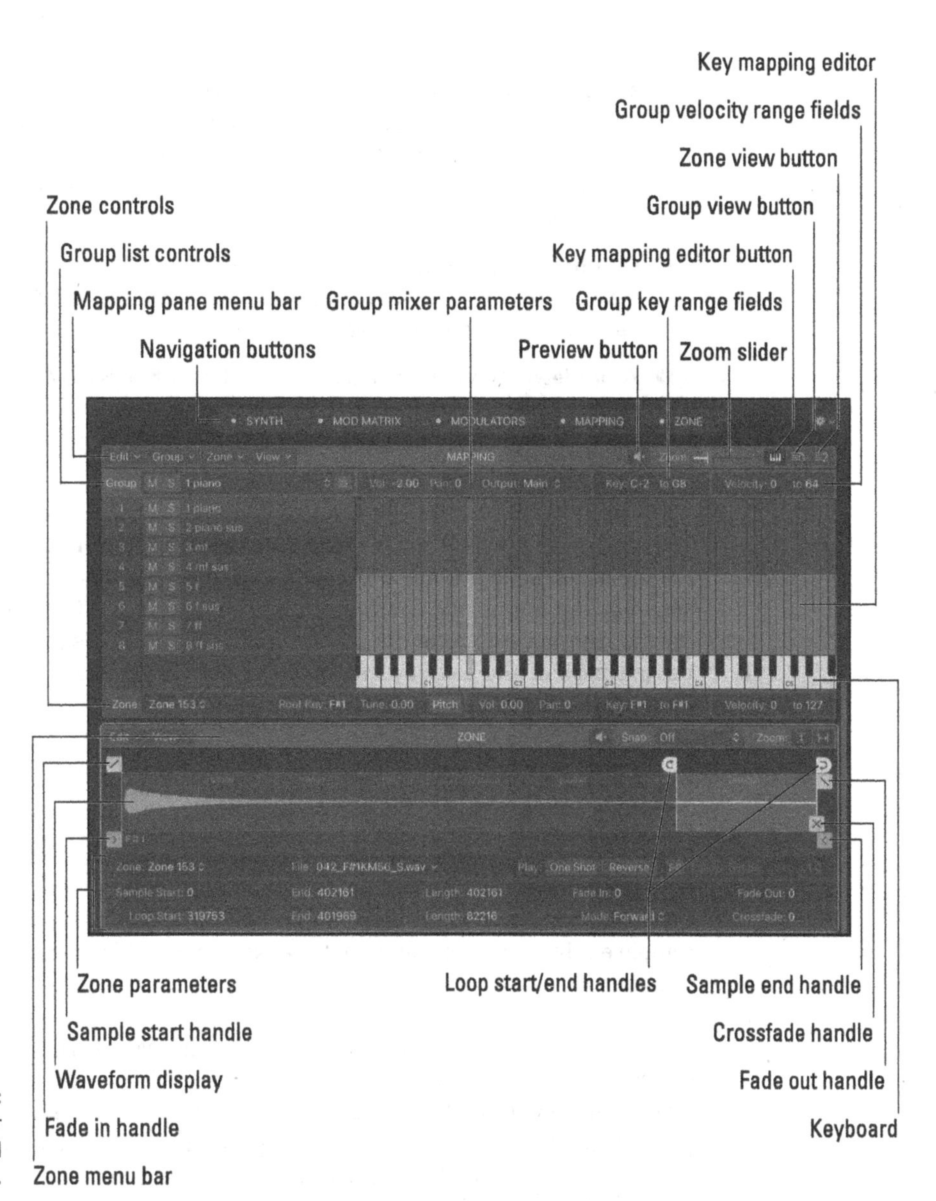

FIGURE 11-10: The Sampler mapping and zone panes.

On the right side of the menu bar is a preview button to play the selected note, a zoom slider, and three view buttons for displaying key mapping editor view, group view, and zone view. Here is a description of the different views:

- **Key mapping editor view:** Shows a Group menu bar at the top containing group list controls, mixer parameters, key range fields, and velocity range fields. Below the menu bar is the key mapping editor, where you can drag and drop audio files or use the Zone menu to add them. Rectangles define the boundaries of individual zones. The width indicates the key range, and the height indicates the velocity. At the bottom of the editor is the Zone menu bar, which contains the Name pop-up menu, pitch fields, mixer fields, key range fields, and velocity range fields.

- **Group view:** Shows the group parameter editor, which is a list containing the group name, mixer parameters, key range parameters, velocity range parameters, and release trigger.
- **Zone view:** Shows the zone parameter editor, which is a list containing the group name, zone name, audio file, pitch parameters, key range parameters, velocity range parameters, mixer parameters, playback parameters, group assignment, and loop parameters.

The Zone pane is where you edit the zone that's selected in the Mapping pane. At the top of the pane is a menu bar with Edit and View menus, a preview button, a snap pop-up menu, and zoom buttons. The waveform display shows the waveform of the selected zone, its marker handles, and shaded marker areas. Adjust the playback of the zone by dragging the marker handles. Below the waveform display are the zone parameters.

Exploring Quick Sampler

Quick Sampler, shown in Figure 11-11, provides an intuitive and powerful workflow for quickly creating unique instruments containing a single audio file. The Quick Sampler interface is divided into two sections. The upper section provides sample editing functions, and the lower section provides parameters to shape the sample's sound.

At the top of Quick Sampler are buttons to choose the sample playback mode. Classic playback mode plays the sample while you hold a key and stops when you release the key. One Shot playback mode plays the sample at the start marker position and finishes at the end marker position. Slice playback mode divides the sample into slices and maps them to keys. Recorder mode allows you to record samples from any audio signal.

To the right of the playback mode buttons are the sample name field and pop-up menu, where you load audio files and perform file-handling tasks. The Snap pop-up menu lets you snap your edits to the nearest beat, transient, or zero crossing. To the right are vertical and horizontal zoom buttons and an Action pop-up menu with commands and functions available to the current playback mode.

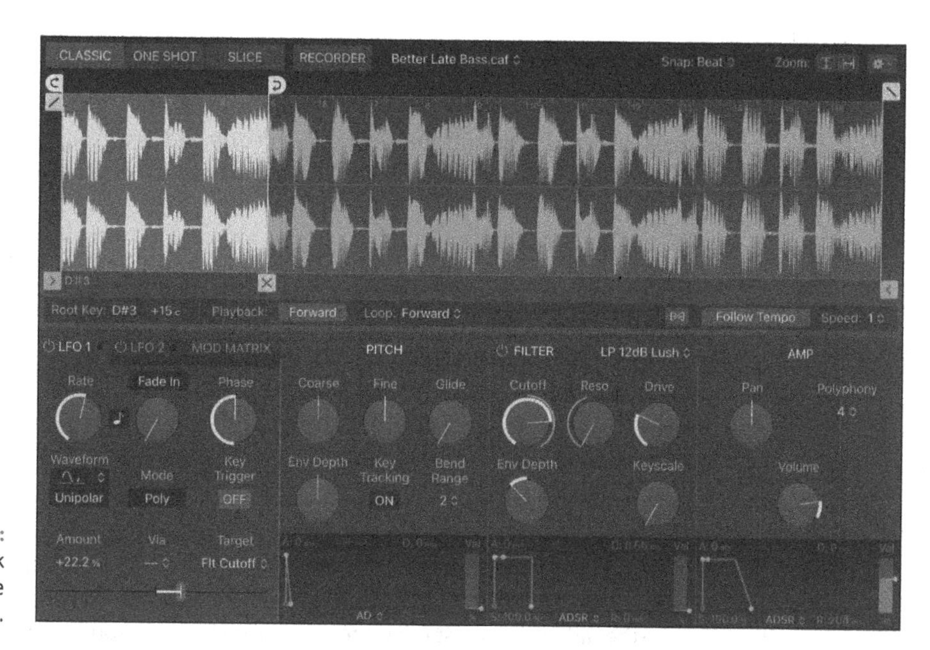

FIGURE 11-11: The Quick Sampler software instrument.

TIP

Control-click the waveform display to see a menu of commands relevant to the selected playback mode. When editing slice markers, this shortcut menu will save time and simplify your workflow.

The lower section contains controls that allow you to shape the sound of the sample. There are two LFO controls and Mod Matrix, Pitch, Filter, and Amp controls. The LFO controls allow you to modulate sample parameters. The Mod Matrix control allows you to create custom modulation routings with more than 20 sources and destinations. The Pitch, Filter, and Amp controls give you precise control over the sample's sound character.

TIP

You can use Quick Sampler also as an effects processor. Simply route the audio signal you want to process into Quick Sampler and then tweak the parameters until you get the sound you're looking for. When combined with the powerful modulation options, this opens up a world of creative possibilities.

Manipulating Samples with Alchemy

Alchemy is a powerful sample-manipulation synthesizer and features many different types of synthesis. Sound sources can be virtual analog synth waveforms, Sampler instruments, and audio files. You can resynthesize audio (that is, you can take any recording and manipulate it as you would any synth patch). Drum loops, vocals, and samples are all fuel for your sound design engine.

You can have up to four sound sources in an Alchemy preset, and you can morph between the sound sources to create complex sounds that change over time. Sound sources can be filtered and shaped independently and as a group. Alchemy includes a full-featured arpeggiator and effects section to bring depth and interest to your patches.

Exploring the Alchemy interface

At the top of the Alchemy interface, shown in Figure 11-12, is the name bar, where you manage your presets and choose from three types of views:

- **Browse:** The preset browser is where you recall presets. You can sort presets by categories and tags and rate the presets you like the most or the least.
- **Simple:** Select the simple view to show only the Performance section. You get access to the Transform pad and X/Y pads, where you morph between snapshots of your preset. You also get several control knobs and menus to further adjust the most important parameters of your sound. The performance controls are common to all views.
- **Advanced:** The advanced view is where you will do most of your programming. The top half of the view shows you different parameters depending on the source mode you've selected on the left side of the interface. If Global mode is selected, the display shows the master parameters of all four available sound sources. If A, B, C, or D mode is selected, the individual sound source parameters are displayed. If Morph is selected, the display shows controls for morphing between different sounds and parameter settings. The lower half of the view is the modulation area, where you set target parameters to be controlled by a list of user-programmable modulation sources.

Designing and resynthesizing sounds

To begin designing a sound, select Advanced view in the name bar. Choose your sound sources in the Sources area, as shown in Figure 11-13. If Global or Morph mode is selected, you see all four sound sources and their master controls. If A, B, C, or D mode is selected, you see advanced parameters for only the selected sound source. Each sound source can have three filters in addition to the two global filters, which affect all sound sources.

The Modulation section includes a modulation rack where you choose modulation sources and their targets. Almost every Alchemy knob can become a modulation target, and you can even modulate the modulators, giving you the flexibility to realize your wildest sound design dreams.

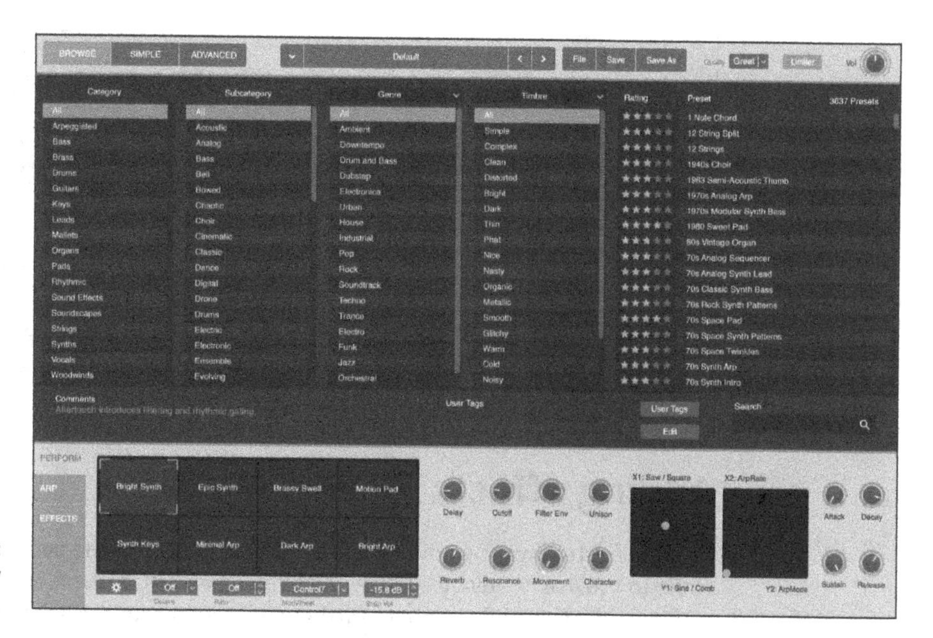

FIGURE 11-12: The Alchemy interface.

FIGURE 11-13: The Alchemy advanced view.

TIP

Alchemy has too many features to cover in this book, so it's worth diving into the manual to get detailed descriptions of each parameter and its functions. Choose Help ➪ Logic Pro Instruments to open an in-depth manual with some Alchemy tutorials to help you get comfortable designing your own cutting-edge sounds.

Using the arpeggiator

Alchemy includes an arpeggiator that allows you to generate complex *arpeggios* (notes played one after another as opposed to simultaneously) from the chords you play. Click the ARP button in the performance section to view the arpeggiator subpage, shown in Figure 11-14. Note that the arpeggiator button is available only in browse or advanced view, not in simple view.

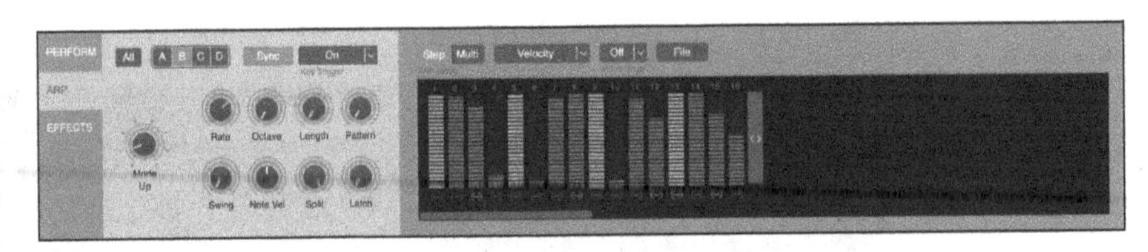

FIGURE 11-14: The Alchemy arpeggiator.

Rotate the Mode knob to turn the arpeggiator on and off, and choose the arpeggio's note order. Click the All, A, B, C, or D button to choose the sound source for the arpeggiator. On the right side of the interface is the arpeggiator sequencer, where you can build sequences that control various sound parameters on each step of your arpeggio.

Adding effects

To add effects to your sound, click the Effects button in the performance section while in browse or advanced view. Alchemy gives you several effects processors, such as distortion and reverb, to further shape your sound. Click the Main, A, B, C, and D buttons, as shown in Figure 11-15, to choose global effects for the entire instrument or independently for any or all four sound sources. Add effects in the effects slot area below the buttons by clicking the effects slot pop-up menu and choosing the desired effect. The effect and its parameters are added to the area on the right, where you can adjust them manually or modulate them using the modulation area of Alchemy.

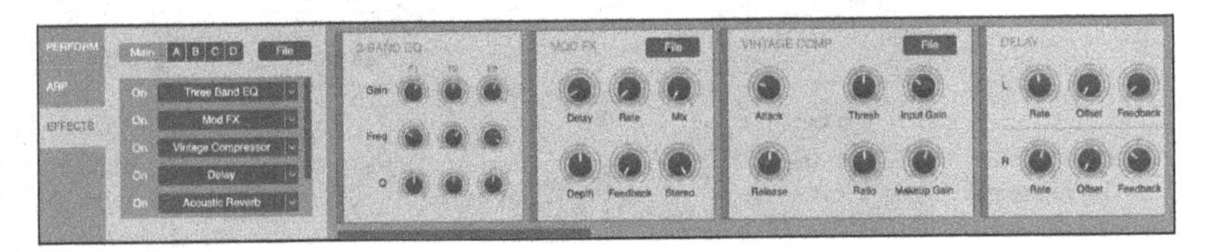

FIGURE 11-15: The Alchemy effects section.

TIP

When an Alchemy parameter is the modulation target, you see an orange arc on the knob. When the knob has a green arc, the parameter is a morphing target. The colored arcs give you a quick visual cue of parameters that the patch may rely on. This feature is useful when you design complex patches with lots of modulation and morphing.

TIP

Many of the presets on the Logic Pro instruments have their overall volume set to zero, which means it neither boosts nor cuts the signal. Consequently, the software instruments can be very loud when you first load them. It's a good idea to find the overall volume of these instruments and turn it down, so you aren't in danger of clipping the signal. Also, you will get a better balance between the other tracks in your project if the instruments aren't too loud.

With all these software instruments to play with, you can have an endless amount of fun. One of the joys of playing synths is how they can surprise and delight you with their unique personalities. Just like getting to know a good friend, listen to your synth and let it tell you its story. In time, you and your synths will become close musical partners.

IN THIS CHAPTER

» Creating an orchestral template

» Loading and playing orchestral patches

» Defining a productive composing workflow with screensets

» Playing samples with expression and articulation

Chapter 12

Conducting a Virtual Orchestra

Imagine standing in front of a full 50-piece orchestra, conducting your own score. The music builds, section by section, into a glorious peak. You pull the intensity down gracefully, and a lone trumpet plays a short, plaintive motif over a bed of lush strings. The audience is moved to tears. And then a pregnant pause before a slow roar of applause overtakes the concert hall. Flowers are thrown onto the stage, and the crowd begins chanting, "Encore!"

You may have to imagine the applause and chanting. But Logic Pro will provide the orchestra.

Orchestral instruments will never go out of style. An orchestra is Hollywood's best friend. Plus, Logic Pro shines at film scoring. You learn how to import video in Chapter 8. In this chapter, you discover how to set up an orchestra and score your own soundtrack.

A video isn't required to compose with an orchestra. But if you'd like to practice scoring to video, an online search for royalty-free videos will give you something to import into your project.

One big reason why Logic Pro is great at orchestration is its music notation capabilities. The notation displayed in the score editor is separate from the MIDI events, so the score can look perfect even if the playing isn't. You can view the score layout in Logic Pro and print lead sheets for the performers and the complete orchestral score for the conductor. If you read music, you'll find that the score editor is a great place to input music, especially if you're transcribing a score from sheet music. The Project Gutenberg online library has scores you can input for practice or fun.

Building an Orchestral Template

Logic Pro comes with an enormous selection of sampled orchestral instruments. These sophisticated samples are available as patches in the library. Most of the patches include the Space Designer plug-in, which re-creates acoustic environments, so you have exceptional control over the sound of your orchestra.

Note: You may already have an orchestral template installed with Logic Pro (choose File ⇨ New From Template ⇨ Project Templates ⇨ Orchestral). For some users, the orchestral template isn't included in the installation, even when they download the additional content (as you discover later in this section). I want to teach you how to create an orchestral template so that you can create your own personalized template and be encouraged to experiment with the orchestra. This template, along with a video guide, is available at `https://logicstudiotraining.com/chapter12`.

Setting up an orchestral template will save you time when inspiration strikes. Create the template once and then use it over and over again. To begin, create a new empty project:

1. **Choose File ⇨ New from Template (⌘-N).**

   The template chooser window opens, as shown in Figure 12-1.

2. **In the template chooser, select the Empty Project template.**
3. **In the Details area, select the project settings you want.**

   If the Details area is closed, click the disclosure triangle to reveal the project settings.

4. **Click the Choose button.**

   The new tracks dialog appears.

5. **Select the Software Instrument option and enter the number of tracks you want to create.**

   For this example, enter 32 in the number of tracks field. You'll be able to fit a full orchestra on 32 tracks with a few tracks to spare.

6. **Click the Create button.**

   The main window of your new project opens.

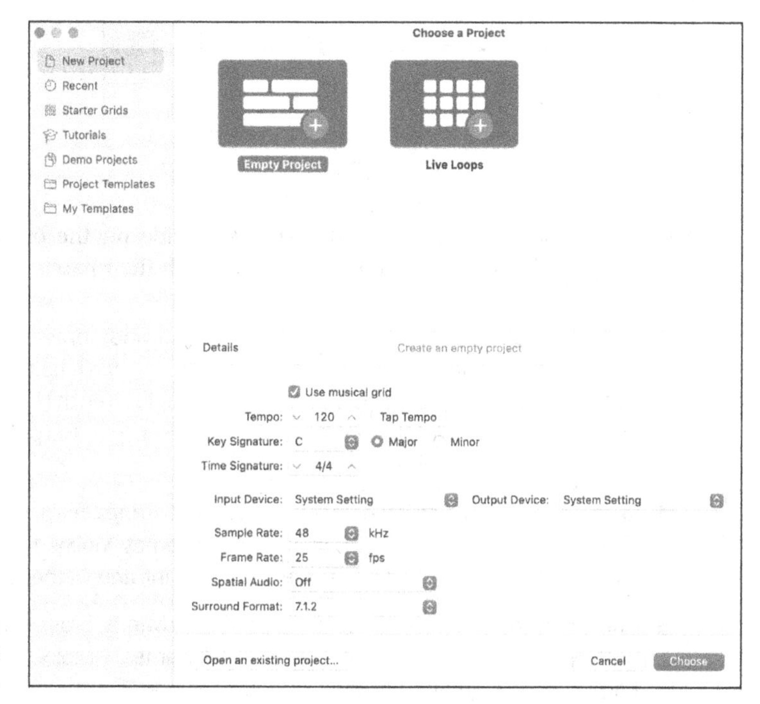

FIGURE 12-1: The template chooser.

Choosing your orchestral instruments

Orchestras come in lots of shapes and sizes. The Logic Pro orchestral patches can deliver the most common variations. And, of course, you're not limited to using acoustic samples exclusively. For example, you can mix an orchestra with electronic instruments or a rock band.

REMEMBER

You must download additional content to take advantage of orchestral patches. Choose Logic Pro ➪ Sound Library ➪ Open Sound Library Manager. Select the Orchestral, Piano, Studio Horns, and Studio Strings check boxes in the Sound Library Manager window, as shown in Figure 12-2, and click Install. After installing your patches, you can find them in the library by choosing View ➪ Show Library (Y).

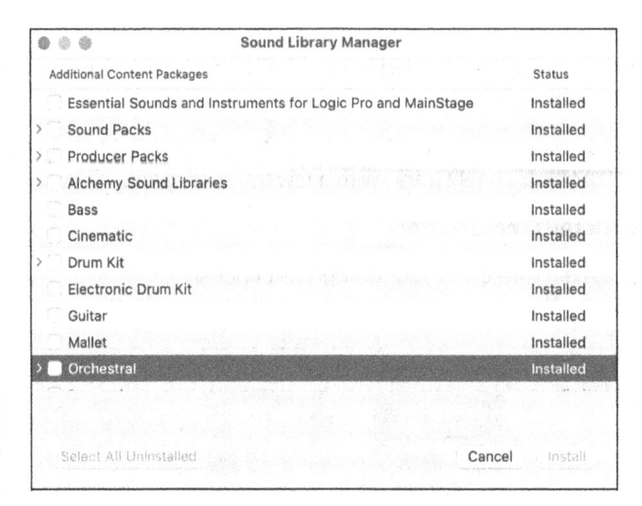

FIGURE 12-2: The Sound Library Manager window.

Before you load your tracks with patches, decide on the orchestra you want. Following is a list of common orchestrations with their patch names:

- **Full orchestra:** A full orchestra consists of woodwinds, brass, percussion, harp, piano, and strings. You can load the following patch names: Piccolo, Flutes, Oboe, English Horn, Clarinets, Bassoon Solo, French Horns, Trumpets, Trombones, Tuba, Timpani, Orchestral Kit, Glockenspiel, Harp, Grand Piano, Violins 1, Violins 2, Violas, Cellos, and Basses.
- **String orchestra:** A string orchestra consists of strings, harp, piano, and percussion. You can load the following patch names: Violins 1, Violins 2, Violas, Cellos, Basses, Harp, Grand Piano, Timpani, and Orchestral Kit.
- **Concert band:** A concert band consists of woodwinds, brass, harp, piano, and percussion. You can load the following patch names: Piccolo, Flutes, Oboe, English Horn, Clarinets, Bassoon Solo, Saxophone, French Horns, Trumpets, Trombones, Tuba, Glockenspiel, Harp, Grand Piano, Timpani, and Orchestral Kit.

TIP

Logic Pro has several grand piano patches, and you can choose between a Yamaha, a Boesendorfer, or a Steinway grand piano. Many orchestral patches also have solo patches in addition to the ensemble patches, such as the French Horn, Clarinet, and Flute. And if you don't want to work with individual instruments and would rather compose with full sections, you can use the Full Brass and Full Strings patch as well as several genre-specific patches such as Romance Strings, Cinema Strings, and Pop Strings.

Choosing appropriate staff styles

Staff styles store information about how a region should be displayed in the score editor. For example, orchestral instruments are traditionally associated with a specific clef type and transposition. Choosing a staff style lets you quickly adjust how a region or entire track looks.

The orchestral patches in your library might already have the appropriate staff style saved with the patch. However, you can easily change the staff style of a track by following these steps:

1. **Select the track.**
2. **Open the inspector by choosing View ⇨ Show Inspector (I).**

   The inspector opens on the left side of the main window.
3. **Open the track inspector pane, shown in Figure 12-3, by clicking the disclosure triangle.**
4. **In the Staff Style drop-down list, select the style.**

   The track will use the new staff style in the score editor.

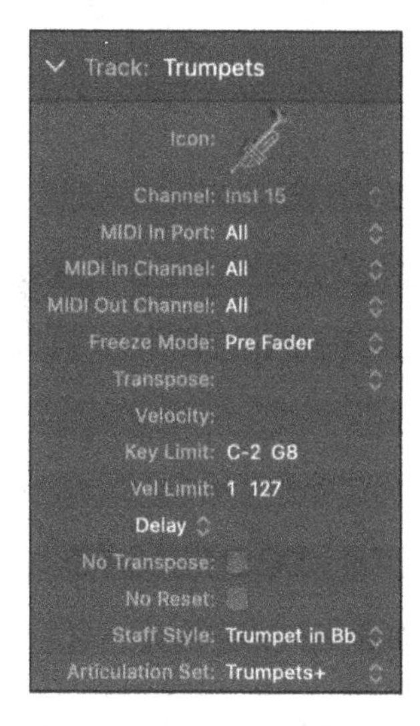

FIGURE 12-3: The track inspector.

A great example of when you might want to change the staff style is with the Trumpets patch. The staff style that loads with the Trumpets patch is transposed for Bb trumpet, which means the trumpet notation is transposed up two semitones. If you compose and orchestrate in the score editor, you might want to switch between the instrument's staff style to a style with no transposition. You can compose more quickly when you don't have to continually do the mental transposition in the score editor. To edit the staff style, follow these steps:

1. **Select the track.**
2. **Choose View ➪ Show Editor (E).**

   The editor window opens at the bottom of the tracks area.
3. **Click the Score tab to display the score editor.**
4. **On the score editor menu, choose Layout ➪ Show Staff Styles.**

   The staff styles window opens, as shown in Figure 12-4.
5. **In the Transpose field, drag your cursor up or down to change the value.**

   The score editor instantly updates to the new transposition for all tracks that use the same staff style.

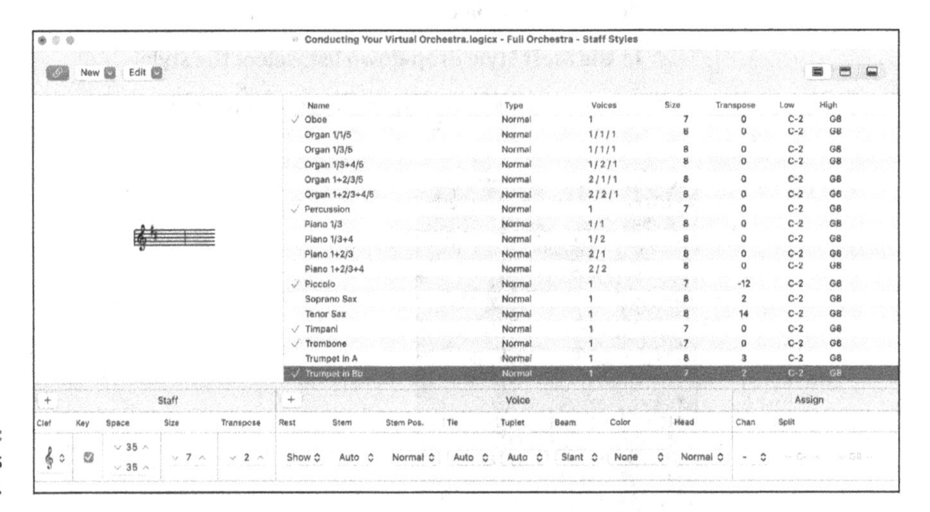

FIGURE 12-4: The staff styles window.

Saving your orchestral template

With all your patches loaded, you could save your project as a template by choosing File ➪ Save as Template. But you can do more to make this template a composer's playground. Because your work as a composer comprises many tasks, you can create screensets dedicated to specific workflows. Here are some screenset ideas to help you compose faster:

- **Choose instruments and input parts.** At the beginning of a project, you spend a lot of time auditioning different instrument patches. Open the library (Y) to make patch selections fast. The tracks area will already be open, enabling you to record new regions and select tracks. If you also open the editor (E), you'll be able to quickly input notes into the piano roll, score, or step editor.
- **Perform parts with smart controls.** When you record, you need to capture your best performance. Open the smart controls (B) to have immediate control over the different articulations of these orchestral patches.
- **Set the project key and tempo.** If you're composing to video, you need regular access to the global tracks to time your music to the video. Open the score editor in a new window (choose Window ➪ Open Score Editor) and display the global tracks (G). This view gives you a score for referencing your music and the global tracks for inputting tempo and key changes (for details, see Chapter 4).
- **View the score editor in linear view.** See the entire score scroll in front of you. Open the score editor in a new window (⌘-5), and choose View ➪ View Mode ➪ Linear View on the score editor menu. Input notes in the score editor by using the pencil tool.
- **Edit MIDI.** Adjusting note lengths and velocities is fast and easy in the piano roll editor. A giant piano roll editor in the tracks area makes MIDI editing simple. (To find out how to edit MIDI, see Chapter 15.)
- **Engrave music.** Like the linear view, the score editor looks beautiful in page view, as shown in Figure 12-5. Open the score editor in a new window and select View ➪ View Mode ➪ Page View. From this view, you can add ornaments and text from the part box in the inspector (I) and print your score for the musicians and conductor.
- **Produce the final mix.** You adjust the levels of the individual instruments in the mixer. Open the mixer in a new window (⌘-2) and make it as big as possible. You learn how to use the mixer in Chapter 16.
- **View video.** If you have a video imported into your project, choose View ➪ Show Movie Window. Watch the movie and play what you see. Sound simple? Logic Pro does its best to make it so.
- **Use project notes.** If you work with a team, documenting your work is a must. Open the project notes by choosing View ➪ Show Note Pads (Option-⌘-P). In the project notes, create a log of important changes, instructions, lyrics, or anything else you want to remember later.

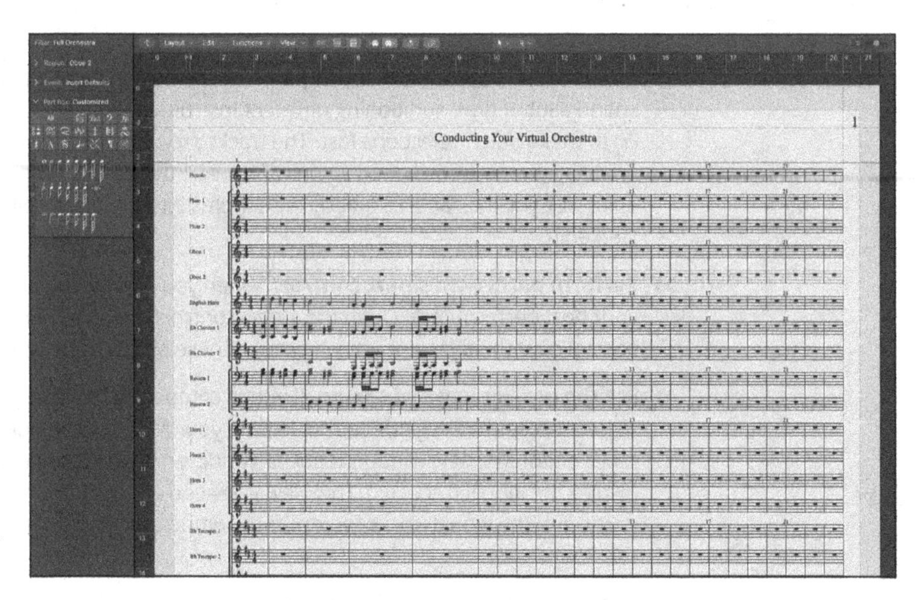

FIGURE 12-5: Score editor page view.

When you have these screensets set up the way you like, choose Lock on the screenset menu. After the screensets are locked, you can change the view as much as you like, but your original settings will be recalled when you choose the screenset.

TIP

You've already learned how to save your project as a template, but you can also save your entire orchestra as a patch. First, select all the tracks and create a summing track stack (as described in Chapter 4). Save the track stack as a patch, and you can load your orchestra into any project.

Performing Your Orchestra

Your orchestral patches come with several velocity levels and different articulations to help you create realistic sounding performances. The patches also load with smart controls designed to make your performances expressive and dynamic.

Select a track and open the smart controls (press B), as shown in Figure 12-6. Note the Legato and Staccato buttons in the clarinet smart controls; these switch between two articulation samples. You can quickly assign these buttons to your MIDI controller; for details, see Chapter 3.

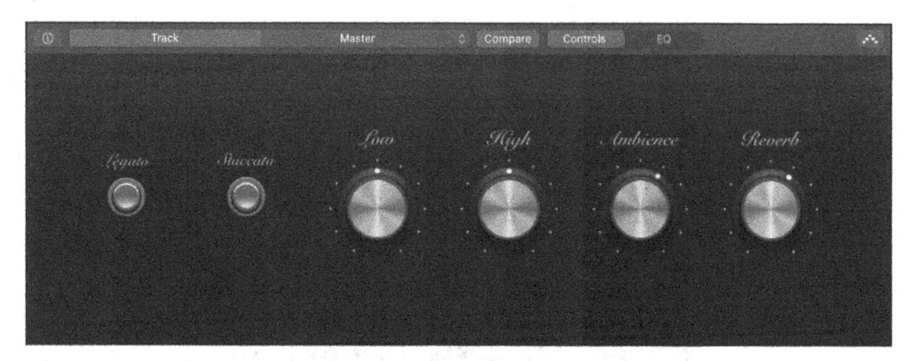

FIGURE 12-6: The woodwind smart controls.

Sampler instruments are often a surprise to play. You never know how many different samples or velocity levels you can experiment with. Designers of great sample instruments give you instantly playable instruments, and the Logic Pro patches deliver on that promise with their intuitive design. But if you want to know exactly what your instrument can do, there's no better place than the Sampler mapping pane. A good patch on which to perform a sample autopsy is the Orchestral Kit:

1. **Select the track that has the Orchestral Kit patch loaded.**

   You should already have a track with the Orchestral Kit patch loaded if you followed the instructions in the "Choosing your orchestral instruments" section.

2. **Open Sampler by clicking the center of the instrument slot in the inspector (I).**
3. **Click the Mapping pane navigation bar shortcut button at the top of the Sampler interface.**

   The Mapping pane scrolls into view.

Note how some of the keys have multiple velocity levels, as shown in Figure 12-7. Depending on how hard you hit your MIDI controller, different samples will play. Each sample turns yellow as you play it. Watching the velocity of a multi-sampled Sampler instrument as you play is a great way to get to know the feel of your MIDI controller. Working with the instrument's preset velocity will help you to play a sampled instrument more expressively.

Visit `https://logicstudiotraining.com/chapter12` to download a template with the natural ranges of the orchestral instruments written in the track notes.

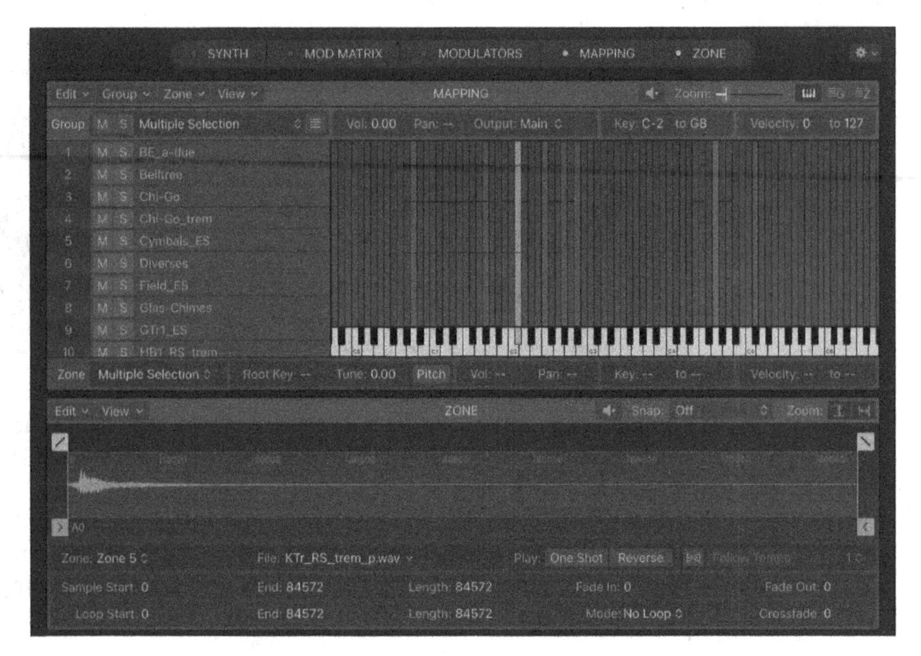

FIGURE 12-7: The Sampler mapping pane.

Playing with Studio Strings and Horns

Logic Pro includes a set of multi-sampled string and horn instruments called *studio instruments.* You find them in the library (press Y to open the library) under the Studio Horns and Studio Strings menus. These instruments include solo and grouped instruments with multiple articulations to help you create realistic string and horn parts. They also feature *auto voice split,* which automatically assigns chordal notes to different instruments. For example, when you play a chord with the Studio Strings instrument, low notes are played by the bass, high notes are played by violins, and middle notes are played by the violas and cellos.

Following are descriptions of studio instruments parameters, as demonstrated by the Studio Strings, shown in Figure 12-8:

- **Articulation pop-up menu:** Select instrument articulations.
- **Preset pop-up menu:** The currently selected preset is displayed in the preset pop-up menu at the top of the interface, below the Articulation menu. Use this menu to choose different groups of instruments or solo instruments.
- **Auto Voice Split button:** Automatically assign notes you play to different instruments. Turning it off will still assign notes to other instruments, but splits will be based on specific key ranges.

» **Dynamics via CC button:** Control the instrument's timbre and dynamics using MIDI continuous controller messages. Additional parameters for choosing controller sources are available in the extended parameters area.

» **Last Played Articulation field:** See the most recently used articulation.

» **Parameter knobs:** Adjust the volume and other parameters. The knobs are on the right side of the interface; the ones you see depend on the selected preset.

» **Extended Parameters:** Display extended parameters by clicking the disclosure triangle at the bottom of the interface. The parameters you see depend on the selected preset.

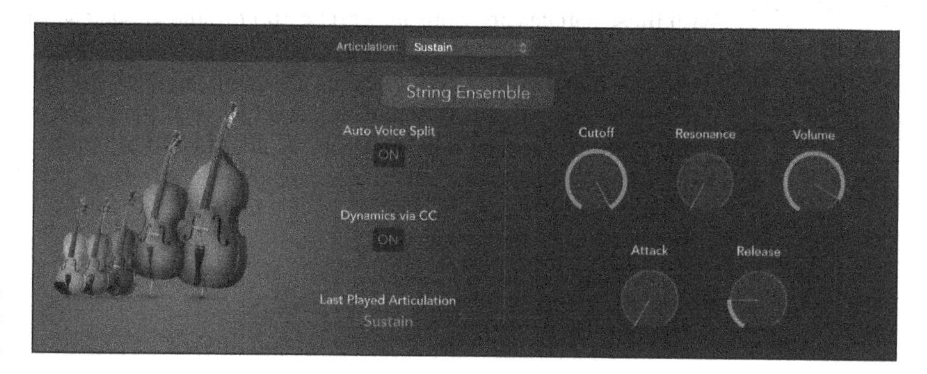

FIGURE 12-8: The Studio Strings interface.

Traveling the World Instruments

As a composer, using instruments from other cultures allows you to create authentic scores and satisfy artistic and client needs. In addition to the full orchestra you already have, you can create a project alternative that includes groups of world instrument patches. To create a new project alternative, do the following:

1. **Choose File ⇨ Project Alternatives ⇨ New Alternative.**

   The new alternative dialog opens.

2. **Enter your new alternative name and click OK.**

   A new project alternative is created, duplicating the current project settings.

Here are several groups of patches you could add to the full orchestra:

» **India:** Indian Bansuri Flute, Indian Shehnai Oboe, Indian Sitar, Indian & Middle Eastern Kit

» **Africa:** African Kalimba, African Kit, South African Singers, South African Voice Effects

» **Medieval:** Celtic Hammered Dulcimer, Celtic Harp, European Folk Kit, Medieval Lute, Celtic Tin Whistle, Medieval Recorder, Irish Bouzouki

» **Asia:** Asian Kit, Chinese Dizi Flute, Chinese Erhu Violin, Chinese Guzheng Zither, Chinese Kit, Chinese Ruan Moon Guitar, Chinese Xiao Flute, Indonesian Gamelan Gongs, Indonesian Gamelan, Japanese Koto, Japanese Shakuhachi Flute, Taiko Drums, Tibetan Singing Bowls

REMEMBER

You need to download the World instruments from Sound Library Manager, as you learned previously in the "Choosing your orchestral instruments" section.

With these patches and your new orchestral template, you can follow in the footsteps of John Williams and pull the heartstrings of movie watchers. You can learn from the masters and input scores of composing giants Bach, Mozart, and Beethoven. You can add a string orchestra to your next rock song or record an Indian raga. There's so much you can do, and now that you have your orchestra saved as a template and a patch, you can add it to any project that needs a virtual orchestra.

4 Arranging and Editing Your Project

IN THIS PART . . .

Arrange tracks and regions in the tracks area and make global changes to your arrangement.

Get to know the audio and MIDI editors.

Adjust the pitch and timing of your recordings and view your MIDI as notation.

IN THIS CHAPTER

- » Arranging tracks and regions in the tracks area
- » Making global changes to your arrangement
- » Arranging techniques to create interest
- » Creating groove templates to enhance timing

Chapter 13
Arranging Your Music

When you arrange music, you organize instruments and parts until they sound and feel good and capture your listener's attention. Logic Pro gives you a structured and flexible environment to make your arranging skills develop and blossom. You get instant feedback on your changes and tools to make improvements quickly. With Logic Pro as your arranging partner, you'll find that it does most of the work, leaving you to enjoy the results.

In this chapter, you discover how to make your grooves feel better and tighter. You learn how to systematically build an arrangement that will keep your listeners interested. And you find efficient ways to enhance your arrangement and create parts that stick.

Working in the Tracks Area

After all your tracks are recorded, it's time to arrange them. You'll do most of your arranging work with regions in the tracks area, deciding which regions to play and when and where to play them.

Recording more material than you need to build a well-organized arrangement is a common practice. Expert arrangers will tell you that a big part of arranging is deciding what not to play. With Logic Pro's visual tracks area, you should have no challenges organizing your regions into pleasing patterns that stimulate and entertain.

REMEMBER

You must select Enable Complete Features in the Advanced Settings pane to get the most out of Logic Pro and the information in this chapter. Choose Logic Pro ⇨ Settings ⇨ Advanced and then select Enable Complete Features.

Using the ruler

The ruler is the place your eyes look when you need to know where you are in the project. The ruler shows you the position of the playhead and regions in the tracks area, and it quickly lets you know how much of the project you're seeing in the main window. You use the ruler also to guide you as you move regions in your arrangement.

In Chapter 2, you find out how to set the ruler to display bars and beats or clock time from the General pane of Project Settings (Option-P). But you can have the best of both worlds and view bars and beats as well as clock time by displaying the secondary ruler. To show the secondary ruler, choose View ⇨ Secondary Ruler (Control-Option-Shift-R) on the tracks area menu bar. A second ruler will appear, showing either bars and beats or clock time, depending on your current selection in the project preferences.

A third ruler that's useful to have available is the marquee ruler. Choose View ⇨ Marquee Ruler on the tracks area menu bar. A small band of empty space, called the *marquee strip,* will appear below the main ruler. The marquee ruler shows the location of current marquee selections. You can drag the marquee selection forward or backward in your project or adjust the edges of your selection.

By default, the marquee tool is set as your ⌘ -click tool. This means you can quickly make selections using the marquee tool by ⌘ -dragging across the regions you want to select. The marquee tool is also great for setting the playback position because playback always begins from the left side of your marquee selection. If your cursor is at the bottom of the main window, it can be faster to make a quick marquee selection to set the playback position instead of moving your cursor to the top of the tracks area and clicking in the ruler to set the playback position.

TIP

An advanced and precise way to navigate your project is to use the Go to Position dialog, shown in Figure 13-1. Press / or choose Navigate ⇨ Go To ⇨ Position. When the dialog opens, type the bar number in the New field and click OK or press Return. This method is my preferred way of navigating Logic Pro because the key commands are so much faster than using a trackpad or mouse.

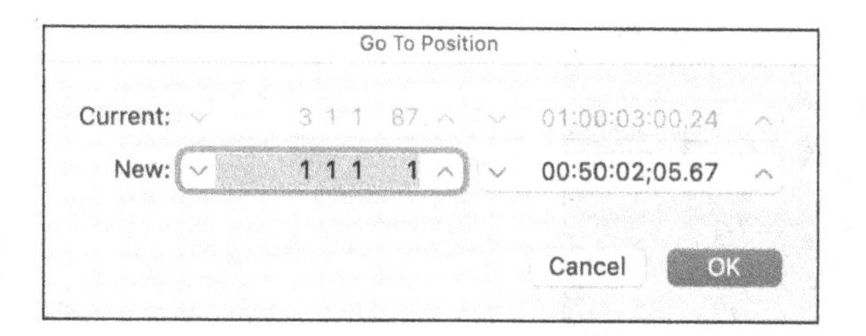

FIGURE 13-1: The Go to Position dialog.

Scrubbing the arrangement

You can drag your cursor along the ruler to audition your project, a process known as *scrubbing*. Scrubbing is useful when you need to find a precise spot for an edit. In the transport, click the pause icon (shown in the margin). If the pause icon isn't visible, customize the control bar as described in Chapter 3, and select Pause on the transport menu. Drag your cursor along the bottom part of the ruler, forward or backward, and the playhead will scrub the MIDI at the speed of your cursor.

By default, only MIDI is scrubbed. If you want your audio regions to scrub with your MIDI, choose Logic Pro ➪ Settings ➪ Audio, click the Editing tab, and select the Scrubbing with Audio in Tracks Area check box. If an audio region is selected in the tracks area, only the selected audio will be scrubbed along with the MIDI.

Although not enabled by default, you can set up key commands to Scrub Rewind and Scrub Forward in your project. You find out how to create key commands in Chapter 3.

Investigating the region inspector

The inspector has a lot of arranging power. And because a lot of arranging is done with regions, the region inspector area is an excellent arranging tool. Display the inspector (I) and click the disclosure triangle to display the region inspector area, shown in Figure 13-2. When you select a region or multiple regions in the tracks area, the region inspector is updated based on the selection. If you want to select all the regions and edit them simultaneously using the region inspector, click the track header, and all the regions on the track will be selected. Then you can edit them as a group from the inspector.

The parameters available in the region inspector depend on the type of selected track. All region inspectors have an additional disclosure triangle to hide and show more parameters. The benefit of editing the regions in the inspector is that the edits are nondestructive. You can always return to the region's original state by undoing the inspector changes.

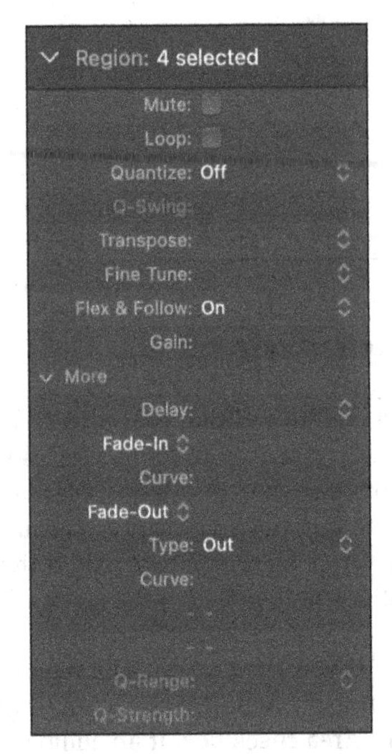

FIGURE 13-2: The region inspector.

Here's a description of some of the available parameters:

- **Mute:** Mute the selected regions.
- **Loop:** Loop the selected regions indefinitely or until they reach the next region on the track.
- **Quantize:** Correct the timing of notes. Choose the quantize value on the pop-up menu and all your notes will snap to the nearest division.
- **Q-Swing:** Determine the amount of shuffle feel to add to your quantize value. A setting between 50% and 75% will give you a natural swing feel.
- **Transpose:** Change the value in the Transpose field to transpose MIDI regions and Apple loops.
- **Gain:** Raise or lower the volume of audio regions and Apple loops. The main downside of using the Gain field to lower the volume of regions is that the audio waveform gets smaller in the tracks area and audio editors. It's better to use this Gain parameter to raise the level of tracks that were recorded too low.
- **Delay:** Adjust the timing of your regions. An excellent use for the Delay field is with orchestral choir patches. Because choir patches have a slow attack time, delaying the region by a negative value pulls the region forward in the project. You can get a natural timing of the choir with the rest of the tracks by

adjusting the delay to compensate for the slow attack. If you want a region to have a laid-back feel, the Delay setting is perfect for this kind of experimentation.

» **Fade In/Out:** Fade in and out audio regions. Using the region inspector, you can quickly fade multiple tracks, such as drums, simultaneously. You often need to fade into the beginning of audio regions to minimize room noise that's present before the part begins. For example, you may need to remove room noise from live drums. With the region inspector, you can select an entire group of drum tracks and fade them in and out.

Investigating the track inspector

Below the region inspector is the track inspector, as shown in Figure 13-3. You use the track inspector to edit the currently selected track. To edit multiple tracks simultaneously, select them by Shift-clicking the track headers.

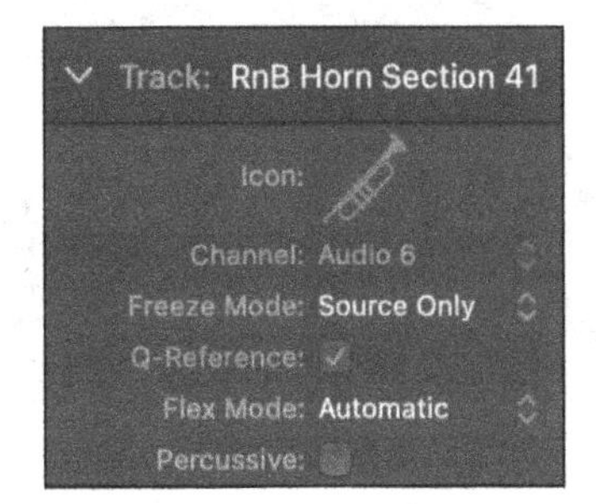

FIGURE 13-3: The track inspector.

REMEMBER

You can select a region without selecting the track that the region is on, so make sure you select the track and not just a region on the track.

Like the region inspector, the track inspector shows you different parameters depending on the type of track selected. Here's a description of some of the parameters available:

» **Icon:** Control-click the track icon to choose an icon on the pop-up menu, as shown in Figure 13-4.

» **Channel:** Display the channel strip type and number. The channel parameter is for reference only and isn't editable.

» **Freeze mode:** Freeze a track to reduce processing power on the track by temporarily turning the track and all its audio effects into an audio file. (In Chapter 3, you find out how to freeze tracks in the track header.) Freeze mode parameter can be set to Source Only, in which the track is frozen without Effects plug-ins, or Pre Fader, in which the track is frozen with all Effects plug-ins.

- **Q-reference:** On audio tracks, use the track's *transients* (peaks in volume) as reference points during the quantization of an edit group. For more on editing audio, see Chapter 14.
- **Flex mode:** Define how the audio will be processed while flex editing. You can discover more about flex editing in Chapter 14.
- **MIDI channel:** On MIDI tracks, choose the MIDI channel for the track's output.

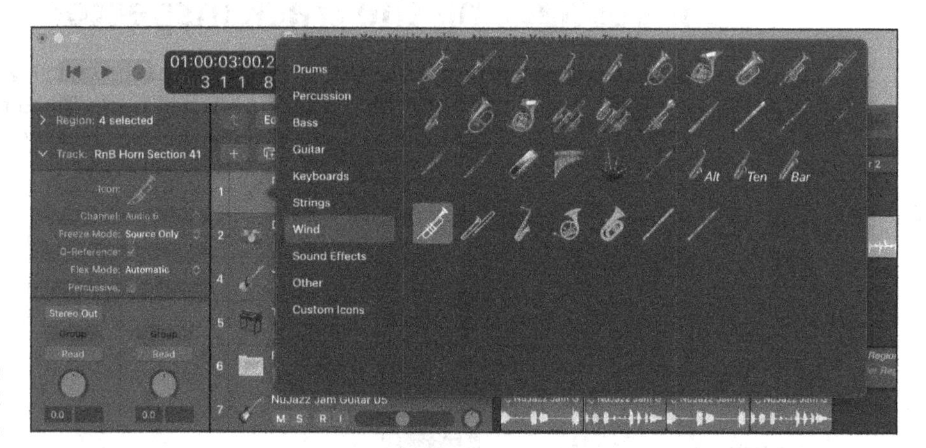

FIGURE 13-4: The track icon menu.

Showing Your Global Tracks

In Chapter 4, you find out about global tracks, which affect all your project's tracks. Global tracks are helpful when you're arranging because they help you create markers for quick navigation and important sections of your project.

To view the global tracks, choose Track ➪ Global Tracks ➪ Show Global Tracks (G). The global tracks appear at the top of the tracks area, as shown in Figure 13-5. If you don't see a global track you're looking for, choose Track ➪ Global Tracks ➪ Configure Global Tracks (Option-G) and select the global tracks you want to view.

TIP

You can resize and reorder global tracks to customize your workspace. Drag the track headers up or down to reorder them. To resize a global track, drag the track header dividing line up or down.

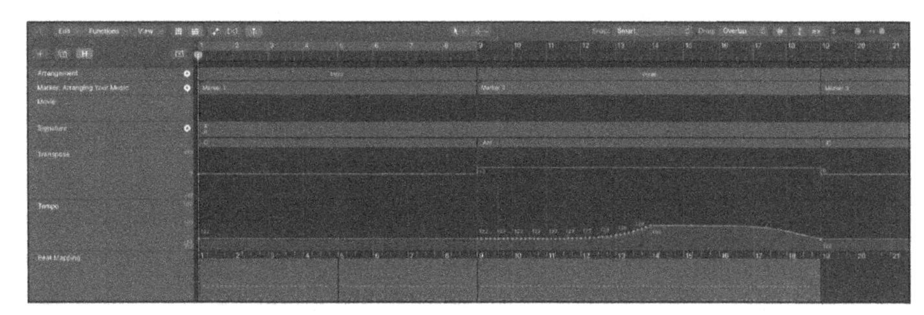

FIGURE 13-5: The global tracks.

Creating and naming markers

Markers are excellent arrangement tools. They allow you to navigate your project and select regions quickly. You can also rename markers to help you identify sections in your project. You can view and edit markers in several ways:

- » Choose Track ➪ Global Tracks ➪ Show Marker Track or press Shift-⌘-K to display the marker track.
- » Choose View ➪ Show List Editors (D) and then click the Marker tab to view the marker list editor, as shown in Figure 13-6.
- » Choose Navigate ➪ Open Marker List to view the marker list in a new window.

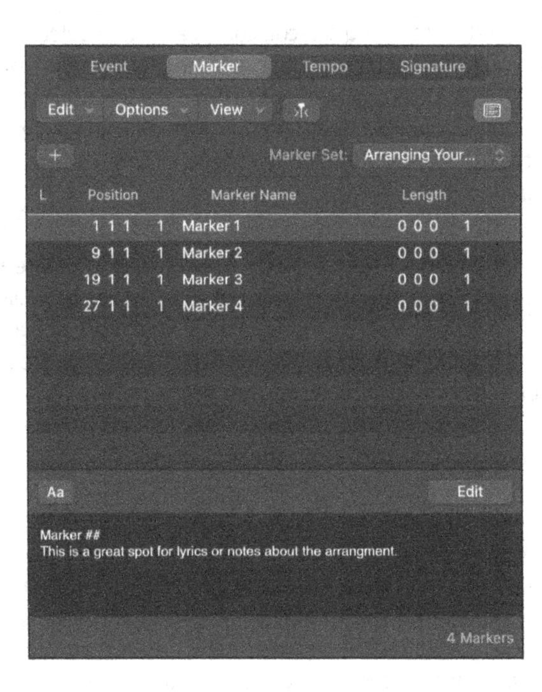

FIGURE 13-6: The marker list editor.

To create and edit markers, do one of the following:

- » Click the plus sign icon in any of the marker editors to add a new marker at the current playhead position.
- » In the global marker track, use the pencil tool and click the location where you want to create a new marker.
- » In the marker list editor or marker list window, choose Options ➪ Create.
- » At any time, press Option-apostrophe (') to create a new marker at the current playhead position.

When you create a new marker, the default behavior is to round the marker to the nearest beat. If you want to create a marker without rounding, choose Navigate ➪ Other ➪ Create without Rounding or choose Options ➪ Create without Rounding from the marker lists.

Moving, renaming, and deleting markers is a straightforward process:

- » In the global marker track, drag the edges of the marker to change its start and end points. Double-click the marker or use the text tool to rename it. You can delete selected markers by pressing Delete. Control-click the marker to access a pop-up menu with more marker commands.
- » In the marker list editors, you can cut, copy, paste, and delete markers on the Edit menu. You can adjust the position and length of markers by typing in the marker list. Double-clicking a marker will open the Edit Marker Text area at the bottom of the editor window. Use this area to rename markers and add text notes to your marker. This spot is excellent for lyrics or notes about the arrangement.

You can also store alternate marker sets. If you need to create markers at a few places in your project to make edits later but don't want to mess up your current arrangement of markers, you can create a new marker set and even give it a descriptive name. From the marker list editor, choose New Set on the Marker Set drop-down menu. To create a new marker set from the global tracks, click the Marker title in the track header and choose Marker Sets ➪ New Set on the pop-up menu. You can also rename sets from these menus.

TIP

If you have an extended keyboard with an extra number pad, press the number keys to instantly move the playhead to the corresponding marker number. You can also navigate markers using the Next Marker and Previous Marker key commands: Option-period (.) and Option-comma (,). Whenever you wish you could go to the same places in your project over and over again, create a new marker set and a few markers. Then you can zip to the stored locations quickly.

Here are some more uses and tips for markers:

- Use the Go to Marker Number key command (Option-/) to open a dialog where you can type the marker number and press Enter to instantly take the playhead to the marker.
- Markers make it fast and easy to set locators. Use the Set Locators by Marker and Enable Cycle key command (Control-Option-C). You can explore other key commands by opening the Key Commands window (Option-K) and typing *marker* in the search field.
- You can color your markers by choosing View ⇨ Show Colors (Option-C). Click a color in the palette, and the selected marker will be colored.
- The marker track can be resized to view all the text in your markers. Put your cursor over the bottom edge of the marker track in the track header and pull down when your cursor becomes the resize tool.

Creating key signature and time signature changes

The signature global track does double duty, handling the key signature and the time signature. Choose Track ⇨ Global Tracks ⇨ Show Signature Track. The top half of the track shows the time signature, and the bottom half shows the key signature. You can also view and edit the signature track from the signature list editor. Choose View ⇨ Show List Editors (D) and then click the Signature tab to view the signature list editor, shown in Figure 13-7.

You use the pencil tool to create signature changes in global tracks. Click the top half of the signature track to create a time signature change. The Time Signature window opens, as shown in Figure 13-8, and you can choose the number of beats per bar and the note value that equals one beat. You can also define how you want the beat divided. For example, a 7/4 time signature divided as 2+2+3 can be entered as 223 in the Beat Grouping field. The beat grouping feature affects the display of notation in the score editor.

You create key signature changes by clicking the bottom half of the signature track. The Key Signature window opens, as shown in Figure 13-9. Choose the key, Major or Minor, and whether to disable double flats (bb) or double sharps (x) in the score editor.

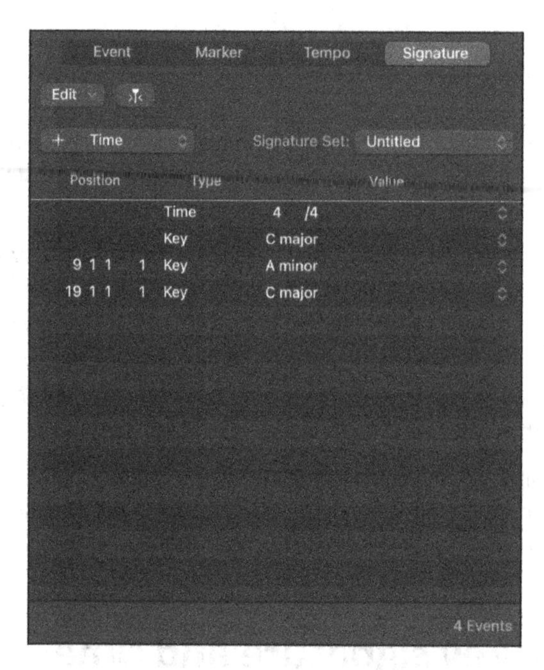

FIGURE 13-7: The signature list editor.

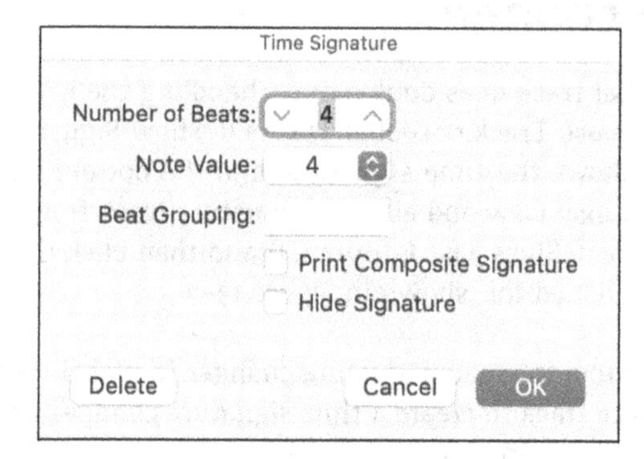

FIGURE 13-8: The Time Signature window.

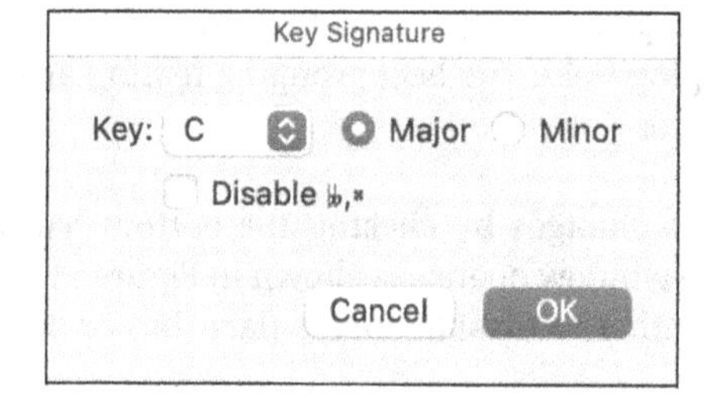

FIGURE 13-9: The Key Signature window.

To edit signature changes, double-click the signature change in the signature track. You can edit your signature changes also in the signature list editor. Just like with the marker track, you can create signature sets. Click the Signature name in the track header and choose Signature Sets ➪ New Set on the drop-down menu. Or use the signature list editor to create, delete, and rename signature sets on the Signature Set drop-down menu.

REMEMBER

Key signature changes don't change the pitch of the audio or MIDI regions you've recorded. They change only the display of the regions in the score editor.

Creating tempo changes

Most projects have a single tempo. But when projects need tempo changes, it's nice to know that Logic Pro makes the job simple and effective. Choose Track ➪ Global Tracks ➪ Show Tempo Track (Shift-⌘-T). The tempo track behaves similarly to the other global tracks. But unlike the other tracks, the tempo track can handle more edit points and create smooth curves for natural feeling tempo changes.

Use the pencil tool to create a tempo change in the tempo track. A solid tempo control point will appear on the tempo line along with a hollow tempo curve automation point that you can drag left and right or up and down to create tempo curves between two points, as shown in Figure 13-10.

FIGURE 13-10: The tempo track.

Change the tempo resolution, change the quantization, and create tempo sets by clicking the Tempo name in the tempo track header and making a choice on the pop-up menu. Another way to adjust your tempo is with the tempo list editor, as shown in Figure 13-11. Press D to open the list editor and then click the Tempo tab.

In the tempo list editor, click the Additional Info button to see all tempo points used to create your tempo curves. Because a curve can contain hundreds of tempo changes, it's helpful to hide that information when viewing your tempo changes. Choose Options ➪ Tempo Operations to open the tempo operations window, as shown in Figure 13-12.

The tempo operations window can create complex tempo curves or scale existing tempo changes. Explore the Operation menu to see the other tempo operations available. Because the tempo operations window has an Undo button, it's a great place to try out your tempo changes until you get it just right.

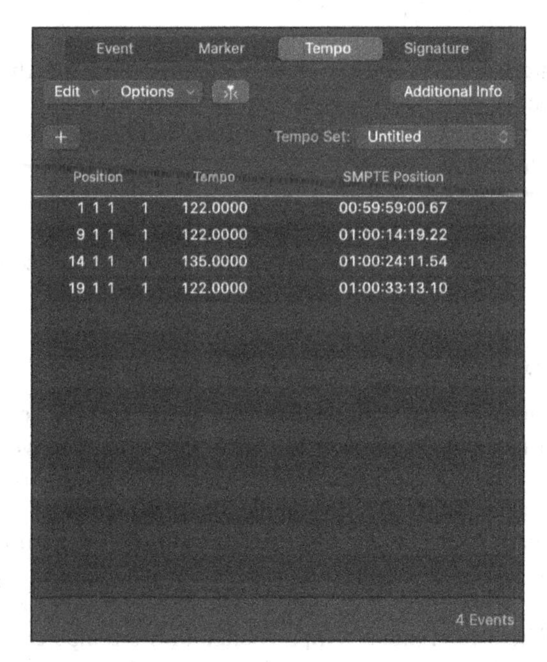

FIGURE 13-11: The tempo list editor.

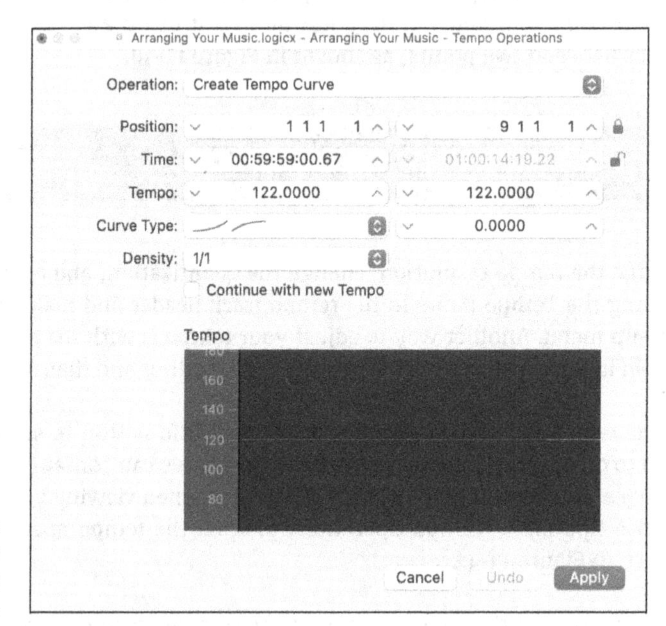

FIGURE 13-12: The tempo operations window.

REMEMBER

Tempo changes don't change the playback speed of audio regions. A tempo change only adjusts the starting point of audio regions in the project timeline. It's always a good idea to plan your tempo changes before you begin recording.

Creating arrangement markers

Arrangement markers are similar to regular markers but can be used to rearrange your project globally. To view the arrangement track, choose Track ➪ Global Tracks ➪ Show Arrangement Track (Shift-⌘-A). You create arrangement markers with the pencil tool. By default, arrangement markers are eight bars long, but you can drag the edges of the markers to resize them. Click the name of the marker to choose a different preset name on the pop-up menu, or choose Rename to create your marker name.

You can select arrangement markers and drag them to new positions on the arrangement track. All regions within the arrangement markers will be moved to the new position. You can select multiple arrangement markers by Shift-clicking them, and you can copy arrangement markers, including the regions within them, by Option-dragging the marker to a new location. All arrangement markers and their content will be shifted to the right to make room for the new arrangement. You can replace markers by ⌘ -dragging an arrangement marker on top of another.

Creating transposition points

The transposition track transposes the key of your Apple loops and MIDI regions globally. Audio regions, drum loops, and Apple loops with undefined keys are unaffected by the transposition track. If you don't want a software instrument or MIDI track to be transposed, select No Transpose in the track inspector.

To open the transposition track, choose Track ➪ Global Tracks ➪ Show Transposition Track (Shift-⌘-X). You create transposition control points with the pencil tool. Drag the control points up or down with the pointer tool to adjust the level of transposition. Delete selected control points by pressing Delete.

REMEMBER

Decide quickly if you're going to transpose individual regions or transpose the track globally. If you change the transposition of a single region and then decide to transpose the entire project from the transposition track, you may end up transposing that region more than you intend.

TIP

If you want the ultimate control, it's best to transpose individual regions from the region inspector, but that process can be time consuming. If you're just trying something out and want to transpose your project temporarily, the transposition track is the right tool for the job. It works great for trying out harmonic modulations in songs or when you're finding the correct key for a singer at the beginning of your project.

Beat Mapping Your Arrangement

If you didn't record to a metronome or you're importing audio that was recorded in a different session or other software, you'll need to align your ruler to the imported tracks. For that task, you can use the beat-mapping track. Beat mapping makes the project tempo follow the audio or MIDI regions in your project. After you beat map your project, you can do things such as overdub with the metronome, quantize regions, loop regions, and view MIDI in the score editor.

To show the beat-mapping track, shown in Figure 13-13, choose Tracks ⇨ Global Tracks ⇨ Show Beat Mapping Track (Shift-⌘-B).

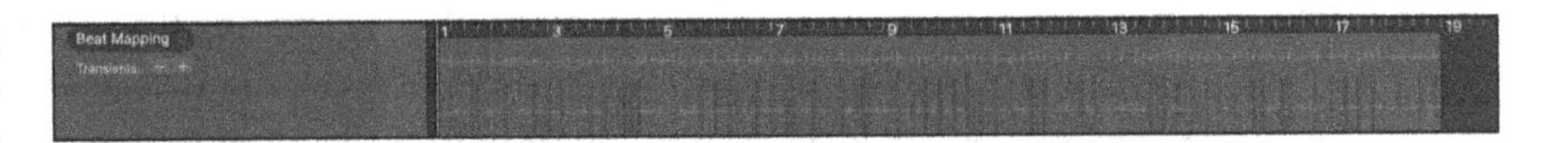

FIGURE 13-13: The beat-mapping track.

Beat mapping MIDI regions

Only some projects require a metronome. Sometimes, you want to play freely, known as *tempo rubato.* But when the recording is completed, you can use beat mapping to adjust the ruler to fit the performance. Beat mapping makes editing easier and makes it possible to view your MIDI recordings in the score editor. Here's how you beat map MIDI regions:

1. **Select the MIDI regions.**

   The beat-mapping track displays the contents of the selected regions.

2. **Click the position in the beat-mapping ruler that you want to map to a note.**

   A vertical line appears in the beat-mapping track.

3. **Drag the vertical line to the note.**

   The tempo adjusts, and the ruler position and note are aligned.

4. **Repeat Steps 2 and 3 with other ruler positions and notes.**

TIP

You might want to beat map the ruler to a place in a region that doesn't have a note. To do this, Control-drag the vertical line to the place on the region you want to map.

Beat mapping audio regions

The process for beat mapping audio regions is similar to the one for MIDI regions, except you have to analyze the audio beforehand. To analyze the audio for beat detection, follow these steps:

1. **Select the audio regions to be analyzed.**
2. **Click the Beat Mapping title in the global tracks header, and then choose Analyze Transients on the pop-up menu.**

   The audio is analyzed for peaks in volume, known as *transients.* Transients appear as blue lines in the beat-mapping track.
3. **Click the position in the beat-mapping ruler that you want to map to a transient.**
4. **Drag the vertical line to the transient.**

   The tempo adjusts to the audio region.

Automatically beat-mapping regions

As mentioned at the beginning of this chapter, Logic Pro does much of the arranging work for you. It's good to know how to beat map audio and MIDI regions, but you can also let Logic Pro automatically beat map your regions. Simply do the following:

1. **Select the regions to beat map.**
2. **Click the Beat Mapping title in the global tracks header, and then choose Beats from Region on the pop-up menu.**

   The Set Beats by Guide Region(s) window opens, as shown in Figure 13-14.
3. **Choose the note value.**

   The note value sets the beat map resolution.
4. **If an audio region is selected, select the algorithm.**

   Select the Tolerating Missing or Additional Events check box to map only the most important transients. Select the Use Exactly All Existing Events check box to map every transient.
5. **Click OK.**

   Transients appear as blue lines in the beat-mapping track, and the tempo is adjusted to the region.

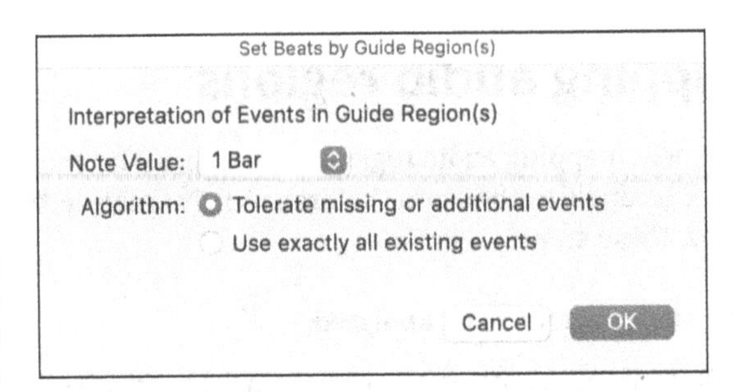

FIGURE 13-14: Set Beats by Guide Region(s) window.

Arranging Regions in the Tracks Area

Arranging in Logic Pro consists mostly of organizing your regions so they make sense to your ear. You can also apply arranging techniques to a single region in terms of harmony, such as string or horn arranging. But the tracks area is the broad stroke palette for you to paint your sonic masterpiece. Here are some tips to help you begin arranging:

- **Create contrast.** Creating contrast in your arrangement adds energy, forward motion, and interest. If you want your arrangement to sound big, make it small first. Create contrast through the use of different combinations of instruments in your song sections. Moving from dense to sparse, intense to relaxed, and unstable to stable will keep your listener's attention. Look for ways to provide refreshing contrast in your song form, rhythm, harmony, and melody.
- **Arrange your song form.** Your song form can be arranged and rearranged. At the least, your song will probably have two different sections. Use your song form as an arrangement tool. For example, change the order of your sections using the arrangement markers to see if they sound better.
- **Add and release tension.** You can simplify the job of all harmony to tension and release. Every time you play the root chord, you release the tension. (In the key of C, for example, the root chord is C.) Every time you move away from the root chord, you build tension. Pay close attention to where you place the root chord in your song and find ways to build and release tension. In terms of contrast, you can adjust the length of your chord progression, add or subtract chords from sections, and start each section on different chords.
- **Provide consistency and contrast.** The big driver of pop music is rhythm. (I use the term *pop music* loosely, as in the opposite of unpopular music.) Listeners want consistency and variation, in other words, contrast. Give the listener consistency by keeping the rhythms similar in the A sections, and give the listener contrast by doing something different in your B sections. In songwriting, for example, keep the rhythms of your verses consistent but

contrast the chorus rhythm from your verse. Note that the term *rhythm* includes the rhythm of your harmony and melody and the rhythm of the drums and other rhythmic instruments.

» **Rearrange sounds.** Your instrument choices can complement and contrast each other. You learn how to create an orchestral template in Chapter 12, and the different orchestral sections are a great example of complement and contrast. Brass is metallic and bright, woodwinds are nasal and soft, strings are warm and lush, and percussion is sharp and punchy. All those sounds can be rearranged in many wonderful combinations to create interest for your listener.

Selecting regions

The bulk of arranging work is done on regions, so you'll do a lot of selecting in the tracks area with the pointer tool. You open the tool menu by pressing T. Note that each tool has a key command associated with it, as shown in Figure 13-15. After you open the tool menu by pressing T, you can press the key command for the tool you want. Pressing T twice selects the pointer tool. These shortcuts make selecting regions quick and easy.

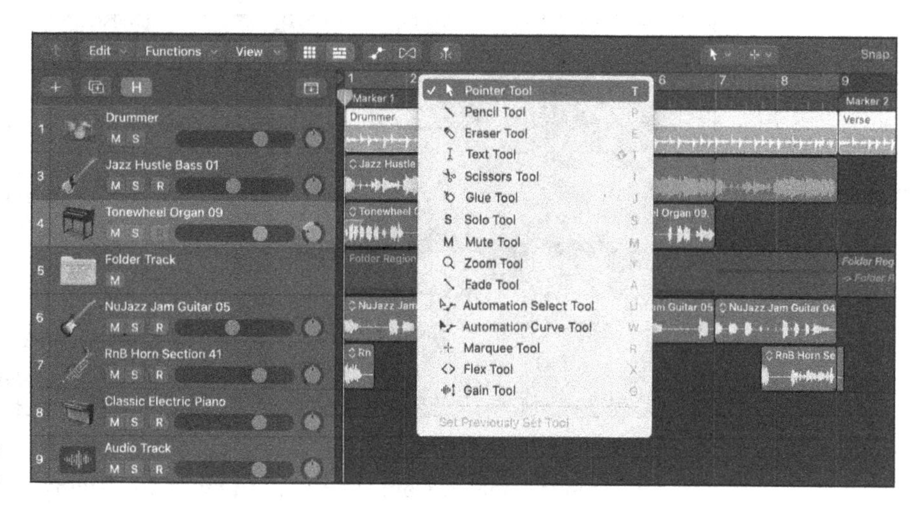

FIGURE 13-15: The tool menu.

TIP

You can select more than one region by Shift-clicking regions one by one or by dragging your cursor around the regions you want to select. You can select all the regions on a track by clicking the track header.

You can select regions in more ways by using the Edit menu. To select all the regions in your project, choose Edit ⇨ Select ⇨ All. If you Control-click an empty spot in the tracks area, you have more selection choices, such as Select Overlapped Regions/Events, which is helpful if you need to select regions that are underneath other regions. If you Control-click a region, another menu of selection commands appears in a pop-up window, such as Select ⇨ Invert Selection. You can also navigate and select regions by using the arrow keys.

Selecting parts of regions with the marquee tool

Sometimes, you'll want to select only parts of regions. For example, a live drum performance could contain a dozen tracks, each with a single region the length of the entire song. When you want to select only a portion of those regions, you use the marquee tool.

To select a portion of a region or group of regions, select the marquee tool on the tool menu (T) and drag your cursor around the area you want to select. The selection is outlined and brighter in color, as shown in Figure 13-16. In addition to selecting regions, your marquee selection is used for playback. As long as the selection is in place, playback begins at the left of the selected regions.

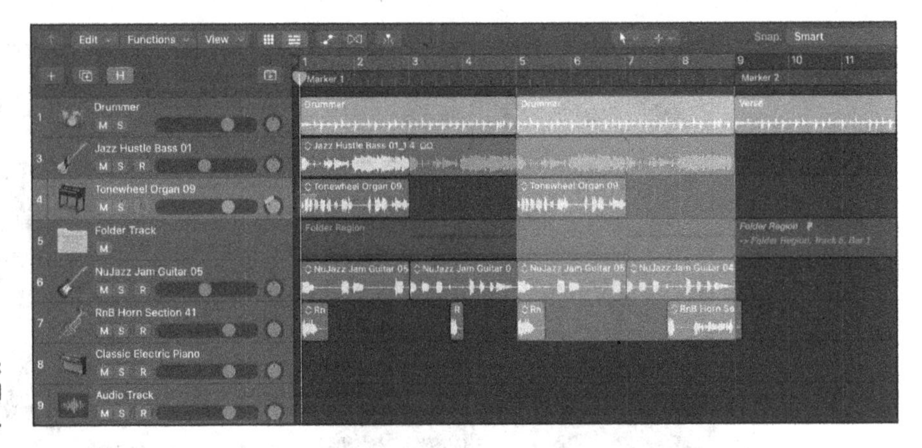

FIGURE 13-16: A marquee tool selection.

You can remove a region from the selection by Shift-clicking the region. If the marquee ruler is displayed, you can move the selection left or right by dragging in the ruler. To display the marquee ruler, choose View ⇨ Marquee Ruler on the tracks area menu bar.

Marquee selections have many uses. You can apply edits to your selection, and you can also navigate by using your selection. For example, pressing Z will zoom to fit your selection. You can set your locators by the marquee selection (for use in cycle mode) by pressing ⌘ -U.

Logic Pro doesn't take selecting objects for granted. If you take the time to practice some alternate methods of selecting regions besides simply clicking regions individually with the pointer tool, you'll be rewarded with a fast and productive workflow, which means less time clicking and more time creating.

TIP

Here's an excellent reason to colorize your tracks. If you color groups of tracks the same color, you can choose Edit ➪ Select ➪ Equal Colored Regions (Shift-C) and instantly select them all.

Moving regions

After you select regions, you can drag them anywhere in the tracks area. However, keep in mind that you can drag a region to a different track type, but the region will play only if it's moved to the same track type.

You copy regions by Option-dragging them to a new location. Change the length of a region by dragging its bottom corner to the new length. You loop a region by dragging its top corner to the right as far as you want it to loop.

A great key command to memorize is the nudge command. Press Option-right arrow to nudge the selected region to the right by the nudge value. Press Option-left arrow to nudge the region to the left. Set the nudge value by Control-clicking a region and choosing Move ➪ Set Nudge Value To on the pop-up menu. You can set the nudge value to a variety of lengths, such as bars and beats, milliseconds and samples, or video frames. Press the semicolon key (;) to move the selected region to the playhead position.

Copying and pasting works the same as in many other Mac applications. You make a selection and copy it by pressing ⌘ -C or choosing Edit ➪ Copy. You paste your selection on the selected track at the playhead position by pressing ⌘ -V or choosing Edit ➪ Paste.

TIP

Don't forget to deselect your regions when you're finished making changes. Deselecting your regions is a safety measure against accidentally editing or deleting regions without realizing it. The key command to deselect all is Shift-Option-D. You can also deselect regions by clicking an empty spot in the tracks area or choosing Edit ➪ Select ➪ Deselect All.

Soloing and muting regions

Soloing regions is less of an arrangement tool and more of a mixing or editing tool. You solo regions or tracks to listen to the details of the recording. Sometimes, editing or mixing in solo mode is useful for silencing all distractions and hearing minute details. To play regions in solo mode, select the solo tool on the tool menu (T) and click-hold the region to play it without any other tracks.

The mute tool is much more useful in an arranging workflow. You want the capability to silence regions without committing to deleting them. Expert arrangers know it's as important to leave things out as it is to put things in. It's common to record more than you need and then mute what you don't need. Muting is a considerable part of arranging.

Select the mute tool on the tool menu (T) and click the regions you want to mute. Muted regions appear dark gray in the tracks area. To unmute a region, click the region again using the mute tool.

You can also mute regions in the region inspector. Press I to open the inspector and then click the disclosure triangle to open the region inspector, as shown in Figure 13-17. Select the Mute check box to silence the region. Notice how the mute tool updates the mute check box in the region inspector and vice versa.

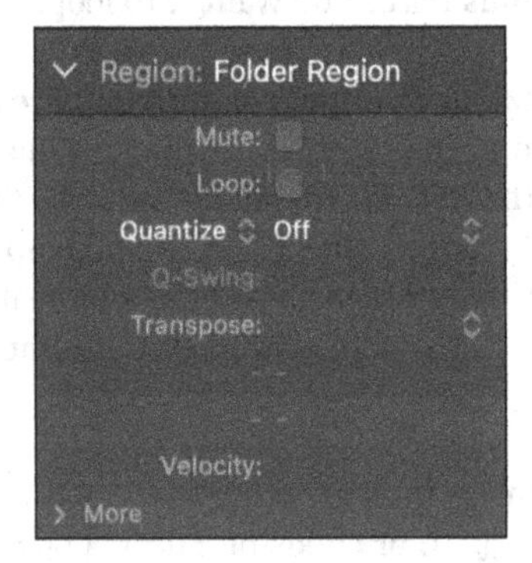

FIGURE 13-17: The region inspector.

TIP

Two useful key commands to remember are Control-M to mute regions and Shift-M to select muted regions. For example, if you have a few regions muted but not selected, press Shift-M to select the muted regions and then press Control-M to unmute the regions.

Time-stretching regions

If you import a MIDI or audio file that doesn't fit your project tempo perfectly, you can stretch or shrink the region to match the tempo. To time stretch a region, Option-drag the lower-right or lower-left edge of the region. The region will be compressed or expanded. If you're time-stretching an audio region, you can choose Edit ⇨ Time Stretch ⇨ Time Stretching Algorithm and select one of the following options:

» **Universal:** The default algorithm and a good option for most time stretching.

» **Complex:** A good option for full mixes and complex sounds.

» **Percussive:** A good option for drums and percussive sounds.

» **Legacy Algorithms:** A submenu that contains algorithms from previous Logic Pro versions. Check this submenu for more options if the previous algorithms aren't giving you the desired result.

Demixing MIDI regions

Standard MIDI files contain more information than just note values. These files contain tempo changes, marker names and positions, individual track names, and the MIDI events such as notes, time positions, and MIDI channels. The two MIDI file formats are 0 and 1. Format 0 contains all the data in a single track, and format 1 contains multiple tracks. Depending on the type of MIDI file you import, you could be importing multiple MIDI channels on a single track. In situations like this, you can demix the MIDI events in a region by MIDI channel:

1. **Select the MIDI region.**
2. **Choose Edit ➪ Separate MIDI Events ➪ By Event Channel.**

   A new track is added to the tracks area, and a MIDI region is created for each channel containing data.

A similar situation is called for when you have a MIDI region containing a drum pattern and you want each drum on its own track for individual control. Choose Edit ➪ Separate MIDI Events ➪ By Note Pitch. Each note that contains data will be added to a new track and region. You can also separate MIDI events by Control-clicking the region and choosing MIDI ➪ Separate by Note Pitch or Separate by MIDI Channel on the pop-up menu.

Fading and crossfading audio regions

When you organize audio regions in the tracks area, sometimes you need to crossfade between two regions smoothly. A crossfade makes the transition between two audio files sound more natural. And sometimes, you need to fade in and out of audio regions to eliminate noise. Guitar players who love distortion know what I mean.

Select the fade tool on the tool menu (T), and drag over the beginning of an audio region to fade in or drag over the end to fade out. A fade is created on the region, as shown in Figure 13-18. To crossfade between two regions, drag over the end of one region and the beginning of another. With the fade tool selected, you can adjust the fade length by dragging it in the region.

FIGURE 13-18: Audio region crossfade.

Stripping silence from audio regions

Sometimes you'll record a take the entire length of a song but record audio only some of the time, for example, when recording background vocal tracks. You might be recording only a few notes every fourth bar, but you leave record on for the entire length of the song. You can quickly strip the silence from the long region to create many smaller regions. Control-click the region and choose Split ➪ Remove Silence from Audio Region (Control-X) on the pop-up menu. The Remove Silence window opens, as shown in Figure 13-19, and you can adjust the settings that determine the number of regions. Click OK, and your big region becomes many regions.

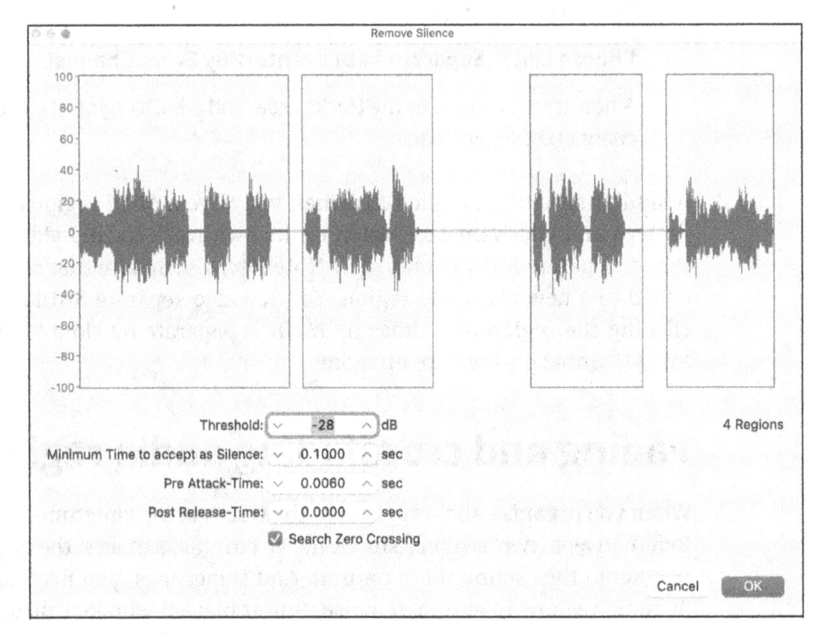

FIGURE 13-19: The Remove Silence window.

Creating Folder Tracks

Logic Pro can help you organize your arrangement and experiment with different arrangements by using folders. A folder is a container. It looks just like a region, but it can contain other regions. You can also have folders within folders. You might use folders to:

- Organize tracks and song sections
- Nest parts of an arrangement
- Nest alternate versions of a part

TIP

If you want to organize your tracks, another option is a folder track stack. You find out how to create track stacks in Chapter 4. Track stacks make it easy to open and close folders of tracks, which simplifies your workspace.

Packing and unpacking folders

You can select as many regions as you want and put them into a folder. If the selected regions are on the same track, the folder is added to the same track as the regions. But if the selected regions are on different tracks, a new folder track is created. To create a folder, do the following:

1. **Select the regions you want to include in the folder.**
2. **On the tracks area menu bar, choose Functions ⇨ Folder ⇨ Pack Folder.**

   A folder track is created containing the regions, as shown in Figure 13-20, and the regions are removed from the tracks area.

FIGURE 13-20: A folder track.

A folder looks like a region in the tracks area. A folder doesn't have a channel strip, so a folder track does not have audio controls. Double-click the folder to open the folder contents and display the tracks area of the folder. To leave the folder, click the leave folder icon (shown in the margin) on the tracks area menu bar or double-click an empty spot in the tracks area.

TIP

You can drag a folder region to an audio or MIDI track type and the corresponding contents of the region will play through the track's channel strip. So if you drag a folder that contains MIDI to a MIDI track type, the MIDI regions play through the track's channel strip. Likewise, a folder containing audio regions can be played on an audio track.

To unpack the contents of a folder, do the following:

1. **Select the folder region.**
2. **Choose Functions ➪ Folder ➪ Unpack Folder to Existing Tracks or Unpack Folder to New Tracks.**

   The regions contained in the folder are added to the tracks area.

Adding and removing regions

To add regions to an existing folder, drag them onto the folder at the location you want them to play. The region is moved into the folder along with its track. To remove regions from the folder track, open a second main window and drag the regions from the folder to the main tracks area:

1. **Press ⌘-1 or choose Window ➪ Open Main Window.**

   A second main window opens in a new window.
2. **Double-click the folder region to view its contents.**

   The tracks area displays all the regions contained in the folder.
3. **Drag the regions from the folder to the tracks area in the first main window where you want them to play.**

   Your region is moved to the main tracks area and is no longer contained in the folder.

Creating alias folders and regions

Region aliases and clones are references not to science-fiction movies but to regions in the tracks area. You can create a region that references an original region, allowing you to edit the original and have the alias change too. These regions are referred to as *aliases* if they're MIDI regions or *clones* if they're audio regions. You can also create folder aliases. Aliases are great for parts that repeat because you have to edit only the original region, and all the aliases will be updated with the new edit.

To create an alias of a region, press Shift-Option while dragging the region to a new location. If you make an alias of a MIDI region or folder region, the alias region looks different than the original, as shown in Figure 13-21. The name is in italics and the region has an alias icon. Cloned audio regions look the same as the original, which makes them less desirable than MIDI or folder regions.

FIGURE 13-21: A region alias.

TIP

Audio clones have limited cloning capability because only the start and end points of the original region are cloned. You can get around this limitation by packing the audio regions into a folder and then making an alias of the folder. Then all the edits you make to the regions in the original folder will be cloned in the folder alias.

Using Groove Templates

Another important part of arranging is creating parts that feel good together. Logic Pro gives you two amazing ways to make audio and MIDI tracks follow the same groove:

- **Groove track:** You can set one track in your project as the groove track and select other tracks to follow the timing of the groove track. Control-click anywhere on the track list and choose Track Header Components ⇨ Groove Track. Click the track number of the track that defines the groove. A second column of check boxes appears in the track header, as shown in Figure 13-22. Select the check box to follow the groove track.
- **Groove template:** You can create custom quantized grids based on an audio or MIDI region. You apply the grid to other regions so they get the same feel.

FIGURE 13-22: The groove track.

TIP

You can make groove templates as long as you like, but it's usually best to create short ones between two and four bars. Because you can create multiple groove templates in a project, you can model the groove from several different instruments and musical parts and use it throughout your project to enhance the timing and feel.

Creating a groove template

If you have a drummer track in your project, selecting a drummer region for your groove template is a great choice. But if you have a region that you think defines the groove better, select it. Better yet, create groove templates from both regions. Nothing is holding you back from creating groove templates from every region in your project. To create a groove template:

1. **Select the region you want to model for the groove template.**
2. **On the Flex Mode pop-up menu in the track inspector (I), select a flex time mode, as shown in Figure 13-23.**

   Because the transients in the audio region are used to quantize the region, either Slicing or Rhythmic flex time mode is a good choice.
3. **On the Quantize pop-up menu in the region inspector, select Make Groove Template.**

   The groove template is added to the quantize menu, named after the region.
4. **Select the regions you want to quantize with the groove template.**
5. **On the Quantize pop-up menu in the region inspector, select the new groove template.**

   The timing of the region matches the groove template.

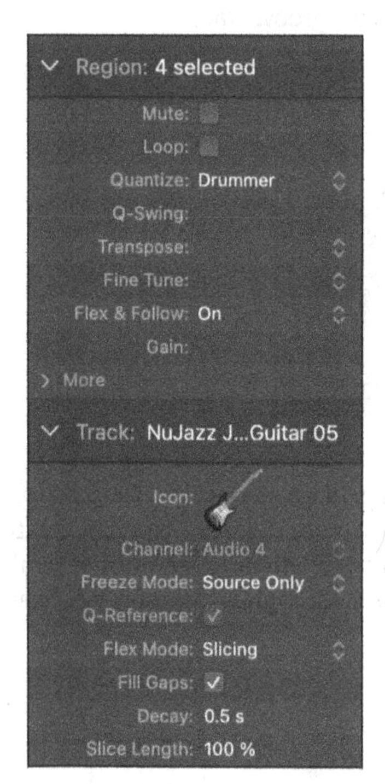

FIGURE 13-23: The track and region inspector.

REMEMBER

If you delete the region on which a groove template is based, the groove template is deleted too. After a groove template quantizes a region, you can create a second groove template from the quantized region. This safeguards your groove template from being lost if the original region is deleted. You can also manually remove groove templates by using the region inspector Quantize menu.

Importing groove templates from other projects

Because groove templates are based on regions in your project, any media you import can be used as a groove template. You can collect regions with good grooves and save them in a project template, as you learn in Chapter 2. Start your next project with this template, and all your grooves will be ready to load.

TIP

Groove track and groove templates can be used together. You can create groove templates from the tracks that follow your groove tracks. You can now delete the original groove track because the tracks that were following the groove track are now following a groove template.

Importing third-party groove templates

You can import third-party groove templates by choosing File ➪ Import ➪ Other and navigating to the file. Logic Pro supports standard MIDI files and DNA Groove Template files. Because groove templates are basically MIDI events that are used to set quantize values, any imported MIDI file can become a groove template. Also, because you can quantize audio using flex time, as you learn in Chapter 14, even your imported audio can become a groove template. Groove templates are unbeatable at getting the right feel in your arrangement.

Now that you can build an arrangement in the tracks area, you can listen to how your favorite artists arrange their music and duplicate what they do in Logic Pro. You can import audio you love and make templates to tighten up your groove. Your command of folders and global tracks will make you more productive.

If you put Logic Pro to work for you, your arrangements will feel great and sound solid. Even the greatest musicians can benefit from Logic Pro's timing enhancements and workflow shortcuts. Don't feel odd or embarrassed if you find yourself saying, "Logic Pro, I couldn't have done it without you." I'm almost certain Logic Pro thinks the same about you.

IN THIS CHAPTER

» Getting to know the audio editors

» Creating composite takes

» Fixing pitch and audio timing

» Repairing audio glitches

Chapter 14
Editing Audio Tracks

Imagine you are back in the time before personal computers and want to edit audio. You would have to use a razor blade on tape — with no undo command. The process wouldn't be creative or quick, and you wouldn't edit analog audio just to experiment.

Fast-forward to the present. Music is in high demand in this high-speed world. You don't have the luxury of taking months to finish a project — your internet fans wanted it yesterday. Logic Pro makes editing audio so simple and powerful that you can meet the demands of your fans and clients without breaking a sweat — or cutting your fingers on a razor blade.

In this chapter, you discover how to create the perfect take with Logic Pro's quick swipe comping. You find out how to get the timing of your parts just right and tune up the pitch. And you learn how to get the best sound out of your audio and even create sounds that you didn't record in the first place.

Knowing Your Audio Editors

Logic Pro provides you with several ways to edit audio. You can do audio editing right from the tracks area, as you learn in Chapter 13. In this section, you discover the specialized Logic Pro audio editors.

The audio track editor

The audio track editor, shown in Figure 14-1, is the default Logic Pro audio editor. It works well for most audio editing needs. You can open the audio track editor in any of the following ways:

- » Double-click an audio region.
- » Choose View ➪ Show Editor with an audio region selected.
- » Press E with an audio region selected.

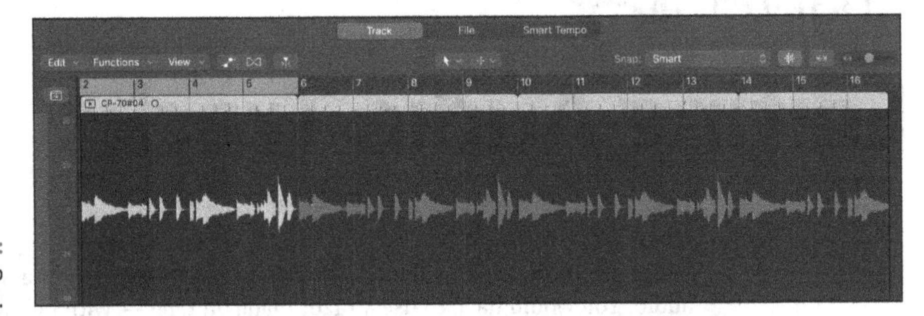

FIGURE 14-1: The audio track editor.

TIP

When you're starting out with Logic Pro, it's a good idea to display Quick Help. Click the Quick Help icon in the control bar or choose Help ➪ Quick Help. Quick Help shows you the name and function of anything your cursor hovers over. The quick help window is displayed at the top of your inspector or in a floating window that follows your cursor when you choose Help ➪ Quick Help Follows Pointer. Quick Help is useful when you can't remember what all the nameless icons do in the menu bar and other areas.

The audio file editor

The audio file editor, shown in Figure 14-2, is used for sample-level audio editing. It works well for audio repairs, such as removing pops and clicks. You can open the audio file editor in the following ways:

- » Double-click an audio region, and then click the File tab at the top of the audio editor.
- » Choose View ➪ Show Editor with an audio region selected, and then click the File tab at the top of the audio editor.

» Choose Window ➪ Open Audio File Editor to open the editor in a new window.

» Press Option and double-click an audio region in the project audio browser.

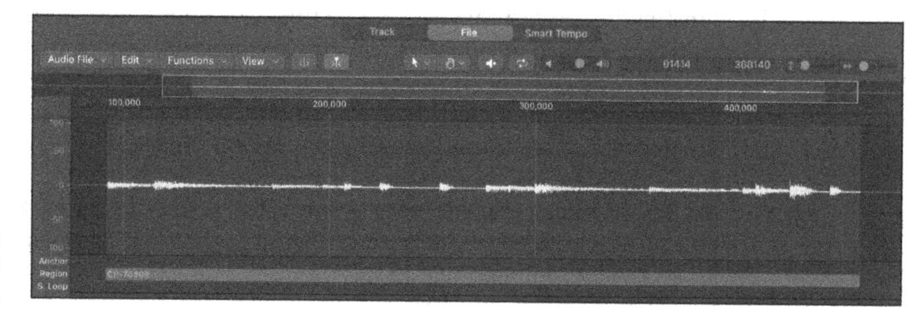

FIGURE 14-2: The audio file editor.

WARNING

The audio file editor is considered destructive because the original audio file on your hard drive is modified. You can still undo your edits, but working with copies of your audio files is always safer.

The smart tempo editor

The smart tempo editor, shown in Figure 14-3, is used to edit the beat markers and tempo data of audio files that contain tempo information. You can open the smart tempo editor in any of the following ways:

» Double-click an audio region, and then click the Smart Tempo tab at the top of the audio editor.

» Choose View ➪ Show Editor with an audio region selected, and then click the Smart Tempo tab at the top of the audio editor.

» Choose Window ➪ Open Smart Tempo Editor to open the editor in a new window.

» Choose Edit ➪ Tempo ➪ Show Smart Tempo Editor.

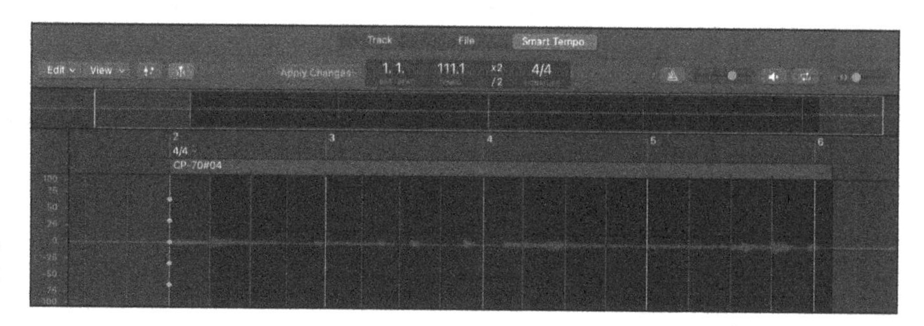

FIGURE 14-3: The smart tempo editor.

Creating the Perfect Take with Quick Swipe Comping

Creating the perfect take used to be time consuming. But since Logic Pro automatically records take folders, creating a composite take (a *comp take*) is almost too easy. You can quickly create a comp take by swiping on takes in a take folder, as you learn in this section. Quick swipe comping is another example of how Logic Pro does the heavy lifting, allowing you to focus on creating.

REMEMBER

Enable Complete Features must be selected in the Advanced Settings pane to access the additional audio editors. Choose Logic Pro ⇨ Settings ⇨ Advanced and select Enable Complete Features.

Comping takes

A *comp take* is studio-speak for a composite take, a single recording made from many recordings. Comping used to require a lot of copying and pasting — as well as razor blades and tape before digital audio came along. As you learn in Chapter 6, Logic Pro puts your recorded takes on multiple lanes of a single track, as shown in Figure 14-4. Each lane enables you to drag through the region to hide or display audio from the take and build a final composite take. For example, with a vocal performance, you can take one word or even just part of a word from a single take and improve your singer's performance. (Just don't tell the singer, or you might bruise their ego.)

To create a single take from multiple takes, do the following:

1. **In the upper-left corner of the take folder, click the disclosure triangle to see the individual take lanes.**
2. **Preview an individual take by clicking it and playing your project.**

   If you have many takes, it's a good idea to take notes with a pen and paper or with the Logic Pro notepad (Option-⌘-P). Pay attention to what sounds good and what doesn't.
3. **Use the pointer or the pencil tool to drag over the areas of the take you want in your comp.**

   Dragging over one take automatically deselects the same area of the other takes and updates the current comp.

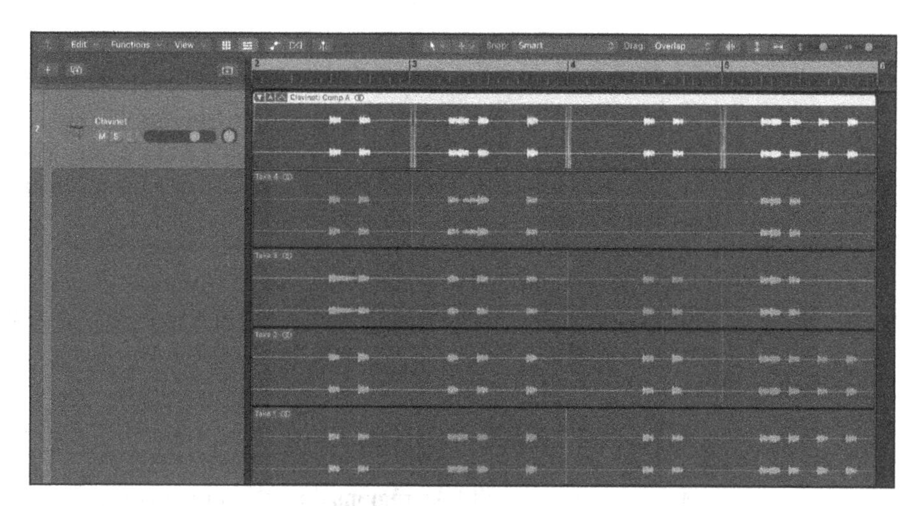

FIGURE 14-4: A take folder.

In addition to dragging selections within takes, you can edit the comp in more advanced ways:

- » Shorten a take section without extending the selection of other takes by Shift-dragging the take section. Shortening a take section leaves silence between the two adjacent sections in the comp.
- » Move the selection left or right by placing your cursor in the center of the section and dragging when the cursor becomes a two-headed arrow.
- » Alternate sections between takes by clicking an empty section of a take. The take you click will become active, and the formerly active take will become inactive. Use this method to audition sections of takes quickly.
- » Remove a section from the comp by Shift-clicking the section. To remove the entire take from the comp, Shift-click the take header.

Creating alternate comps

When you create a comp that you like, it's a good idea to duplicate it for safekeeping. You might also want to try out different ideas. Follow these steps to create alternative comps:

1. **To the right of the take folder disclosure triangle, click the current comp indicator.**

   A pop-up menu appears that lets you select takes and gives you commands for your comps and take folder.

2. **In the pop-up menu, choose Create New Comp or Duplicate Comp.**

   Your new or duplicate comp appears in the menu list, where you can select and edit it. The Create New Comp command is available only if you haven't created a comp. After you've created a comp, the Duplicate Comp command is available.

After you've duplicated your comp, you can use the pop-up menu to select between the two comps and rename the comp. You can also delete the comp or all comps except the current one.

Editing take regions

You might want to edit take regions directly but not edit your comp. For example, you can change the start and end points of a take region or split the region in two. You can turn quick swipe comping on and off by clicking the take folder's quick swipe comping icon (shown in the margin). When quick swipe comping is on, you can create and edit comps by dragging in the take regions, as described previously.

When quick swipe comping is off, the icon shows a pair of scissors, and you can edit the regions directly. You can also do the following to your take regions when quick swipe comping mode is off:

- Select an individual take by clicking it.
- Change start and end points by dragging the lower edges of the take region.
- Slice a take at a comp section borders by Control-clicking and choosing Slice at Comp Section Borders in the pop-up menu.
- Trim takes at active comp section borders by Control-clicking and choosing Trim to Active Comp Section Borders in the pop-up menu.
- Move a take within its lane by dragging it left or right.
- Move a take to a different lane by dragging the take to the new lane.

Packing and unpacking take folders

It's unnecessary to record multiple takes on a single track to use a take folder. You can create a take folder from any audio you have in your project. You can also remove takes from a take folder so that they play independently as regions on their own track.

To pack audio regions into a take folder, follow these steps:

1. **Select the regions you want to pack into a take folder.**
2. **Control-click a selected region, and choose Folder ⇨ Pack Take Folder (Control-Option-⌘-F) in the pop-up menu.**

   The regions are placed into a take folder on the topmost track.

You can do either of the following to unpack a take folder:

- Control-click the take folder header, and then choose Folder ⇨ Unpack Take Folder to Existing Tracks (Control-⌘-U) or Folder ⇨ Unpack Take Folder to New Tracks (Shift-Control-⌘-U).
- Click the current comp indicator to the right of the take folder disclosure triangle, and then choose Unpack, Unpack to New Tracks, or Unpack to New Track Alternatives.

In addition to packing and unpacking take folders, you can do the following in the take folder menu:

- **Flatten:** The currently active comp replaces the take folder. Each take section is its own region with crossfades added to the regions for smooth transitions between the regions.
- **Flatten and merge:** The currently active comp replaces the take folder. Each take section is merged into a single audio file with only one region.
- **Export the active comp to a new track:** The currently active comp is exported to a new track, leaving the original take folder unchanged. Each take section on the new track is in its own region with crossfades added to the regions.
- **Move the active comp to a new track:** The currently active comp is exported to a new track, but the original take sections are removed from the take folder.

TIP

You can create a vocal double effect, a common music production technique, by creating different comps and exporting them to new tracks. If you have several usable takes in a take folder, creating multiple comps and exporting them makes vocal doubling easy. And if the exported comps aren't perfectly in tune or the timing is off, you can adjust them by using flex pitch and flex time, as you're about to learn.

Time Traveling with Flex Time

When flex time was added to Logic Pro in 2009, it was one of those gasp-at-the-possibilities moments. If you've ever been on a deadline and the recording wasn't going well, you probably wished you had something like flex time. You've come to Logic Pro at the right time, as you're about to find out.

Flex time allows you to adjust the timing and groove of performances. You can fix mistakes, experiment, perfect important downbeats, and so much more. To enable flex time on a track in the tracks area, do the following:

1. **Choose Edit ⇨ Show Flex Pitch/Time (⌘-F).**

   Flex icons appear on the track headers, as shown in Figure 14-5.

2. **Click the flex icon on the audio track you want to edit.**

   The flex menu appears on the track header.

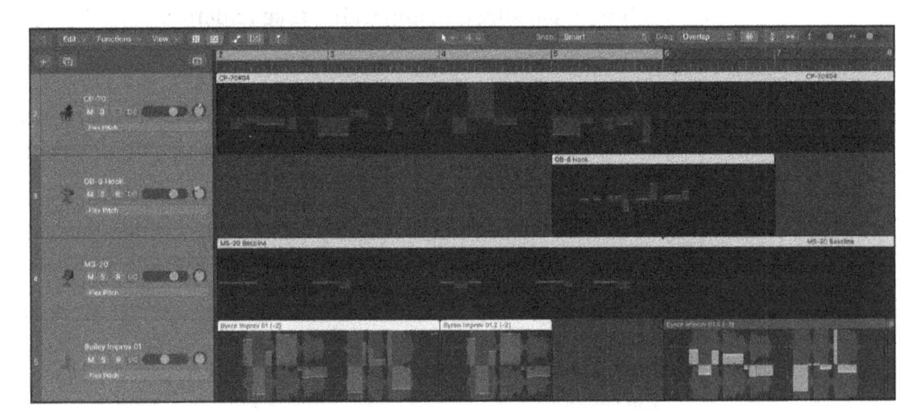

FIGURE 14-5: Tracks with flex mode enabled.

You can enable flex time also from the audio track editor:

1. **Open the audio track editor by double-clicking an audio file or by selecting an audio file and choosing View ⇨ Show Editor (E).**

   The audio track editor opens in the main window.

2. **Click the flex icon on the audio track editor menu bar.**

   The flex menu appears on the menu bar to the right of the flex icon.

Choosing flex time algorithms

After you've turned on flex mode, you can select an algorithm in the flex mode menu, which appears in the following places, to determine how your audio will be altered:

- Track inspector (see Figure 14-6)
- Track header
- Audio track editor

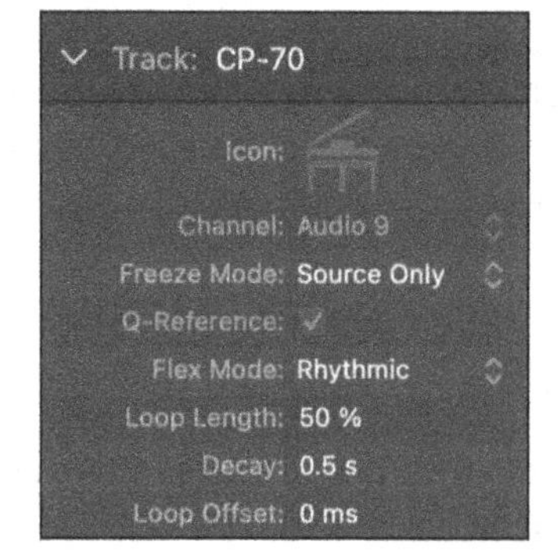

FIGURE 14-6: The track inspector flex parameters.

You can also have Logic Pro analyze your audio and automatically choose the best algorithm for the audio material.

Depending on your chosen flex algorithm, different parameters will be available in the track inspector. You can choose between the following flex algorithms and parameters:

- **Automatic:** Logic Pro analyzes your audio and selects between monophonic, slicing, and rhythmic algorithms.
- **Monophonic:** Used for monophonic audio, such as bass lines, lead sounds, and vocals. Click the Percussive check box if the audio is a percussive monophonic sound, such as a guitar or pitched percussion.
- **Slicing:** Used for percussive audio, such as drums and percussion. Audio isn't expanded or compressed with this algorithm, so moving audio can create gaps. Set the Decay time to compensate for gaps in the audio. Click the Fill Gaps check box to disable the decay time and remove any gaps. Set the Slice Length to shorten each slice by a percentage and adjust the gaps between slices.

» **Rhythmic:** Used for rhythmic audio, such as guitars and keyboards. Set the Loop Length to adjust the length of the loop used at the end of the slice to fill the gap. Set the Decay value to adjust the decay of the loop. Set the Loop Offset to move the loop forward in time, preserving the attack of the transients that follow.

» **Polyphonic:** Used for polyphonic audio such as chords and full mixes. The polyphonic algorithm is the most processor intensive of all the algorithms. Click the Complex check box to add more internal transients to the flex time processing.

» **Speed (FX):** Used for special effects, especially percussive material. This algorithm has no additional parameters.

» **Tempophone (FX):** Used for special effects based on granular synthesis techniques. Adjust the Grain Size used to compress and expand the audio. Adjust the crossfade length to set the quality of the effect. Higher values produce softer effects, and lower values produce hard effects.

Using flex markers

You adjust the timing of your audio with flex markers. After you enable flex time editing, your audio is analyzed using the chosen algorithm and transient markers are added to your audio, as shown in Figure 14-7. You can use these markers to help you see where you need to adjust the timing.

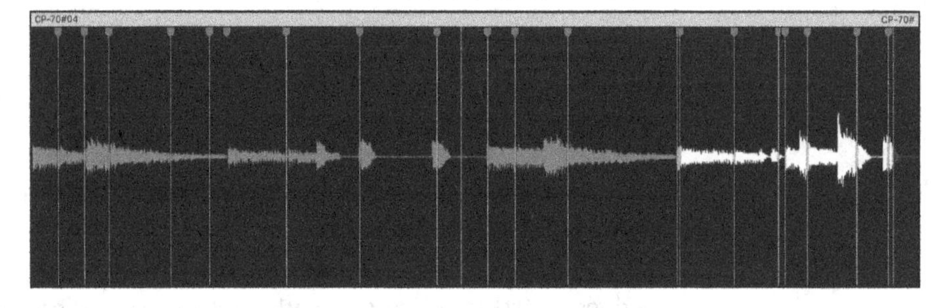

FIGURE 14-7: Transient markers.

Flex time editing can be done in the tracks area and the audio track editor. Follow these steps to adjust the timing of your audio tracks with flex time:

1. **In the track header, click the flex icon to enable flex time editing.**

   Your audio is analyzed and transient markers are added to the region.

2. **Choose a flex time algorithm.**

3. **Click the spot you want to adjust.**

   A flex marker is added to the track. You can use the transient markers to guide your editing, but you can also add flex markers to points without transient markers.

   If you click the lower half of the region, multiple flex markers are added. In addition to the place you click, a marker is added to the left and right transients, making the flex editing more precise.

4. **Drag the marker to the left or right to adjust the timing.**

   The audio is compressed or expanded to fit the edit.

TIP

You can connect a flex marker to a transient in a region on an adjacent track, allowing you to tighten the timing between two separate tracks. Drag a flex marker up or down to a transient on another region until the line between the two regions turns yellow, and then release. Your track will be adjusted to the transient.

Using the flex tool

The tool menus in the tracks area and the audio track editor contain a flex tool to use on regions when flex editing isn't shown. To use the flex tool on an audio region, do the following:

1. **Open the tool menu (T) and choose the Flex Tool.**
2. **Drag the audio waveform to adjust the timing.**

TIP

Remember the marquee tool in Chapter 13? It's an excellent tool for flex editing. Create a selection with the marquee tool, and then drag the selection with the flex tool to adjust the timing. You can also click the selection with the pointer tool to create flex time markers from the selection and the closest transients. If you click in the upper half of the region, flex markers are added to the four positions: the edges of the selection and the nearest transients outside the selection. If you click in the lower half of the region, flex markers are added to three positions: the selection edges and the place you clicked.

Tuning with Flex Pitch

It's a good thing Logic Pro users were given seven years to come down from the glee of flex time before Logic Pro announced flex pitch. We needed to catch our breath.

Editing pitch isn't anything new, and a few companies have done a pretty good job implementing it. But flex pitch makes editing pitch simple thanks to its intuitive workflow and the fact that it's built into Logic Pro. To enable flex pitch on a track in the tracks area, follow these steps:

1. **In the tracks area menu bar, click the flex icon to enable flex editing.**

   Flex icons appear on the track headers.

2. **In the track header, click the flex icon.**

   Flex editing is now enabled on the track.

3. **On the Flex Mode menu in the track header or the track inspector, choose Flex Pitch.**

   The regions on the track display the flex pitch editing bars, as shown in Figure 14-8.

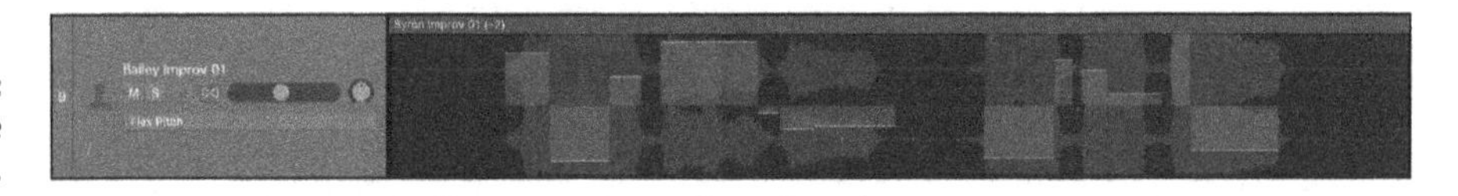

FIGURE 14-8: Flex pitch in the tracks area.

TIP

Because flex time and flex pitch use different algorithms, switching back and forth between the two might trigger Logic Pro to ask you how to handle the switch: reapply, revert, or suspend your current edits to move forward. To preserve the edits you've made so far, it may be a good idea to finish editing the timing, bounce the region in place, and then edit the pitch, or vice versa. Bouncing the region in place renders your edits to a new audio file and places it at the same position in the timeline as the original. Select the region and choose File ➪ Bounce ➪ Regions in Place. And whenever you're using elastic audio editing techniques such as flex time or flex pitch, it's a good idea to bounce your edits in place when you're finished because you never know when a software update might change an algorithm and thus change your edits.

Editing pitch in the tracks area

After you enable flex pitch on a track, the region in the tracks area displays horizontal bars above or below each note. These bars show how close or far away the pitch is from the center pitch, indicated by the horizontal center of the region. If a bar is above the center, the pitch is sharp. If a bar is below the center, the pitch is flat. To edit the pitch, drag the bars closer to the center.

TIP

You can speed up the editing process by Control-clicking the audio region and choosing Set All to Perfect Pitch on the menu. If you don't like your edits, you can start fresh by choosing Set All to Original Pitch.

You can adjust two flex Pitch parameters from the track inspector:

- **Formant Track:** *Formants* are bands of frequencies that determine the sound quality of vowels. Drag up on the Formant Track field to increase the tracking interval.
- **Formant Shift:** This parameter adjusts how formants are affected by changes in pitch. You can get some cool vocal effects by setting the Formant Shift value to extremes. Positive values give you squeaky Mickey Mouse effects, and negative values give you deep-voiced Barry White effects. The formant shift parameter is fun to play with and helps you get more natural sounding flex pitch edits.

Editing pitch in the audio track editor

You can edit pitch also in the audio track editor. The flex pitch presentation in the audio track editor is similar to the pitch presentation in the MIDI piano roll editor, and you have more control over the edited sound. To use flex pitch in the audio track editor, do the following:

1. **Double-click an audio region to open the audio track editor, or select the region and press E.**

   The audio track editor opens at the bottom of the main window.

2. **On the audio track editor menu bar, click the flex icon.**

   Flex editing is enabled on the track.

3. **On the Flex Mode menu on the audio track editor menu bar, choose Flex Pitch.**

   Flex pitch data is added to each note, as shown in Figure 14-9.

Editing pitch event hot spots

In the audio track editor, flex pitch creates a rectangle for each note. The notes correspond to the piano keys to the left of the editor. Drag the rectangle left, right, up, or down to adjust both time and pitch. You can also split the note with the scissors tool to create two notes or merge notes by using the glue tool. You can reset the pitch by Control-clicking the note and choosing Set to Original Pitch.

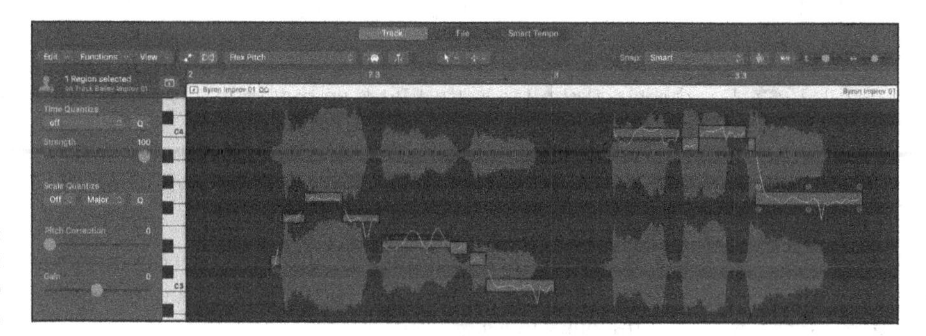

FIGURE 14-9: Flex pitch in the audio track editor.

As you hover your cursor over each rectangle, six hot spots become available so you can edit the following flex Pitch parameters:

- **Pitch Drift:** The left and right hot spots at the top of the rectangle control the pitch drift at the beginning and end of the note. Use these parameters to allow the pitch to drift into or out of the note. These parameters are useful for keeping a natural feel to the pitch because singers often scoop into notes and fall out of notes.
- **Fine Pitch:** Use the top center hot spot to fine-tune the pitch.
- **Gain:** The bottom left hot spot controls the note's volume.
- **Vibrato:** The bottom middle hot spot controls the pitch vibrato.
- **Formant Shift:** The bottom-right hot spot controls how formants are affected by the pitch shift.

TIP

Have you ever wished you could sing a melody and have it turned into MIDI or notation? Flex pitch lets you do that. After you've finished flex pitch editing, choose Edit ➪ Create MIDI Track from Flex Pitch Data on the audio track editor menu. A new software instrument track is created and the flex pitch data is added to a MIDI region as MIDI notes.

Quantizing the pitch and scale of a region

With flex pitch enabled in the audio track editor, Logic Pro can automatically quantize a region's pitch, scale, and time. Choose View ➪ Show Local Inspector to view the quantization parameters. At the top of the inspector is the pitch correction slider, which adjusts the amount of pitch tuning. Below the pitch correction slider are the Time Quantize parameters, which enable you to choose the note value and strength of quantization. The Gain slider adjusts the volume of the region. The Scale Quantization parameters allow you to choose the scale's root note and type. All common scale modes are available, along with other popular scale types and some more esoteric scales.

TIP

Flex pitch can do more than just fix pitch mistakes or inconsistencies. You can also use it to create new parts, such as vocal harmonies. In many cases, the parts you come up with will be perfectly usable. Sometimes, however, moving a pitch too far can sound unnatural, so adjust the formant parameters for better results. But even if you don't keep the part, you can experiment and find out what will make your song sound stronger.

Editing Audio in the Audio File Editor

You probably won't need to use the audio file editor often because the audio track editor will usually be your first choice for editing audio. The audio file editor is reserved for more precise and permanent edits to audio files. You can open the audio file editor, shown in Figure 14-10, in the following ways:

- Choose View ⇨ Show Editor or double-click an audio file, and then click the File tab in the audio editor window.
- Choose Window ⇨ Open Audio File Editor (⌘-6) to open the audio file editor in a new window.

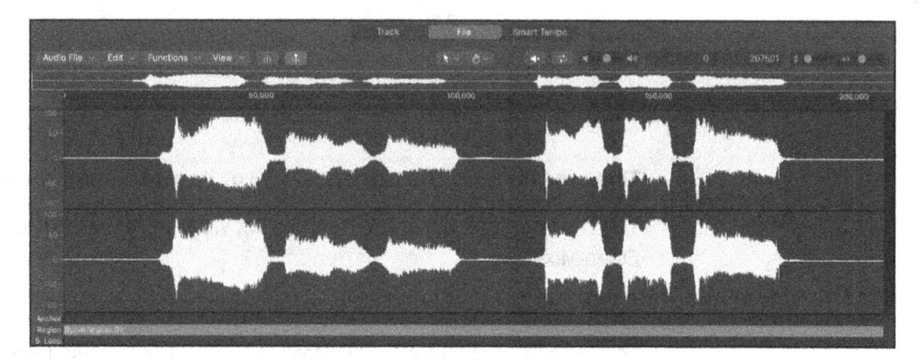

FIGURE 14-10: The audio file editor.

It's worth repeating that the audio file editor is destructive, meaning it changes your audio permanently. Making copies and backups of files you intend to edit is good practice. To protect your audio, choose one of the following:

- **Choose Audio File ⇨ Create Backup (Control-B).** A duplicate of the audio file is stored in the same location as the original file and given a .dup file extension.
- **Choose Audio File ⇨ Save A Copy As.** A dialog opens where you can adjust the settings and location of the file before you click the Save button.

Navigating and playing audio

Playback in the audio file editor is independent of the project audio. That means the transport controls in the control bar don't control the playback of files in the audio file editor. The audio editor is monitored through the prelisten channel strip in the mixer, shown in Figure 14-11. The prelisten channel strip adjusts the audio settings of files played in the audio file editor and project audio browser.

FIGURE 14-11: The prelisten channel strip.

Follow these steps to view the prelisten channel strip:

1. **Open the mixer in the tracks area (X) or its own window (⌘-2).**

   You can open the mixer also by choosing View ➪ Show Mixer or Window ➪ Open Mixer.

2. **On the mixer menu bar, click the All button to display all channel strips.**

   The prelisten channel strip is located to the right of the last audio channel strip.

You can play audio files in the editor by clicking the prelisten icon (shown in the margin). Control-click the prelisten icon to play audio through the prelisten channel strip or the track channel strip. In addition to the prelisten icon, you can play audio in the editor in the following ways:

- Press Control-Option-⌘-spacebar to play and stop the entire audio file.
- Double-click the ruler to play and stop the audio file.

- Click-hold the waveform overview to start playback at the click location.
- Click the cycle icon to repeat the selected audio continuously.

The audio file editor allows you to navigate to precise points in your audio file. Because the audio file editor is used for detailed editing, it's important to have the capability to navigate around your audio file at the smallest level. To navigate your audio file, do one of the following:

- Click the waveform display to navigate to specific locations in the audio file.
- Choose Edit ➪ Go To and select the audio file location you want.
- Choose Functions ➪ Search Peak to find the loudest point in the audio file.
- Choose Functions ➪ Search Silence to find the first silent point in the audio file.
- Use the vertical and horizontal zoom sliders to adjust how much of your audio file you see in the editor window.

Selecting audio

To edit audio in the audio file editor, you need to select the audio in the editing window. The audio file editor gives you precise ways of selecting audio, down to the individual samples. You can select portions of your audio file in the following ways:

- Drag your cursor over the area you want to select.
- Choose Edit ➪ Select All (⌘-A).
- Choose Edit ➪ Select All Previous (Shift-Control-Option-left arrow).
- Choose Edit ➪ Select All Following (Shift-Control-Option-right arrow).
- Choose Edit ➪ Region ➪ Selection to select only the audio that is used in the region.

In addition to these basic ways of selecting parts of your audio file, you can use transient markers to make selections. Transient markers can make selecting and editing audio much faster. To select audio using transients:

1. **On the menu bar, click the transient editing mode icon (see the margin).**

   The audio is analyzed for transients, and markers are added.

2. **Double-click between two markers or use one of the commands on the Edit ➪ Set menu.**

The start and end points of your selection are shown in the info display on the menu. You change the format of the information by selecting one of the following modes from the View menu:

- » Samples
- » Min:Sec:Ms
- » SMPTE Time
- » Bars/Beats

Editing audio

In the audio file editor, you can do basic editing, such as copying and pasting, or more advanced editing, such as fading in and out and removing pops and clicks. Use the audio file editor instead of the audio track editor to accomplish more detailed, permanent edits.

From the Edit menu, you can copy, cut, paste, and delete. Copied audio is put on the clipboard for pasting to a new location. Cutting audio pulls all the audio after the cut forward and places the selection on the clipboard. Pasting audio occurs at the clicked location and makes room for the paste by pushing back all the audio that follows. Deleting audio is similar to cutting, except the selected contents aren't copied to the clipboard.

After you have selected some audio, you can choose several editing commands on the Functions menu. Here are some common editing functions:

- » **Silence:** Remove all sound. Be careful when you silence audio because too much silence can make an audio recording sound unnatural. In many cases, it's better to lower the gain by using the Change Gain function. Silencing audio is better at the beginning or end of audio files or when you need to remove an unwanted artifact.
- » **Reverse:** Make the sound play backward. You'll love this command if you've ever wanted to create a reverse cymbal crash sound to swell into a section. It's also useful for making clean versions of explicit vocal parts.
- » **Remove DC Offset:** Center the waveform and remove potential cracks and pops that DC (direct current) offset can produce. Use this function if your audio hardware introduces DC offset into your audio. DC offset is clearly visible on your audio waveforms by a vertical shift in the waveform display.
- » **Trim:** Remove everything except your selected audio.

The pencil tool is probably the most helpful tool for removing pops and clicks from your audio. To use the pencil tool, you'll need to zoom in closely on the area you want to edit. If you're not zoomed in enough, clicking the waveform with the pencil will automatically change it to the zoom tool, a nice hint that you're not zoomed in enough. After you're zoomed in enough, drag from left to right to draw in the new waveform.

TIP

You can set Logic Pro to edit your audio in an external editor:

1. **Choose Logic Pro ➪ Settings ➪ Audio.**

   The Audio Settings window opens.

2. **Click the Audio File Editor tab, as shown in Figure 14-12.**
3. **Click the External Sample Editor field.**

   A browser window opens where you can navigate to your external sample editor.

4. **Select your external sample editor in the browser, and then click Choose.**

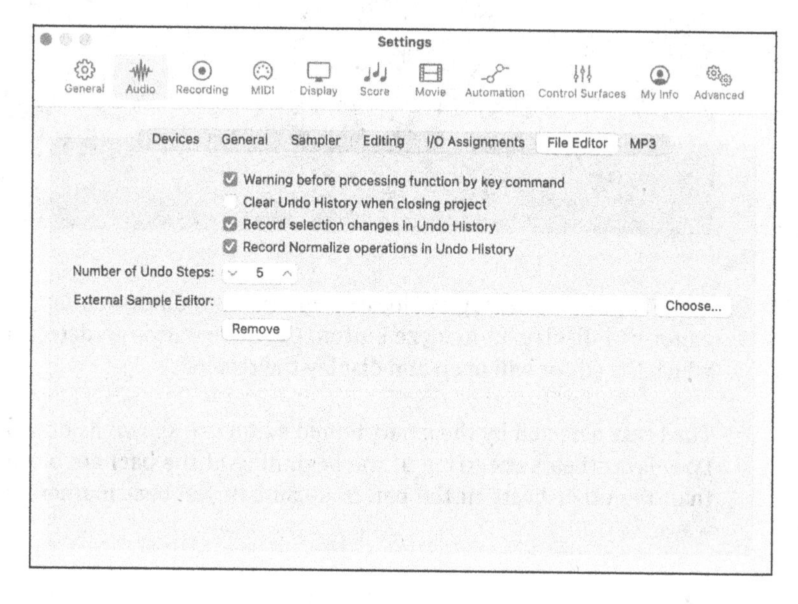

FIGURE 14-12: The Audio Settings window.

After you've set up your external audio editor, select an audio file and choose Edit ➪ Open in (Shift-W) on the tracks area menu bar. Your audio will open in the external audio editor, ready to be manipulated to your ear's content.

Editing Tempo in the Smart Tempo Editor

When you record audio using smart tempo (which you learn about in Chapter 6), tempo information is embedded in the recorded audio file. You can edit this tempo information using the smart tempo editor, shown in Figure 14-13. You can open the smart tempo editor in the following ways:

- » Choose View ⇨ Show Editor, or double-click an audio file and then click the Smart Tempo tab in the audio editor window.
- » Choose Window ⇨ Open Smart Tempo Editor to open the smart tempo editor in a new window.

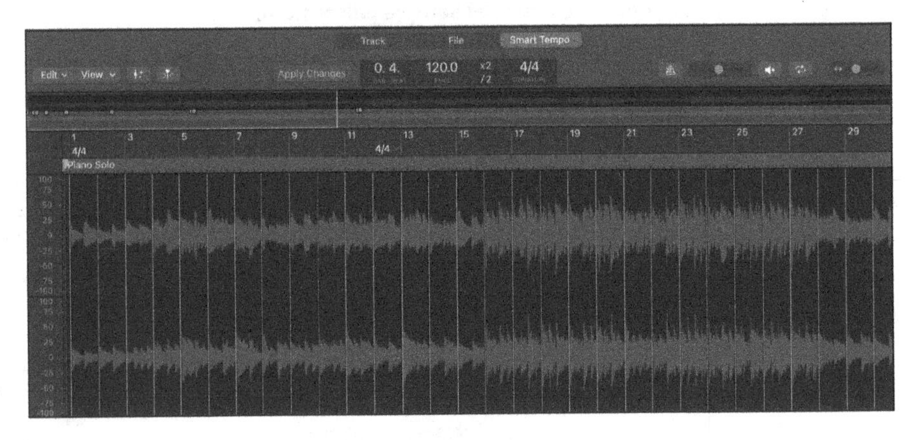

FIGURE 14-13: The smart tempo editor.

If the selected audio region doesn't contain tempo information, the smart tempo editor will display an Analyze button. Click the button to detect the tempo, after which the editor will open and display the results.

The beats detected by the smart tempo editor are shown as orange beat markers. *Downbeats* (beats occurring at the beginning of the bar) are brighter and thicker than the other beats in the bar. You can edit the beat markers in the following ways:

- » Tap the tempo as the audio file plays by pressing the D key to indicate downbeats and pressing T to indicate other beats.
- » Drag the beat markers to the correct location in the audio file.

Beat markers contain several handles that appear as you hover your cursor over the marker. As you hover over each of the handles, help tags reveal the following handle functions:

» **Set downbeat:** Click this handle to set the downbeat. This handle isn't available for beats already set as the downbeat.

» **Move marker:** Drag the move marker handle to move individual beats.

» **Scale selection:** Select multiple beats by dragging between the beat markers and then drag the scale selection handle to move the selected beat markers proportionally.

» **Scale left, move right:** Select multiple beats by dragging between the beat markers and then dragging this handle to scale the selected area and move the beats to the right of the selected area.

» **Scale All:** Drag the scale all handle to scale all beat markers in a selection or in the entire file.

» **Move All:** Drag this handle to move all beat markers.

The smart tempo editor control bar contains several menus and displays to help you edit the file's tempo information. You can change the time signature in the Signature display pop-up menu. You can double or halve the tempo by pressing the x2 or /2 button, respectively. Use the Tempo pop-up menu to choose whether the tempo is constant or variable. Use the Actions menu to perform various actions on the file, such as adapting the project tempo to a region or reverting your edits to the original tempo information.

We've come a long way since the days of tape and razor blades. You can take control of your audio quickly and easily. Your creative process can stretch into the realm of audio manipulation or consist of simply shaping things until you and your clients are happy.

With the capability to create perfect takes and massage the timing and tuning of your parts, you can deliver the best music your imagination has to offer. Your fans want more music. Go give it to them.

IN THIS CHAPTER

» Getting to know the MIDI editors

» Inputting and editing MIDI notes and data

» Adjusting the timing of your MIDI recordings

» Viewing and editing your MIDI as notation

Chapter 15
Editing MIDI Tracks

Logic Pro began as a MIDI sequencer. Even today, it stands alone in its superior handling of MIDI data. MIDI is much more flexible than audio. You can change sounds, notes, velocity, and length, and tweak until you're happy. Yes, Logic Pro can make you happy.

MIDI has many benefits. The small file sizes make it easy to share and use when space is a concern, such as in game music. It's an excellent tool for composition, film scoring, and music notation. Unlike acoustic instruments, MIDI controllers can take the shape of keyboards, drum pads, wind instruments, guitars, and other stringed instruments, but they can sound like anything at all. And MIDI makes music preproduction much easier because you can build songs before you record them.

In this chapter, you take advantage of your MIDI studio. You discover how to input and edit MIDI, learn timesaving editing tips, and make your music look and sound exactly how you want.

Knowing Your MIDI Editors

The piano roll editor, shown in Figure 15-1, is the default Logic Pro MIDI editor. It has the most features and is designed for speed and complete control. Inspired by player pianos, which use grid-based punch cards to make music, this editor is user friendly and your go-to MIDI editor. To open the piano roll editor, do one of the following:

- Double-click a MIDI region.
- Choose View ➪ Show Editor or press E with a MIDI region selected.
- Choose Window ➪ Open Piano Roll or press ⌘ -4 to open the piano roll editor in a new window.

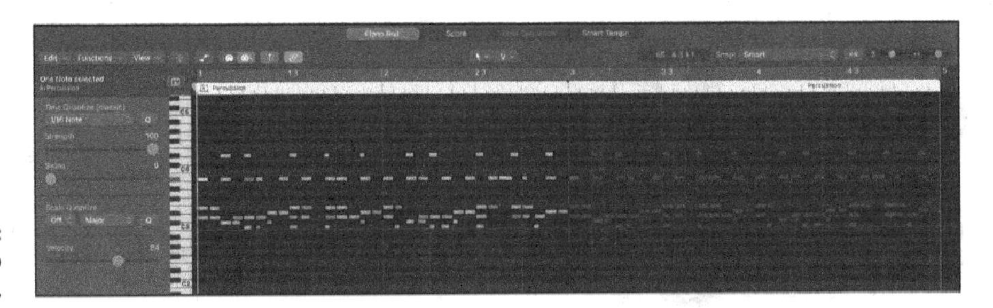

FIGURE 15-1: The piano roll editor.

The score editor, shown in Figure 15-2, is the choice for notation and traditional music representation. It is excellent for composition and scoring, and is a requirement for getting your music played by other professional musicians. If you enjoy reading music, you'll appreciate the score editor. To open the score editor, you can do any of the following:

- Double-click a MIDI region to open the MIDI editors, and then click the Score tab.
- Choose View ➪ Show Editor with a MIDI region selected, and then click the Score tab or press N.
- Choose Window ➪ Open Score Editor or press ⌘ -5 to open the score editor in a new window.

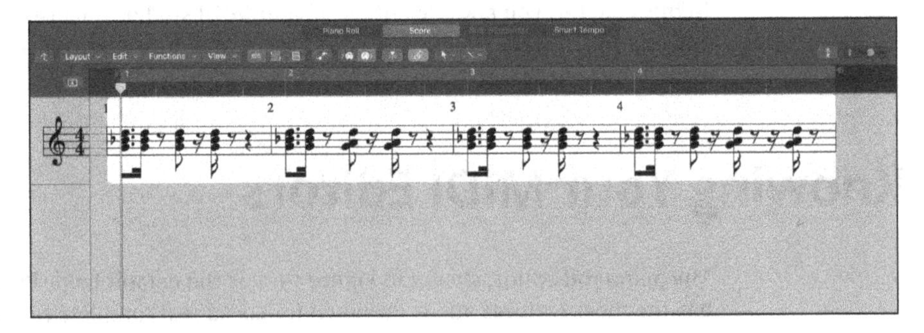

FIGURE 15-2: The score editor.

The step sequencer, shown in Figure 15-3, works well at building musical patterns, especially drum patterns. It mimics hardware step sequencers, in which you build a sequence of notes by using a grid. Even though it's grid-based, like the piano roll editor, the step sequencer has a different layout that allows you to quickly edit MIDI event types other than notes, such as velocity or MIDI controller data. The step sequencer works only with pattern regions in the tracks area or pattern cells in the live loops grid. *Pattern regions* and *pattern cells* are MIDI regions that have pattern data. To create a pattern region, do the following:

- Control-click an empty area of a software instrument track and choose Create Pattern Region from the shortcut menu.
- Control-click a MIDI region and choose Convert ➪ Convert to Pattern Region.

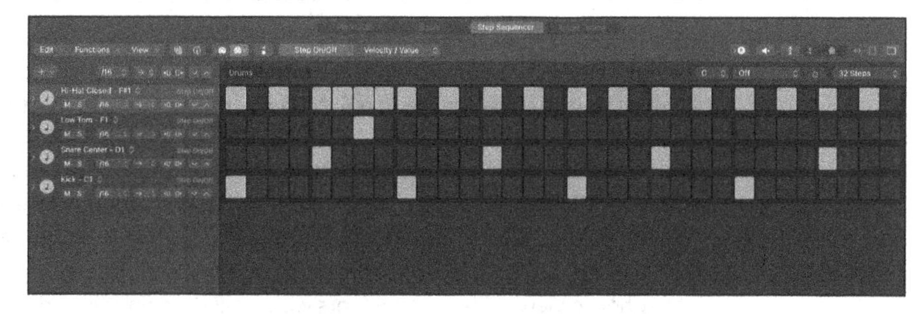

FIGURE 15-3: The step sequencer.

To open the step sequencer, do the following:

- Double-click a MIDI pattern region to open the MIDI editors, and then click the Step Sequencer tab.
- Choose View ➪ Show Editor with a MIDI pattern region selected, and then click the Step Sequencer tab.
- Choose Window ➪ Open Step Sequencer to open the step sequencer in a new window.

The smart tempo editor, shown in Figure 15-4, is used to edit the beat markers and tempo data of MIDI files that contain tempo information. You can open the smart tempo editor in any of the following ways:

- Double-click a MIDI region, and then click the Smart Tempo tab at the top of the MIDI editor.
- Choose View ➪ Show Editor with a MIDI region selected, and then click the Smart Tempo tab at the top of the MIDI editor.

» Choose Window ➪ Open Smart Tempo Editor to open the editor in a new window.

» Choose Edit ➪ Tempo ➪ Show Smart Tempo Editor.

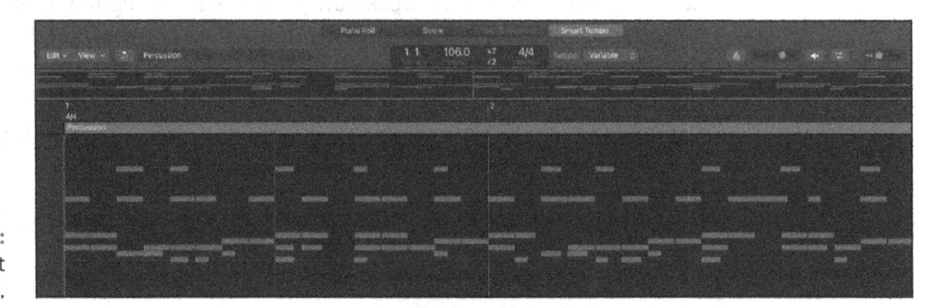

FIGURE 15-4: The smart tempo editor.

If your selected region doesn't contain tempo information, you'll need to click Analyze to initiate smart tempo detection and access the smart tempo editor.

The event list editor, shown in Figure 15-5, might not be the most attractive MIDI editor, but it's probably the most complete editor, listing every MIDI event you record. To open the event list editor, you can do one of the following:

» Choose View ➪ Show List Editors with a MIDI region selected, and then click the Event tab or press D.

» Choose Window ➪ Open Event List to open the event list editor in a new window or press ⌘ -7.

The MIDI environment, shown in Figure 15-6, used to be a more fundamental, and often intimidating, part of the Logic Pro workflow. Now, however, the environment has been pushed into the background as a legacy feature, but you wouldn't be able to sequence without it. The MIDI environment is integral to the input and output of your audio and MIDI. Do one of the following to open the MIDI environment:

» While holding down the Option key, choose Window ➪ Open MIDI Environment to open the MIDI environment in a new window.

» Press ⌘ -0 to open the MIDI environment window.

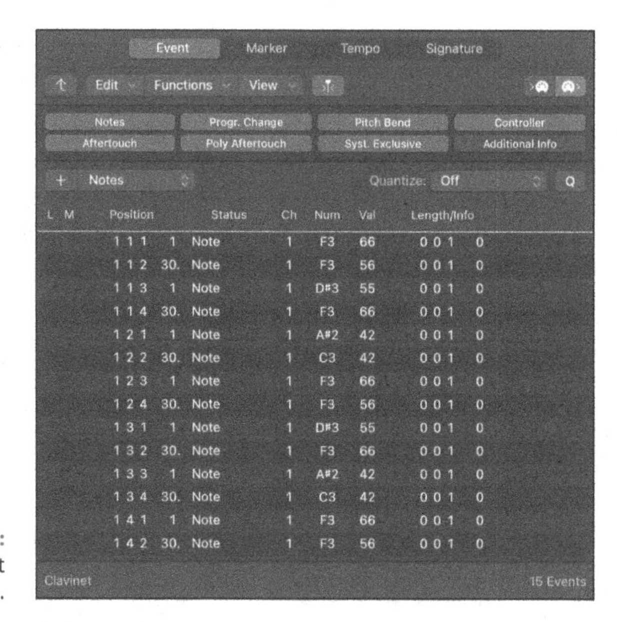

FIGURE 15-5: The event list editor.

FIGURE 15-6: The MIDI environment.

The MIDI transform window, shown in Figure 15-7, alters your MIDI and can save you from time-consuming and repetitive editing tasks. Whenever you want to automate the transformation of large amounts of MIDI data, use the MIDI transform window. You can open the MIDI transform window in a couple of ways:

- Choose Window ➪ Open MIDI Transform to open the MIDI Transform window in a new window.
- Press ⌘-9 to open the MIDI transform window.

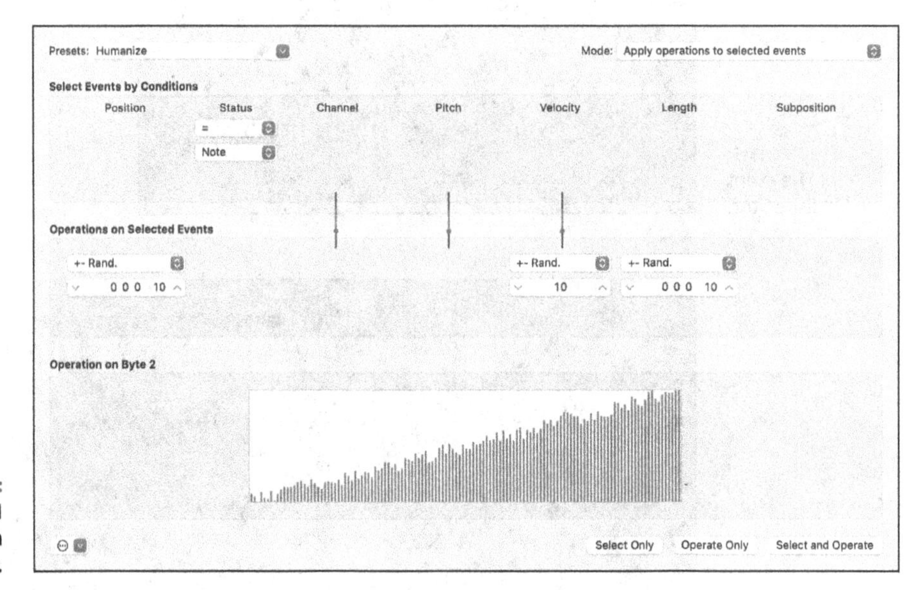

FIGURE 15-7: The MIDI transform window.

REMEMBER

To enable the full MIDI editing features of Logic Pro, you must have Enable Complete Controls selected in the Advanced Settings pane. Choose Logic Pro ➪ Settings ➪ Advanced, and then select Enable Complete Controls.

Editing MIDI in the Piano Roll Editor

In the piano roll editor, MIDI notes are displayed as colored bars on a grid, as shown in Figure 15-8. MIDI notes correspond to a vertical keyboard on the left side of the editor. Like the tracks area, there's a time grid, a ruler, and a menu bar to help you locate your position in the project and make precise edits.

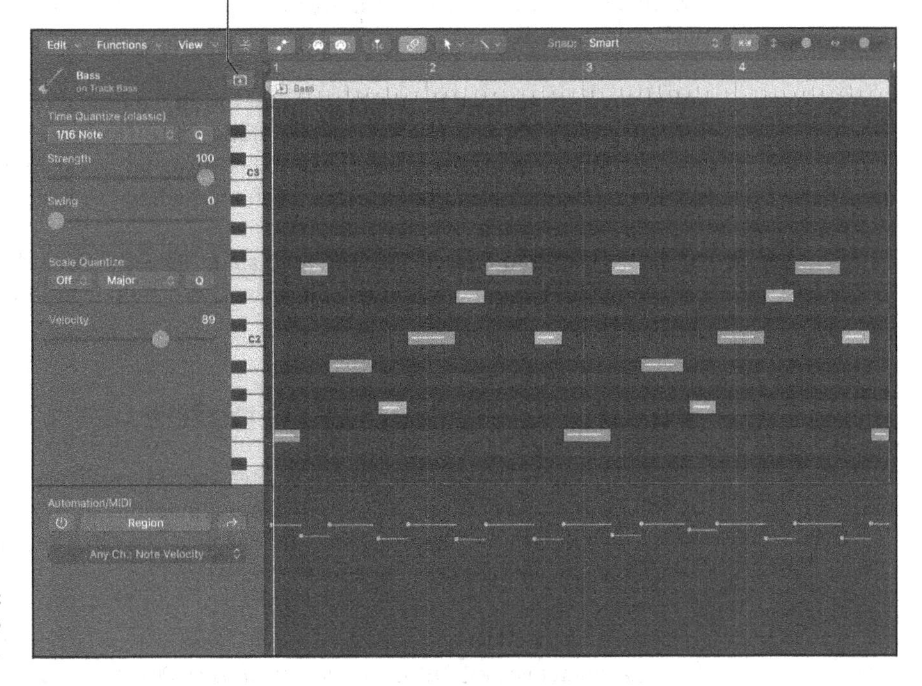

FIGURE 15-8: The piano roll editor.

TIP

You can show the global tracks or just the marker track in the piano roll editor by clicking the show/hide global tracks icon above the piano roll keyboard. Displaying the global tracks is particularly useful when viewing the piano roll editor in a separate window from the tracks area.

Adding and editing notes

In addition to recording MIDI (see Chapter 7), you can add notes directly in the piano roll editor with the pencil tool by clicking the location on the grid where you want to create the note. New notes are created using the same length as the previously created note or currently selected note. You can adjust the length of a note as you create it by holding and dragging the note. Once a note is added, you can play it by clicking it.

Edit MIDI notes just like you edit a region (described in Chapter 4). You can drag the ends of the notes to resize them. Copy notes by Option-clicking and then dragging the note to the new position. Delete notes by selecting the note and then pressing Delete.

The piano roll editor makes selecting notes easy by providing several commands in the Edit ➪ Select menu. You can select the highest and lowest notes and select overlapped notes. You can also move to the previous note (left arrow) and the next note (right arrow).

After you have selected the notes you want to edit, you can move them left or right on the grid by choosing Edit ➪ Move ➪ Nudge Left (Option-left arrow) or Nudge Right (Option-right arrow). You can move them up or down a semitone or an octave by choosing the appropriate option on the Edit ➪ Transpose menu.

REMEMBER

It's a good idea to display Quick Help to aid in identifying all the nameless icons in the menu bar and other areas. Click the Quick Help icon in the control bar or choose Help ➪ Quick Help.

Editing the velocity of notes

Inside each note bar is a line that indicates the velocity level, which determines how loud or soft a note will be played. Longer lines indicate higher velocity. If you hover your cursor over a note, a help tag displays the pitch and the exact velocity level. You can edit the note velocity in these ways:

- » Click a note and drag up or down with the velocity tool. Editing velocity is the parameter you will probably edit second-most, after pitch, so set the velocity tool as your ⌘ -click tool. For details on setting the ⌘ -click tool, see Chapter 3.
- » Select the notes and drag the Velocity slider in the piano roll local inspector. If the local inspector isn't displayed, choose View ➪ Show Local Inspector on the piano roll editor menu.
- » You can set all the selected notes to the same velocity by pressing Shift-Option while dragging any one of the notes with the velocity tool.

TIP

To select multiple notes of the same velocity, select one note, Control-click the note, and then choose Select Equal Colored Regions/Events (Shift-C).

Quantizing notes

Got timing problems? Not anymore. As described in Chapter 14, Logic Pro can quantize your audio, which is pretty amazing. But MIDI data is even more flexible because it won't suffer from audio artifacts after you adjust the timing. You can adjust the quantization of notes in the following ways:

- » Select the notes you want to quantize and use the Time Quantize parameters in the piano roll local inspector. Adjust the Strength slider to make the quantization sound more natural and not too perfect. Use the Swing slider to adjust the level of swing feel.

» Select and click notes with the Quantize tool. The quantize note value is chosen on the Time Quantize menu in the local inspector.

» Select the notes you want to quantize and choose Functions ⇨ Quantize Notes or press Q.

TIP

Quantizing is nondestructive, so you can undo the quantization by choosing Functions ⇨ Undo Quantization or pressing Shift-⌘-Q.

Muting notes

The jazz legend Thelonious Monk once said, "What you don't play can be more important than what you do." If MIDI sequencing were around when he said it, he could have been talking about muting notes. You can be sure that Monk would have loved muting notes. To mute and unmute notes in the piano roll editor, do one of the following:

» Use the mute tool to click the note you want to mute or unmute. You can tell when notes are muted because they have no color. You can mute and unmute several notes at once by selecting the notes and clicking with the mute tool.

» Select the notes you want to mute or unmute, and choose Functions ⇨ Mute Notes on/off or press Control-M.

TIP

You can select all the notes of the same pitch by clicking the corresponding key on the piano roll keyboard. You can select multiple notes by Shift-clicking the keys on the piano roll keyboard. Select all muted notes by selecting Edit ⇨ Select ⇨ Muted Notes or by pressing Shift-M.

Using automation

Click the show automation icon in the piano roll editor menu bar, and a lane opens at the bottom of the piano roll editor, as shown in Figure 15-9. This Automation/MIDI lane makes it easy to draw and edit MIDI controller data graphically. (MIDI continuous controller data is information that can be used to control various aspects of sound, such as volume or pitch.) Automation is especially useful for quickly editing the velocity of multiple note events.

In the Automation/MIDI lane, use the pencil tool to click and drag where you want to add MIDI controller data. You can drag in the lane to create smooth lines for multiple controller points to follow. All corresponding MIDI notes in the piano roll editor will be updated with the new controller data. Choose the MIDI data you want to edit using the automation parameter pop-up menu in the Automation/MIDI section of the piano roll editor local inspector. Another common use for automation is editing sustain pedal events.

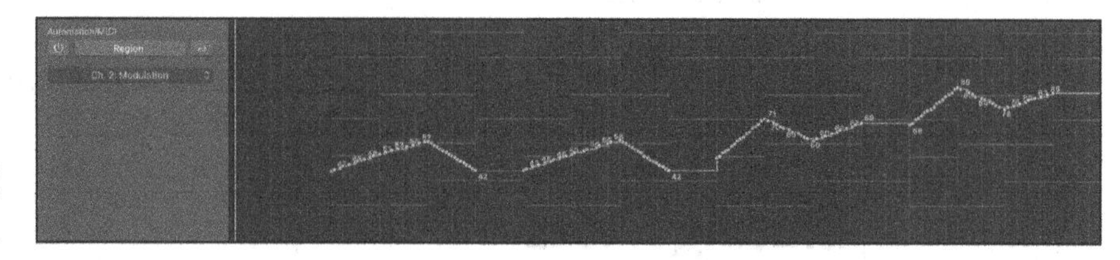

FIGURE 15-9: The Automation/MIDI lane.

Editing MIDI in the Step Sequencer

If you've played with a drum machine or Ultrabeat, you'll quickly get the hang of the step editor. The step editor is unbeatable at programming drum patterns, but you can use it in other creative ways too. It's unparalleled at organizing sounds and other MIDI events rhythmically.

Creating and editing steps

Steps are represented as squares on a grid, as shown in Figure 15-10. Each step represents MIDI events, such as note events or controller data. To create steps, press the Step On/Off button in the menu bar and click where you want the step. You can create multiple steps by dragging horizontally through the row.

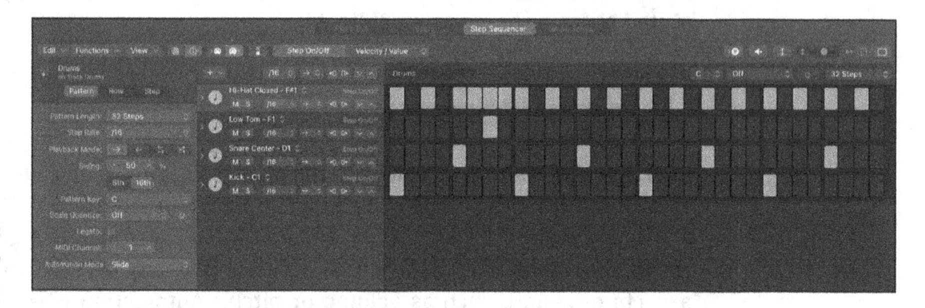

FIGURE 15-10: Steps in the step sequencer.

Edit the value of steps by pressing the edit mode selector button in the menu bar and choosing the parameter you want to edit, then drag the value up or down. You can copy, cut, paste, and delete steps just as you would in the other editors. You can also edit steps by using the Edit and Functions menu in the step sequencer menu bar.

Creating and editing rows

Step sequencer rows can show more than just note events. From the row header's assignment pop-up menu, you can assign a row to show MIDI controller data, such as modulation or velocity. To create a new row, press the new rows button above the row headers. Edit the row parameters in the local inspector (refer to the bottom left in Figure 15-10). If the row inspector isn't shown, choose View ⇨ Local Inspector, and press the Row button. In the local inspector, you can also edit the parameters for individually selected steps and the entire pattern.

TIP

You can use multiple rows for the same instrument. This allows you to quickly create eighth-note hi-hat patterns in one lane and sixteenth-note hi-hat patterns in a separate row.

Creating and saving patterns and templates

Patterns and templates are groups of rows that you can save and recall. Patterns include notes and MIDI events, while templates include empty grids. To create a new pattern or template, choose View ⇨ Pattern Browser (or press the pattern browser button in the menu bar), and then choose Save Pattern or Save Template from the pattern browser actions pop-up menu.

Editing MIDI in the Score Editor

A thorough discussion of the score editor requires several chapters, if not an entire book. The rules of professional music notation aren't set in stone. Music publishers have different notation styles, but one thing is fairly certain: Logic Pro can produce high-quality music engraving quickly.

REMEMBER

The score editor quantizes your notation visually without actually quantizing the underlying MIDI data. This allows you to change how the notation looks without changing how it sounds. For example, you might play a short and fast note (staccato) but want the score to show longer notes with a staccato marking, as is tradition. The power of the score editor is how you can customize the way your music looks.

Creating and editing notes in the score

You can add notes in the score editor using the pencil tool and the step input keyboard. To add notes with the pencil tool:

1. **Select a MIDI region in the tracks area and open the score editor.**
2. **Press I to open the Inspector.**
3. **Choose a note value from the Part Box in the inspector, as shown in Figure 15-11.**

   If you don't see notes in the part box, click either the All button or the notes icon in the part box.
4. **Click the location where you want to add the note using the pencil tool.**

   A new MIDI note is added to the region. You can also drag notes to the score directly from the part box.

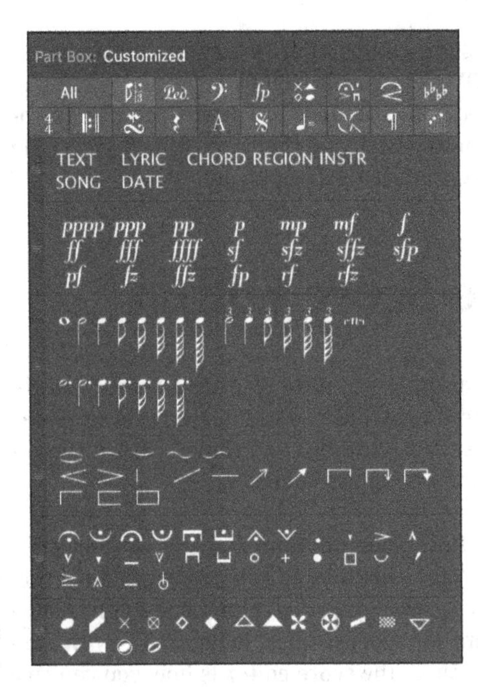

FIGURE 15-11: The part box.

To add notes using the step input keyboard:

1. **Select a MIDI region in the tracks area and open the score editor.**
2. **Choose Window ⇨ Show Step Input Keyboard or press Option - ⌘ - K.**

The step input keyboard opens, as shown in Figure 15-12.

3. **Choose a note value from the step input keyboard.**

4. **Move the playhead to the position you want to add the note.**

5. **Click the note you want to add on the step input keyboard.**

   A note is added to the region.

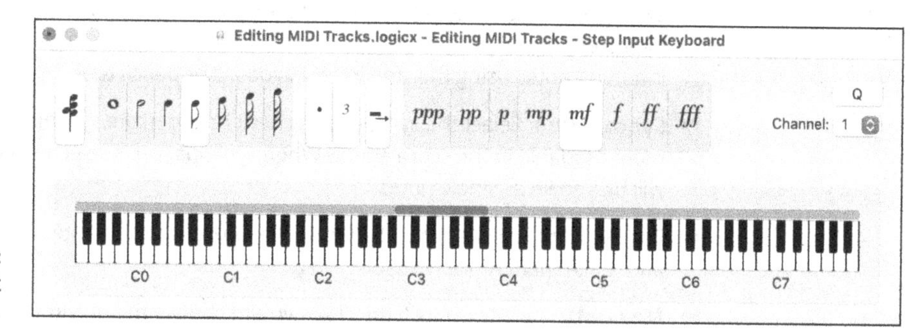

FIGURE 15-12: The step input keyboard.

Using the pointer tool, you can select and move notes left or right in the bar or up and down to change the pitch. You can change a selected note's length and velocity from the event inspector, as shown in Figure 15-13. You can also quantize the MIDI note event, as opposed to editing only the visual quantization, using the Time Quantize parameters in the score editor local inspector. To view the local inspector, choose View ⇨ Show Local Inspector.

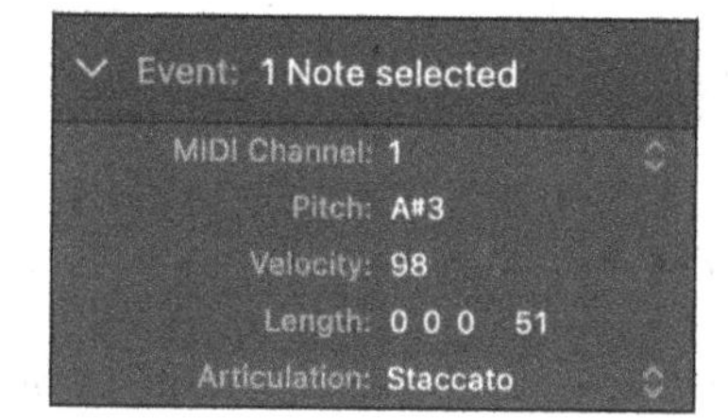

FIGURE 15-13: The event inspector.

Investigating the score region inspector

Every region in the tracks area can have different display properties in the score editor. You change these properties in the score editor region inspector, as shown in Figure 15-14. Here's a description of the region inspector parameters:

- **Style:** Choose the staff style for the selected track. Staff styles determine how the MIDI data is displayed in the score editor. For example, you can choose a bass staff style for a bass track or a piano style for a piano track. The staff style will determine the clef, transposition, and other display parameters.

- **Quantize:** Choose the minimum note value that is displayed in the score editor. The quantize value only determines the display in the score editor and doesn't edit the actual MIDI data.
- **Interpretation:** Select the interpretation check box to make the score more readable instead of presenting the MIDI data precisely as it's sequenced. For example, shorter note values like sixteenth notes are often rounded up when interpretation is selected. It's a good idea to deselect interpretation when inputting notes using the step input keyboard or pencil tool.
- **Syncopation:** Select the syncopation check box to make syncopated rhythms more readable in the score editor. When syncopation is selected, tied notes will be shown as single notes.
- **No Overlap:** Select the No Overlap check box to shorten notes that overlap and make them more readable.
- **Max Dots:** The Max Dots field allows you to choose the maximum number of dotted notes in the score.

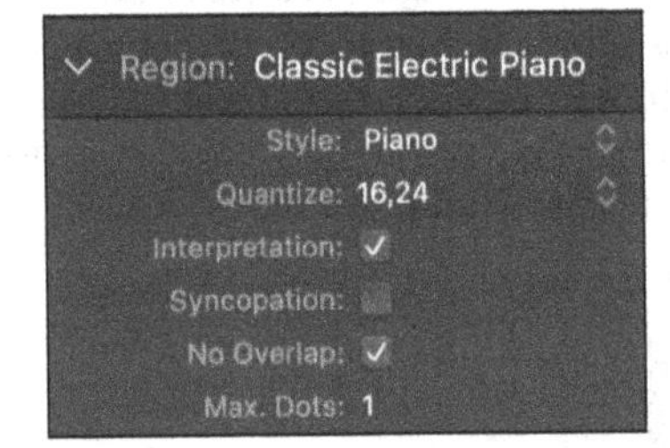

FIGURE 15-14: The region inspector.

Adding items from the part box

You can add a lot more than just notes to the score editor. To make your scores more meaningful to the reader, you can add symbols, text, chords, and much more. For example, you can add accents to notes or change the note heads themselves. To add symbols to notes, drag the symbol you want to add from the part box onto the note. To add symbols to multiple notes, select the notes and then drag the symbol from the part box onto any one of the notes.

Many of the symbols can be added to any location in the score. For example, dynamic marks can be added to tell the player how loud or soft to play the part. Simply drag the dynamic symbol you want to add from the part box onto the score.

The score editor is a fantastic resource and worth exploring in detail. You can add chords to your score, including guitar tablature. You can also add text and lyrics to produce lead sheets for performing or copywriting. If you want an in-depth tour of the score editor, visit `https://logicstudiotraining.com/chapter15` for a complete score editor video guide.

Editing MIDI in the MIDI Transform Window

The MIDI transform window can save you hours of tedious MIDI editing. Whenever you find yourself clicking a lot, open the MIDI transform window and get ready to be amazed. Just a little planning can make your MIDI editing go faster and get you back to creating music.

To edit MIDI using the MIDI transform window, follow these steps:

1. **Select a region or MIDI events within a region.**
2. **Press ⌘-9.**

   The MIDI transform window opens, as shown in Figure 15-15.
3. **On the Presets menu, choose a preset.**
4. **Click Operate Only.**

   Your selected MIDI is transformed according to the preset.

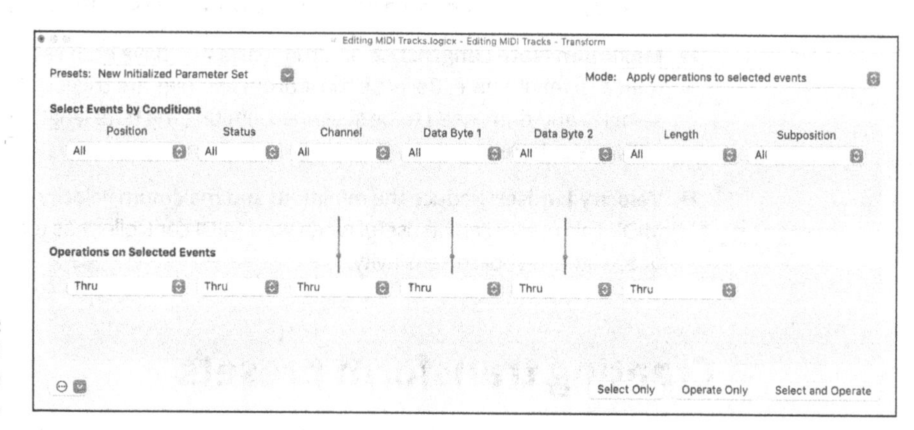

FIGURE 15-15: The MIDI transform window.

In addition to the Operate Only button, two other buttons determine how the MIDI transform window behaves:

- **Select Only:** Click the Select Only button to select MIDI events that meet the criteria in the transform window. Selecting MIDI events before you operate on them is useful when you want to double-check that your transform parameters are set up correctly.
- **Select and Operate:** It's not necessary to select the MIDI events in a region before transforming them. You can tell the transform window to select MIDI that meets your criteria and then transform the MIDI by clicking Select and Operate.

TIP

You can also open the MIDI transform window from the piano roll editor by choosing Functions ⇨ Midi Transform and selecting any function in the list.

Using transform presets

You can choose from several MIDI transform presets on the Presets drop-down menu. Here's a description of some of the presets you might want to consider using:

- **Humanize:** Introduce a range of random velocity, position, and length to your MIDI notes. This preset is perfect for MIDI that has been entered using the step input keyboard or the pencil tool precisely on the grid.
- **Crescendo:** Make your MIDI gradually get louder. This preset is useful when building dynamics into your arrangement. You can easily turn this preset into a decrescendo, the opposite of a crescendo, by reversing the velocity values.
- **Maximum Note Length:** Use on drum parts that have note values longer than a sixteenth note. Because most drum samples are triggered as one-shot samples and don't need to be sustained, limiting the note length makes the MIDI easier to read.
- **Velocity Limiter:** Reduce the minimum and maximum velocity of your MIDI. This preset can be useful when your MIDI controller has too much or too little keyboard sensitivity.

Creating transform presets

You can create your own MIDI transform presets from your own settings. To create a MIDI transform preset:

1. **On the Presets menu, choose Create New Transform Set.**

   A dialog asks if you want to create a new preset or rename the current preset.

2. **Click Create.**

   The MIDI transform window is initialized to the default settings.

3. **In the MIDI transform window, set the conditions and operations.**

   The Logic Pro manual describes the transform window's conditions and operations in detail. Open the manual by choosing Help ⇨ Logic Pro Help.

4. **On the Presets menu, click the preset name and then rename your preset.**

5. **Press Enter.**

   Your new MIDI transform preset is renamed, saved, and added to the menu.

Editing MIDI in the Event List Editor

The event list editor isn't pretty, but it is the most complete MIDI editor. It can often be the best editor when you want to change things directly, without a graphical user interface to interpret your MIDI data.

Displaying events

The buttons at the top of the list editor, shown in Figure 15-16, filter the different event types. Click the buttons to hide or show the corresponding events in the list area.

TIP

If you click the leave folder icon at the top left of the event list editor, you can display all the regions and folders in your project.

Creating and editing events

You can create events by clicking the add event icon (plus sign) above the list area. Choose the event type you want to create by clicking the event type menu to the right of the add event icon. New events are added at the playhead position.

Selected events can be edited on the Edit and Functions menus. You can also edit events in the list area by dragging parameters up and down or double-clicking the parameter and entering the value directly.

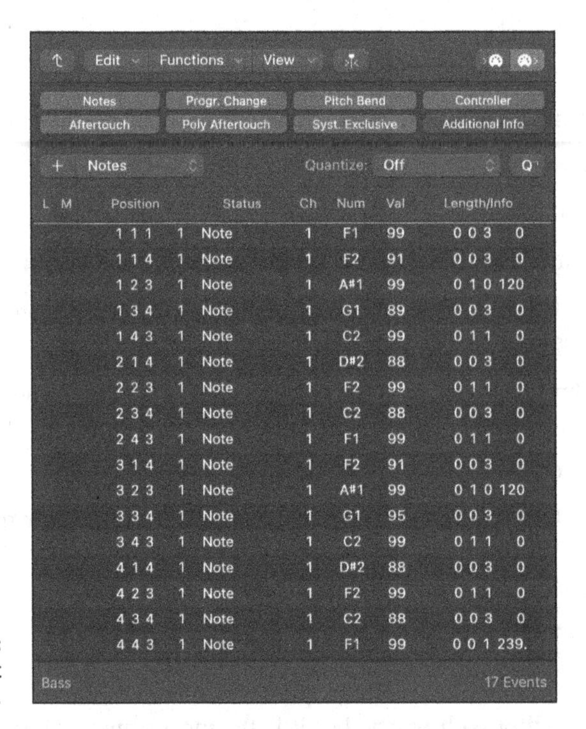

FIGURE 15-16: The event list editor.

TIP

You can create an event that will change screensets in your project at a specific time. (For more on screensets, see Chapter 3.) Choose Meta Event on the event type menu and add an event. Enter 49 in the Num column, and enter the screenset number in the Val column. You might use this meta event to force your project to always begin with a certain screenset, such as a large project Notes window. Or maybe you need to switch between the score editor for reading music and a large global track display for reading lyrics as you record. Another useful meta event is number 52, which automatically stops playback.

Editing Your MIDI Environment

The Logic Pro MIDI environment used to be the main place to build a MIDI project and connect your virtual environment. In Logic Pro, the MIDI environment does its work behind the scenes, leaving you to make music. But if you like to tinker, you'll like the MIDI environment.

The environment contains three basic types of objects:

» **Input objects:** Two input objects are available. The physical input represents the inputs of your MIDI interfaces. The sequencer input represents the Logic

Pro input. You can have only one physical and one sequencer input per project.

- **Instrument objects:** Instrument objects represent the MIDI devices you have connected to Logic Pro.
- **All other objects:** Many other objects are available in the environment that create and control MIDI data and signal flow, such as faders, knobs, and switches.

Every object you see in your project is contained in the MIDI environment. Your software instrument tracks, audio tracks, and outputs all have corresponding objects in the MIDI environment. You don't need to do anything with these objects, but if you want to do things such as transform the MIDI that comes into Logic Pro, you can do that in the MIDI environment.

Exploring object parameters

With the MIDI environment open, you can select objects and adjust their parameters. To view an object's parameters, open the object inspector by pressing I. Every object has a name, an icon, and an Assignable check box, as shown in Figure 15-17. The assignable parameter makes objects available for reassignment in the tracks area. For example, you could reassign a track to a different object in the MIDI environment by Control-clicking the track header and choosing the assignment on the Reassign Track menu.

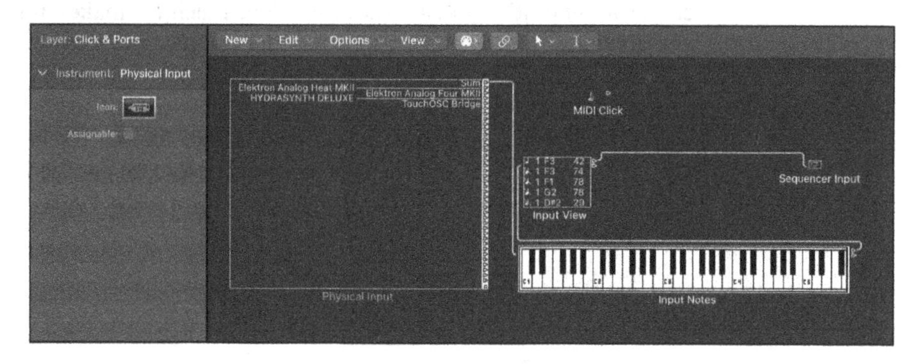

FIGURE 15-17: The MIDI environment object inspector.

Other parameters depend on the object selected. Many of the parameters will be similar to what you've experienced with the other MIDI editors. You can choose MIDI channels, values, inputs and outputs, and more. The Logic Pro manual, available by choosing Help ➪ Logic Pro Help, will give you a list of every MIDI environment object and its parameters.

Viewing environment layers

The MIDI environment puts objects on different layers to help organize objects. You choose the layer on the Layer drop-down menu at the top of the inspector. You can also create, rename, and delete layers by using the same menu. Objects aren't required to be on a specific layer. You are free to move and copy objects between layers in the following ways:

- » To move an object to a new layer, select the object you want to move and hold down Option while selecting the new layer. The object will be moved to the new layer. If the object is connected to another object in the original layer, the connection will be preserved.
- » To copy an object to a new layer, select the object and press ⌘ -C, and then select the new layer and press ⌘ -V. The pasted object will also preserve connections to objects in other layers.

Inserting objects in the environment

You can add new objects by using the New menu. The object you choose will be added to the current layer. You can rename your objects by using the text tool or by editing the name in the object inspector. You can move objects by dragging the object's name or by Shift-dragging the object.

One cool thing that I've taught Logic Pro users to do with the environment is to create a generative music machine. This machine makes music that's always changing and composed in real time. You can use this machine to create unique, ambient soundscapes, add excitement and unpredictable elements to your arrangements, and generate new melodic and harmonic ideas for further composition. The subject is beyond the scope of this book, but if you're interested, visit `https://logicstudiotraining.com/environment` to get the video tutorials.

5

Mixing, Mastering, and Sharing Your Music

IN THIS PART . . .

Understand the fundamentals of mixing, set levels and measure volume, use channel strips in the mixer, apply effects, and elevate your sound.

Automate your mix, record live automation, and automate MIDI events.

Enhance the mix with EQ, control dynamics with compression, and reach competitive loudness with limiting.

Create alternative mixes, back up your projects, and share online.

IN THIS CHAPTER

» Understanding the fundamentals of mixing

» Setting levels and measuring volume

» Using channel strips in the mixer

» Keeping track of your project with track notes

Chapter 16

Mixing Your Project

Mixing is so much fun that a lot of Logic Pro users begin mixing before they begin writing a song. So many cool effects and faders and knobs are available to entertain us. The excitement keeps Logic Pro users hooked and coming back for more. Mixing is taking all your parts, blending them, and delivering the result to your audience in a cohesive whole. Ultimately, mixing is all about controlling and directing the experience of the music.

In this chapter, you discover the fundamental concepts of mixing audio to create an experience. You learn the best practices for setting levels and measuring volume, taking the guesswork out of balancing your tracks. You find out how to overcome the most common challenges of mixing and how to produce professional sounding audio.

Understanding Important Mixing Concepts

Audio engineering confuses a lot of beginners. It's part art and part science. When you understand the core concepts, you can get good mixing results without years of trial and error. In this section, I show you the nine fundamentals of mixing audio. If you visit `https://logicstudiotraining.com/chapter16`, you can download a video guide in which I demonstrate these core concepts visually.

Mindset

The mixing head game can keep beginners from producing quality mixes. What should you focus on? How can you take all your tracks and make them sound good? I believe that if you focus your mind in the right direction, your mix will come together and satisfy your artistic needs and your listener's needs.

When you're mixing a project, the first thing to consider is your audience. Decide the people for whom you're producing this project, as well as the final listeners. Music is a subjective art, and many beginners make the mistake of trying to please the wrong people. If you're working for a client, please the client and the client's audience. Here are some questions to ask yourself and your client:

- **What other music does the audience listen to?** If you know the musical interests of your listeners, you'll have a target sound that you can emulate or complement. Gather a reference list of music that engages your listeners and make notes on how you can achieve a similar or complementary sound.
- **Does the project have a specific concept?** Many music projects need to achieve a specific vibe or concept. If you need to go after an emotion or a distinct sound, knowing the concept will help you set the criteria for success. If more than one person is working on the project, having a specific concept helps you work together and avoid disagreements.
- **Does the project have boundaries or limitations?** For example, a commercial jingle might require exactly 30 seconds of music. A band might have only four pieces with one singer. The project could have a solid deadline. These boundaries keep you from experimenting too far and help sharpen your focus.

TIP

Your first goal is to take the raw tracks and make them sound adequate. After you've achieved a tolerable mix, you can work on elevating it until it sounds really good. Your final step is to stabilize the mix until it sounds polished and professional. But remember that you will never think it's perfect, and that's okay!

Acoustics

The biggest challenge to achieving a stable and professional mix is the acoustics of your listening environment, including the gear you use and the room where you listen. A pair of quality monitors with an even frequency response is important. The room that you listen in can potentially add or subtract frequencies and give you an inaccurate picture of the sound. Here are some suggestions for making the most of your gear and listening environment:

- **Measure your room and calculate the room modes.** Search online for a room mode calculator to do the math for you. You're looking for areas in the frequency spectrum where your room creates buildups of frequencies or gaps in frequencies. This knowledge can help you adjust to the room's limitations.
- **Position your speakers and listening position in an equilateral triangle.** Find the sweet spot where your ears are equally distant between the two speakers, as shown in Figure 16-1. Also, consider placing your speakers away from walls. Having speakers too close to the back wall of your room can create a boomy sound. Finally, consider using speaker stands or isolation pads to reduce unwanted vibrations and to ensure that you get clear, accurate audio playback with no distortion.
- **Consider treating your room with absorptive and diffusive material.** Acoustic foam placed strategically throughout your room can help absorb and diffuse reflections of sound against the walls and give you a much clearer picture of what you're hearing. Furniture and heavy curtains also help break up audio reflections.
- **Listen in multiple locations through multiple audio systems.** If you want a clear picture of what other listeners will hear, play your mix through common consumer systems in addition to your optimized system and location. Consumer earbuds, laptop speakers, and car stereos are some of the most common audio systems that listeners use. Michael Brauer, the engineer for artists such as Coldplay and John Mayer, is famous for doing a big chunk of his mixing while monitoring the audio through a 90s Sony boombox.

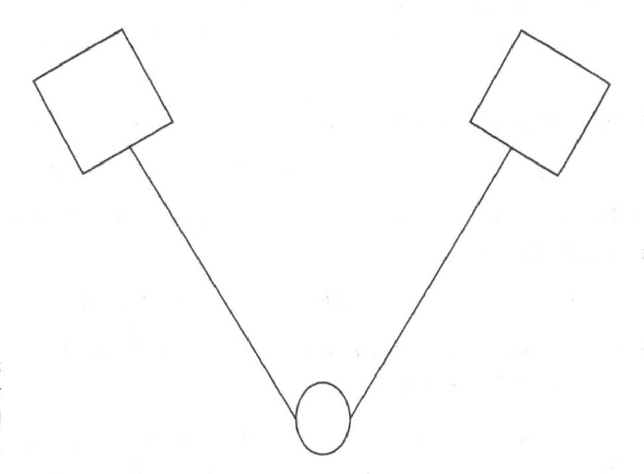

FIGURE 16-1: Ideal speaker placement and listening position.

You don't have to spend thousands of dollars or hire acoustic specialists to significantly improve your acoustics. Spending an hour or two optimizing your room and listening position and calculating room modes will give you a practical advantage as you mix your music.

Balance

With your new mixing mindset and optimized acoustics, you're ready to begin the technical part of mixing. The first step is to adjust the relative levels of your tracks to get an appropriate balance. Different genres of music have different approaches to balance, and it helps to study the genre of music you're mixing to get a feel for the approach. But a general rule is that one track should have the ultimate focus at a time. Soloists and lead vocals are generally more prominent than the rest of the tracks.

A critical component to balance is loudness. Here's what you need to remember about loudness: The human ear doesn't hear all frequencies equally. The ear is more sensitive to frequencies centered around the human voice and less sensitive to lower and higher frequencies. That means if you're listening to music quietly, the vocals will stick out more than the bass guitar or high strings. To compensate for the loudness contour of the human ear, you have to monitor music at a level where the frequencies become more equal.

The film industry has been calibrating the monitoring level to 83 dB SPL for decades because it produces a consistent product and allows audio engineers to create balances that account for the loudness contour of the human ear. Best of all, with today's technology, it's easy to calibrate your monitoring level and make mixing even easier. Here's how to calibrate your monitoring level:

1. **Get a sound-level meter.**

   You can find inexpensive sound-level meters online or at electronics stores. If you have an iPhone or iPad, you can find many free or paid sound-level meters in the app store.

2. **Set your sound-level meter to a C-weighted, slow response.**

   I use my iPhone and the AudioTools app by Studio Six Digital.

3. **Hold or place your sound-level meter where your head would be at your listening position.**

   I like to use a music stand to hold my iPhone while I do the next step.

4. **Play a reference CD or your Logic Pro project, and adjust the monitor gain until the sound level meter reads 83 dB.**

   To adjust your monitor gain, use the master volume knob on your audio interface or amplifier, not the volume of the playback application.

5. **Mark your monitor gain so you can recall this position.**

After you've calibrated the monitoring level of your listening position to 83 dB, you'll have a much easier time getting balances. You don't always have to listen at this level, but whenever you are adjusting the volume of individual tracks, set your monitor gain to the calibrated position.

Panning

Everyone in the music business knows that, for better or worse, the world revolves around lead singers. But in the mix, the world revolves around drummers. The drums and drum machines are giant instruments that take up the low end, the high end, and all the other ends. When it comes to the stereo spectrum, the drums also take up the left side, right side, and center. One of the first questions you ask yourself regarding the panning of your mix is, are the drums heard from the drummer's perspective or the audience's perspective?

Putting your drums in the audience's perspective is the most common technique. You, as the mixer, take the perspective of the audience and pan the drums as you would see them on stage. Kick and snare go in the center, hi-hat goes a little to the right, toms are panned right to left from high to low, and cymbals are spread just as you see them in the Drum Kit instrument, shown in Figure 16-2. The drummer's perspective is the reverse of the audience's perspective and a valid option. You can spread the drums wide or keep them closer to mono. What you do with drums often helps you with the rest of the mix.

FIGURE 16-2: Drum Kit from the audience's perspective.

What if you don't have drums in your mix? Panning is still often associated with the natural acoustic setup of the instruments. An orchestra is panned as you see it on the stage. Electronic dance music treats panning similarly to other pop music, except without extreme panning, which doesn't translate well in a club.

After you've decided on the mix perspective, create clarity and balance from left to right with an equal distribution of rhythmic and frequency activity. Aim for no gaping holes between nine and three o'clock on your pan knobs. You learn exactly how to adjust the pan balance later in the chapter.

Frequency

Panning covers left and right, balance covers front to back, and frequency covers the high and low dimensions of the mix. You don't need to balance each frequency range by making sure they're all equally as loud. Instead, you need to make them present in a way that best represents your vision and the marketplace. Hip-hop must have low end. Pop needs to translate to average systems. Jazz must sound natural, and classical must sound transparent.

If you record your project well, you might not need to use a lot of EQ. Extreme EQ can disturb your balance because you're raising or lowering the volume level of frequencies. It's probably best for beginning mixers to go easy on the EQ and stay subtle. But you will need to use some EQ, and these frequency ranges are generally accepted:

- **Low end:** 20–250 Hz
- **Lower midrange:** 250–1000 Hz
- **Upper midrange:** 1–8 kHz
- **High end:** 8–20 kHz

TIP

The biggest challenge for beginning mixers is getting the low end right. Boosting anything between 20–60 Hz can easily overpower a mix. The low-end workflow I'm about to give you has saved my bass from trouble on many occasions. If the low end is too thick or unclear, mute things that should be in the low end (kick drum, bass guitar, low strings, and so on) and listen for what's left in the mix. If you still hear bass, EQ it out, as you'll learn later in this chapter. Some big culprits to listen for are acoustic guitar, piano, synths, low percussion, and electric guitar. Also, it's best to monitor through speakers with a flat frequency response and consider using a subwoofer to have more precise control of your low-end frequencies.

Depth

Another dimension you can add to your mix is depth, near and far. The difference between balance and depth is that while a lead vocal can be upfront in the mix (balance), it can include an element of depth through the use of reverb or echo, making it also seem far away. If you want sounds to be intimate, leave them dry (without effects). If you want sounds to seem larger than life, add depth.

Your main tools for creating depth in the mix are reverb and delay. *Reverb* is an effect that simulates the reflections of sound bouncing in an enclosed space. *Delay* is an effect that takes the original signal and delays it before playing it back again. Sometimes, delay is used in place of reverb because it gives the impression of space without taking up as much space as reverb. Both effects are commonly used together in a mix.

TIP

The brain can perceive approximately two or three levels of depth at a time. That means you need only a couple of reverbs or delays in your mix.

Effects

Playing with effects is a lot of fun. Just ask a guitar player to use only one effects pedal, and you'll get a look that wonders why you're such a buzz kill. Effects add character to your sound and help instruments create their own sonic space. You'll consistently use the following three major families of effects in your mixing:

- **Modulation effects:** Chorus uses multiple delays for thickness and richness. Phasers use a narrow-band EQ cut that's swept back and forth through the frequency spectrum. Flangers create washy phase cancellations throughout the frequency spectrum. Tremolo and vibrato vary the volume and pitch, respectively.
- **Distortion:** Distortion effects include tube distortion, tape distortion, distortion effect pedals, hard clipping, soft clipping, transformer clipping, noise, record pops, record scratches, and speaker distortion.
- **Filters:** Filtering effects include band-pass filters, filter sweeps, envelope filters, and wah wahs.

Dynamics

How would you like it if everyone yelled at you all the time? Or what if they constantly whispered? We need contrast to feel the effect of loudness and softness. Dynamics are integral to keeping your listener interested and engaged.

Changes in volume are the key to dynamics. Here are three ways you can provide dynamics in your mix:

- **Peaks and valleys:** Your overall mix should have changes in volume. Some genres of music, such as classical and jazz, tolerate more dynamics than others, such as pop or rock. Popular music is often listened to in settings where the need for dynamics is lower, such as in the background of clubs and parties, while classical and jazz are heard in more intimate settings where music is the focus. At the very least, identify the climaxes of the music, and design those areas to be the loudest part of your mix, regardless of the genre.

- **Compression:** Compressor effects are used for effect or to control the dynamics of instruments, groups of instruments, or the entire mix. You discover many types of dynamic control in Chapter 17.
- **Envelopes:** You learn about ADSR envelopes (attack, decay, sustain, and release) in Chapter 11. Envelopes, like compressors, control the dynamics of individual sounds. They help bring out dynamics at a micro level. Synths and samplers have built-in envelopes, but you can also add envelope effects to other instruments, as you learn in the section on using dynamics tools in Chapter 17.

Interest

The final mixing fundamental is interest. All these fundamentals work together, but you could define a workflow around the order in which I've presented them. Set your mindset before you do anything. Set up your acoustics before you set your balances, and so on. Some fundamentals blend together more than others. For example, adjusting your panning helps to set your balance, and your effects and dynamics help your mix sound interesting.

I leave this discussion with three final elements to elevate your mix into something that sounds interesting and keeps your listener's attention:

- **Focus:** Find the main point of interest and focus on it. It could be a lead vocal, a solo, or a riff. Give the listener something interesting to focus on at all times.
- **Uniqueness:** Find a unique element and make it stand out. If you can't find a unique element, find a way to make an ordinary element sound unique.
- **Automation:** Use automation to add dynamics, mutes, panning, and effects, all of which lead to creating interest. You learn about automation in Chapter 18.

Knowing Your Channel Strip Types

You use channel strips to shape your sound and control the signal flow. You have control over the volume, pan, effects, and where the sound is coming from and where it's going. Similar to track types, which you learn about in Chapter 4, different channel strips do different things. Some channel strip types are automatically added to your project, while others you can add for specific purposes.

In this section, you learn about the five different channel strips shown in Figure 16-3 and what they do.

FIGURE 16-3: The channel strips in Logic Pro.

REMEMBER

Enable Complete Features must be selected in the Advanced Settings pane to take full advantage of the mixer and all the information in this chapter. Choose Logic Pro ⇨ Settings ⇨ Advanced, and then select Enable Complete Features.

Audio channel strips

You use audio channel strips (refer to Figure 16-3) for audio tracks. When you create an audio track, an audio channel strip is automatically added to the mixer in your project. Everything you record live, such as guitars or vocals, uses an audio channel strip. Imported and bounced audio also use an audio channel strip.

Instrument channel strips

Software instrument tracks use instrument channel strips (refer to Figure 16-3). Instrument channel strips are similar to audio channel strips except that they use MIDI as the source material instead of audio. Instrument channel strips also allow you to add MIDI effects, which you learn about in the section on adding effects to tracks in Chapter 17.

MIDI channel strips

MIDI channel strips (refer to Figure 16-3) are used to control your external MIDI gear. You can't add audio or MIDI effects to these channel strips, and you can't route audio or save channel strip settings. You can, however, control track automation, mute the track, and send general MIDI settings to the connected instrument, such as volume, pan, and other controller data.

Auxiliary channel strips

Auxiliary channel strips (labeled in Figure 16-3) are used mainly for routing the audio of other track types. You can use auxiliary channel strips as submixes and add effects as a whole. For example, all the tracks in a summing track stack (described in Chapter 4) send their audio outputs to a collective auxiliary channel strip. You can also send a portion of a track's signal to an auxiliary channel strip via effects send to blend reverbs, delays, or other effects with the track's signal.

Output channel strips

All audio is routed through a default output channel strip (refer to the rightmost strip in Figure 16-3) before it goes to your master output. You can add effects to the output channel strip that will affect the entire mix, such as dynamics compression and EQ. If your hardware supports it, you can add additional outputs and assign tracks or submixes to them for outboard processing or additional monitor mixes.

TIP

The inspector has two channel strips, as shown in Figure 16-4. The channel strip on the left shows the selected track's channel strip. The channel strip on the right is dynamic and, by default, shows the output of the channel strip on the left. If the selected track has a send effect, you can click the send slot, and the second channel strip is updated to show the auxiliary channel strip being used to host the send effects. This feature is useful if you're mixing a track and you also want to adjust the sound of the send effects it's using.

FIGURE 16-4: The inspector channel strips.

Using Meters to Visualize Volume and Levels

Loudness. We're kidding ourselves that we can perceive it objectively. Music often sounds instantly better when it gets louder. But you can only turn it up so much before you see diminishing returns (or blow up your speakers). If you don't stage your volume levels with the big picture in mind, you can easily run into a situation where you're constantly seeing red lights at the top of your meters, pulling down faders, and readjusting the balance of your mix. In this section, you discover how to use your meters so you don't trick yourself into thinking that louder is better.

REMEMBER

If you calibrate your monitoring level as I described previously in the "Frequency" section, you'll have a much easier time perceiving loudness, and you won't need to rely on your meters to tell you how loud the music is.

Your ears are the number-one meter of quality. A little bit of ear training can go a long way in polishing your perception of sound. A close second to your ears is the meters on your channel strips. I'd like to give you one rule for setting the levels on your meters: **Peak no higher than between −12 and −6 dBFS on your output channel strip.** All the tracks that sum together will need to be even lower, probably between −18 and −12 dBFS. You will have plenty of opportunities to get your overall mix louder as you go through the mastering process in Chapter 19.

TIP

I have a video about ear training with Logic Pro that I am happy to share with you. Just visit `https://logicstudiotraining.com/eartraining`.

Understanding clipping

The meters in Logic Pro are measured in dBFS, which stands for decibel full scale. 0 dBFS is the top limit, and anything above that is digital clipping. *Digital clipping* means that the signal has gone beyond maximum capacity. The result is distortion in the audio, and not the good kind of distortion that rock guitar players know and love. Digital distortion is an unpleasant sound that you should avoid.

When digital audio came along, 16 bits was the maximum resolution. 16-bit audio accounted for about 96 dB of recordable space, and it was deemed important to record and mix as close to 0 dBFS as possible to keep noise low and use the full digital resolution. Logic Pro is capable of 24-bit recording, so you can record better signals at much lower levels and regain the headroom that was a major benefit of analog mixes. In fact, you'd have to lower your mix by 48 dB to equal a 16-bit recording.

Lowering your mixing levels gives you more headroom and makes mixing easier because you're not always fighting red lights (digital clipping) and having to readjust your balances. Lower mixing levels also make it easier to work with third-party Effects plug-ins or outboard analog gear and raise the final level at the mastering stage.

Choosing pre-fader or post-fader metering

Audio professionals have a common complaint with Logic Pro's meters: They default to post-fader. In post-fader metering, the position of the fader affects the level of the meters. With post-fader metering enabled, the meters will show no level if your channel strip fader is all the way down. Post-fader metering offers visual assistance when you're using a "mix with your eyes" approach. You can see the track level with your eyes on the meter based on the location of your fader. That's not a bad thing, but it's definitely not the whole picture.

Switching to pre-fader metering shows you the level of the signal regardless of the fader's location. Pre-fader metering allows you to see whether a track is clipping, even when your fader has pulled the level all the way down. You'd be surprised how many software instrument presets clip right from the start. You need to know precisely what signal level is hitting the channel so you can make the appropriate decisions regarding adding effects such as compression or EQ.

On analog mixing consoles, many engineers will set the faders to unity gain (the default position) and quickly stabilize track levels using a trim control. You can do something similar by putting Logic Pro's Gain plug-in in the first insert effects slot, setting the basic rough level with the plug-in, and using the channel strip fader to fine-tune the level. You probably won't need to add the Gain plug-in to every track, but it can help if a recorded track is too loud. You learn how to add insert effects in Chapter 17.

Another complaint is that peak level meters, the type Logic Pro and most digital audio workstations use, have almost nothing to do with loudness. For example, two takes from a drummer can sound equally as loud, but one take can peak 4–10 dB higher on a single snare hit. If you raise the other take so that they both peak at the same level, one will sound much louder, maybe even twice as loud. To judge loudness, the ear responds to average levels and not peak levels. Peak meters are great when you need to be concerned with clipping (well, you're always concerned with that), but they don't help you perceive loudness very well. For that reason, professionals often work with VU (volume units) meters, which more closely respond like an ear.

Logic Pro doesn't come with VU metering, so learning to use pre-fader metering will help. Consider searching for a free VU meter plug-in or purchasing one. You can mix without VU meters, but the job is easier with them. To switch to pre-fader metering, follow these steps:

1. **Control-click an empty area of the control bar, and then choose Customize Control Bar and Display.**

   The customization dialog opens.

2. **In the Modes and Functions column, select Pre Fader Metering.**
3. **Click OK.**

   The pre-fader metering icon (shown in the margin) is added to the control bar.

4. **Click the pre-fader metering icon to turn it on.**

TIP

If you want to quickly switch between pre-fader and post-fader metering, set the Toggle Pre-Fader Metering key command. For details, see Chapter 3.

Using the Loudness Meter

The Loudness Meter, shown in Figure 16-5, is a metering plug-in to help you navigate the complicated world of loudness. Use it on your output channel strip to measure the loudness of your entire mix. (You learn how to add plug-ins to tracks in Chapter 17.) Loudness Meter measures your audio signal using the LUFS (loudness units relative to full scale) standard. The LUFS specification aims to normalize the loudness of audio for broadcasting. It was created partly as a response to complaints of excessively loud TV and radio commercials. Streaming services such as Spotify and Apple Music also have LUFS loudness targets so that their end users will have a consistent listening experience.

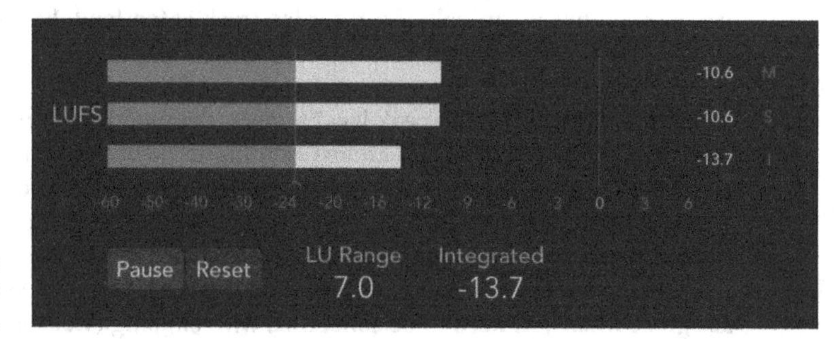

FIGURE 16-5: The Loudness Meter plug-in.

TIP

Loudness targets are continually being updated, so you'll want to research if the platform where you intend to publish your music has requirements or ideal specifications. You're free to publish audio at any loudness level, but publishers such as YouTube and Apple Music may alter the loudness of your audio, potentially compromising the integrity of your mix. And if you ever need to compare the loudness of your mix to another, the Loudness Meter plug-in is an excellent tool for the job. You'll find the plug-in in the Metering folder of your plug-in menu.

Adjusting Channel Strip Controls

The channel strip is where you adjust the volume and sound of a track. You can view a selected track's channel strips in the following ways:

- Choose View ➪ Show Inspector or press I. The channel strip is shown on the inspector. You may have to close the region and track inspector to view all channel strip controls.
- Choose View ➪ Show Mixer or press X. The selected channel strip appears light gray in the mixer. You may need to adjust the size of the mixer to view all channel strip controls.
- Choose Window ➪ Open Mixer or press ⌘ -2 to open the mixer in a new window.

You can adjust multiple selected channel strips simultaneously. To select multiple channel strips, Shift-click the channel strips or make a choice on the Edit ➪ Select menu in the mixer. You can select multiple tracks in the mixer also by dragging along the channel strip names of the channel strips you want to select.

TIP

By default, not all channel strip controls are shown. To show or hide channel strip components, Control-click any track in the mixer or inspector and choose the components you want to show on the Channel Strip Components menu. You can also show and hide channel strip components by using the View menu in the mixer.

Adjusting volume and toggling between levels

Adjust the volume of your tracks by raising or lowering the fader on each channel strip. The default fader state (no boost or cut) is called *unity gain*. You can quickly restore your fader to unity gain by Option-clicking the channel strip.

TIP

Your faders are capable of .01 dB changes. Don't drive yourself crazy with these tiny increments. At least 1 dB of change is required before you reach a just noticeable difference, so moving your fader less than 1 dB isn't worth the effort. Depending on the frequency and complexity of a sound, even a 3 dB change in level is barely noticeable. Save time by moving your faders in increments of 1 dB or more.

Adjusting the pan balance

The pan balance knob is located above the fader and dB readout. Rotate the knob to find the ideal position in the stereo spectrum and Option-click it to reset it to center. Use pan to balance tracks from left to right and to find clarity on individual instruments. You can achieve balance by positioning tracks evenly throughout the stereo field. If the arrangement is balanced, it can dictate the panning. For example, a call and response between two groups of instruments can be panned on opposite sides of the stereo spectrum.

TIP

Your three most sacred positions are center, hard left, and hard right. Vocals, bass, kick drum, and snare are almost always panned center. What you put in the left and right positions will depend on the mixing concept, as described previously in the "Understanding Important Mixing Concepts" section. If two guitar tracks are playing simultaneously, they are often panned hard left and hard right, with other instruments filling in between.

Muting and soloing tracks

At the bottom of the channel strips are two buttons labeled M (mute) and S (solo). Mute a track to disable it from the mix. Solo a track, and all the other tracks' mute buttons blink to indicate they are muted. Sometimes, you can't find a track that was initially soloed because it's hidden or among many other tracks. In that case, press Control-Option-⌘-S to engage the Solo Off for All key command.

TIP

You can put a track into solo-safe mode so it won't mute even if you solo another track. Control-click the solo button of a track to toggle solo-safe mode. A red slash will cross the solo button. Solo-safe mode is useful when you want to hear more than one track at a time when you use solo mode.

Grouping tracks

After you get the balance of a group of instruments, such as drums, just right, you can adjust their volumes together as a group. Good candidates for channel groups are drums, background vocals, doubled lead vocals, and anything that should be adjusted together. To group tracks and choose group settings, follow these steps:

1. **Click the channel strip's Group slot and choose Open Group Settings, or press Shift-Option-G.**

   The group inspector opens, as shown in Figure 16-6.

2. **Select the Groups Active check box.**

   When you want to edit a track individually, deselect the Groups Active check box.

3. **Select the track group check box in the list.**

   A track can be in more than one group at a time. You can also rename the group by double-clicking in the name field and entering the new group name.

4. **Select the track settings you want to adjust as a group.**

   If you don't see the group settings, click the Settings disclosure triangle.

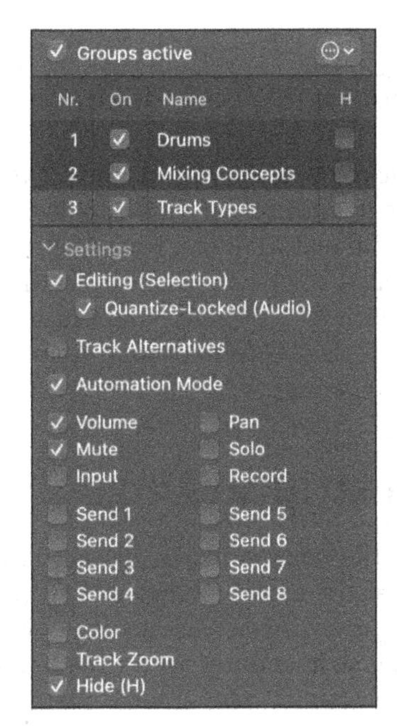

FIGURE 16-6: The group inspector.

Note that you can also edit tracks as a group by selecting the Editing (selection) check box. This option is useful when you edit drum tracks or background vocal tracks as a group. If the grouped tracks are audio tracks and were recorded together, such as drum tracks, select the Quantize-Locked (Audio) check box to ensure that your edits don't adjust the timing of the tracks.

TIP

After you add a track to a group, the group inspector pane is added to the inspector between the region and track inspectors. From the group inspector, you can quickly deselect the Groups Active check box to allow individual track adjustments or make other group adjustments without opening the group inspector window.

Choosing input and output settings

The input and output settings you have available are determined in part by your hardware. While mixing your project, you rarely need to change input settings because you're not recording audio sources. You might want to change your output settings, however. In particular, you can do some creative audio routing by choosing a bus as an output.

Buses are used to route audio to and from auxiliary tracks. To route the output of a track to an auxiliary track, click-hold the output slot and choose a bus on the menu. If a bus is already being used to route audio to an auxiliary track, it will have the name of the auxiliary track in parenthesis. Because you can route many tracks to the same auxiliary track, you can process them as a group. You learn more about auxiliary tracks in Chapter 17.

You can route the output of a track to an output other than the main stereo output if your audio hardware supports it. You can use your additional outputs to route audio to external hardware or additional monitors. You can also choose No Output if you don't want the track to sound at all, or you can select the Surround output if your hardware supports it. You can even select the Binaural output setting, which allows you to pan the track as if it were coming from behind your head. (Wear headphones for this type of panning.)

Selecting channel strip settings

When you set up a channel strip the way you like it, you can save the settings for recall. Click the Setting button at the top of the channel strip and choose Save Channel Strip Setting As. A dialog opens where you can name and save the setting. You can load your saved channel strip settings with the same menu as well as copy, paste, and reset the channel strip.

TIP

You can change channel strip settings by sending MIDI program changes from your MIDI controller. Click the Setting button and choose Save as Performance. Have your MIDI controller send a program change message to change the channel strip setting on the selected track. Only 127 performances can be saved. The MIDI program change value is listed next to the performance name on the Performance menu.

Mixing in Spatial Audio

Spatial Audio is a new way to listen to music that creates a more immersive experience. It does this by using Dolby Atmos to place sounds in different virtual locations around you. This makes it feel like you're in the room with the musicians, which is pretty cool.

Dolby Atmos is an audio format that expands surround sound systems to produce a full three-dimensional sound field that includes height information. In addition, Dolby Atmos isn't reliant on a specific speaker number or configuration for accurate playback, as previous surround formats have been. As a result, any audio device that can reproduce Dolby Atmos may utilize the format, including home theaters and soundbars, computers, headphones, mobile devices, and cars.

Setting up Logic Pro for spatial audio mixing

Logic Pro users are lucky because using Spatial Audio with Logic Pro is a breeze if you have compatible hardware. For example, many of Apple's computers, iPads, iPhones, and AirPods can produce spatial audio. To enable spatial audio in Logic Pro, do the following:

1. **Choose File ⇨ Project Settings ⇨ Audio, set the project sample rate to 48 or 96 kHz, and select Dolby Atmos from the Spatial Audio menu.**
2. **Choose Logic Pro ⇨ Settings ⇨ Audio ⇨ Devices, set the I/O Buffer Size to either 512 or 1024 samples, and click Apply.**
3. **In the mixer (press X), click the Dolby Atmos plug-in on the Master channel strip to open it and select either the Dolby Renderer or Apple Renderer from the Monitoring Format drop-down menu.**
4. **Click the Output slot on a channel strip and select 3D Object Panner from the menu.**
5. **To pan tracks in three-dimensional space, use the 3D Object Panner control on the channel strip or double-click it to open the 3D Object Panner in a plug-in window, as shown in Figure 16-7.**

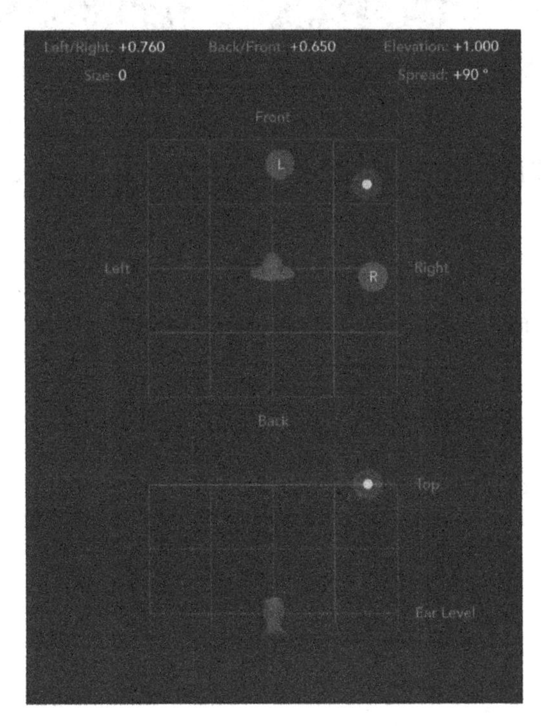

FIGURE 16-7: The 3D Object Panner.

Using the Dolby Atmos plug-in

A Dolby Atmos mix has two different types of tracks: bed tracks and object tracks. First, you route bed tracks to the Surround output and object tracks to the 3D Object Panner via the channel strip output slot. Then, to position the object tracks in 3D space, use the 3D Object Panner. You can plan and visualize your spatial audio mix using the Dolby Atmos plug-in, shown in Figure 16-8.

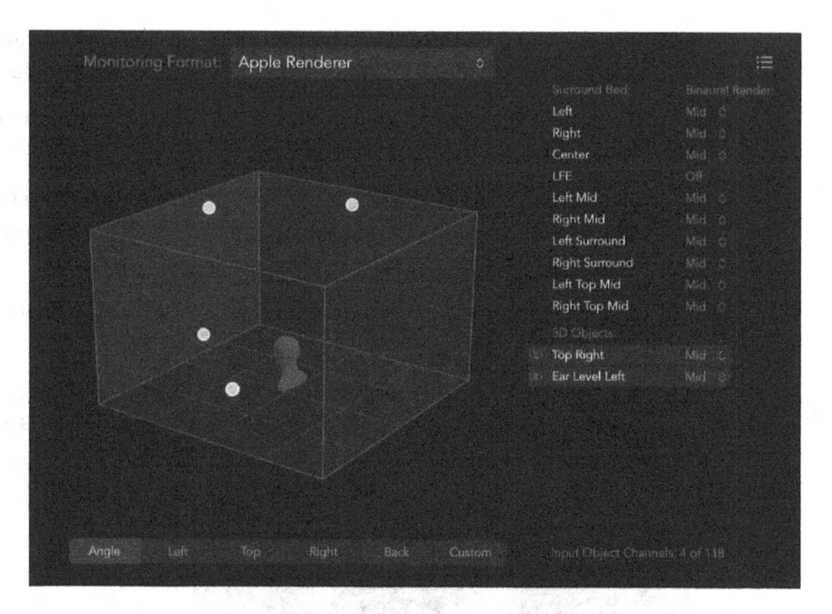

FIGURE 16-8: The Dolby Atmos plug-in.

Monitoring your mix in spatial audio

You'll need hardware capable of reproducing spatial audio to monitor your mix. The good news is that affordable options exist and new hardware is being released regularly. Apple's AirPods and AirPods Pro are impressive at producing spatial audio and have become immensely popular, making them a good choice for mixing.

Once you have compatible hardware, you can choose how to monitor the mix from the Monitoring Format pop-up menu in the Dolby Atmos plug-in. Choose any of the Binaural Apple Renderer formats if you're monitoring with headphones. If you're monitoring through computer speakers, choose the Renderer for Built-in Speakers.

Taking Track Notes

As you mix your project, you make countless changes to each track and its parameters. Mixes can quickly go from stable to unstable without a clear path back to stability. Keeping project and track notes can help you catalog your big milestones and store important reference material. Choose View ⇨ Show Note Pads or press Option-⌘-P to display the note pad, shown in Figure 16-9.

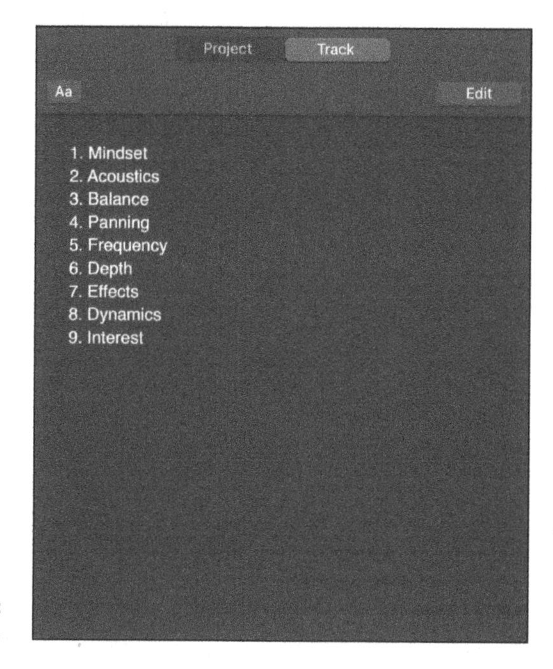

FIGURE 16-9: Track notes.

Project notes are an excellent place to store information about the song arrangement and overall project direction. Track notes are excellent for EQ settings, plug-in settings, and preset names. You can also show track notes at the bottom of every channel strip in the mixer. Control-click a channel strip in the mixer and choose Channel Strip Components ⇨ Track Notes. The project and track notes are great scratch pads.

You learned a lot in this chapter, such as how to use the mixer and measure volume and loudness, as well as some important mixing concepts. Repetition is key to learning, so revisit this chapter again and again as you mix your music. Mixing might initially feel confusing, but you'll get the hang of it with more experience. Don't wait to get experience. Start mixing a project today, and then head over to Chapter 17, where you learn how to shape your mix with effects.

IN THIS CHAPTER

» Applying effects and elevating your sound

» Controlling the flow of signals in the mixer

» Adding depth to your mix with reverb and delay

» Getting your mix competitively loud

Chapter 17

Shaping Your Mix with Effects

Logic Pro comes with many effects, giving you lots of choices. You'll get to know these effects well as you use them again and again. Chapter 16 teaches you important concepts for applying effects. In this chapter, you discover how to add effects and control the flow of audio signals in Logic Pro and how to use the most common plug-ins.

TIP

If you're a laptop artist and you write, arrange, edit, and mix your project as you go, hold your mix loosely until you've settled on the rest. You're going to discover cool sounds by design and by accident, and you should save them as patches so you can return to them later. However, you might want to start mixing from scratch after you've settled on all your parts. Sometimes a clean slate is faster to mix than a full plate.

Adding Effects to Tracks

I don't know many Logic Pro users who get excited about setting track volumes or panning, but they can't seem to get enough of plug-ins and effects. And for a good reason, too, because using effects is where a lot of the fun, experimentation, and magic come into the mix. Logic Pro comes with some of the most amazing sound effects on the market and includes everything you need to make great-sounding music.

Adding audio and MIDI insert effects

Audio insert effects can be added to audio, instrument, auxiliary, and output channel strips. You add Effects plug-ins via the channel strip Audio FX slot. Click an empty effects slot and, on the menu that appears, choose the plug-in you want to add. Insert effects have three clickable areas, as shown in Figure 17-1. You open the plug-in interface by clicking the center of the plug-in slot. You can choose a different plug-in by clicking the right side of the plug-in to open the plug-in menu. You can turn plug-ins on and off by clicking the left side of the plug-in.

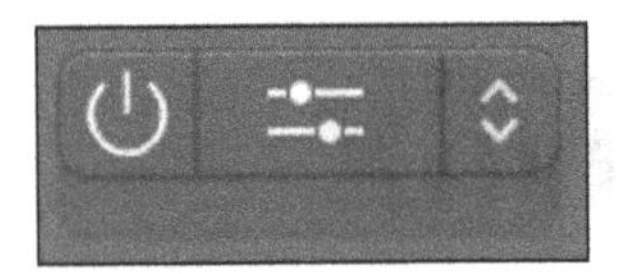

FIGURE 17-1: An insert effect slot.

TIP

Working with plug-ins while mixing can improve your sound immensely, but some plug-ins can add latency and cause peculiar behavior. To accurately determine which ones are causing issues, you can quickly activate or deactivate a plug-in by Option-clicking the plug-in slot on the channel strip. In general, keep the number of high-intensity plug-ins to a minimum and reduce the amount of processing by more resource-intensive plug-ins such as reverbs or click removers. Doing so will help you achieve the highest quality audio results with minimal side effects.

Insert effects are added to the channel strip in series, which means that the audio signal goes through each effect in the order in which it is inserted. You can reorder effects by click-dragging the center of the plug-in to the new location. You can also drag plug-ins to other channel strips. You can copy an effect by Option-dragging the plug-in to a new location.

MIDI effects, as shown in Figure 17-2, can be added only to software instrument channel strips. You add MIDI Effects plug-ins via the channel strip MIDI FX slot. You can replace, move, and copy MIDI Effects plug-ins just like audio Effects plug-ins.

TIP

Choose Help ⇨ Logic Pro Effects to open the online effects manual. Logic Pro comes with 70 audio Effects plug-ins and 9 MIDI Effects plug-ins. The Pedalboard plug-in includes an additional 35 cool-looking stompboxes that you can use also as individual plug-ins — and they're not just for guitar players. Logic Pro has plenty of built-in tone-shaping capabilities.

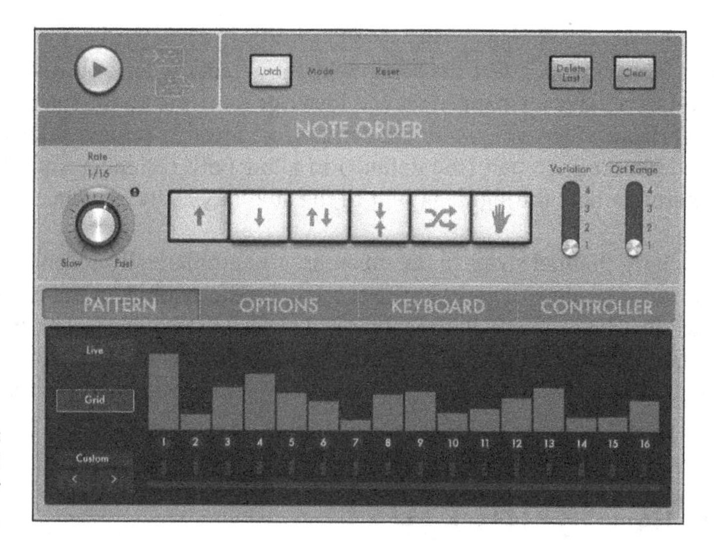

FIGURE 17-2: The MIDI Arpeggiator plug-in.

Adding send effects

While insert effects are serial in the signal flow (one goes through the next), send effects are parallel in signal flow (one goes beside the other). Send effects are useful when you want to send a portion of the signal through an effect, but you want the original sound to remain untouched. For example, you can send a bit of a snare or vocal through a reverb and mix the reverb with the original sound. Reverbs, delays, and compressors are commonly used as send effects. The auxiliary channel that hosts the effects will often have other insert effects on it, such as EQ, compression, or dynamics.

To add a send effect to a channel strip, follow these steps:

1. **Click the channel strip send slot and choose a bus on the menu.**

   The send slot routes the signal through a bus to an auxiliary channel strip. If the bus is empty, an auxiliary channel strip is created and added to the mixer.

2. **Drag up or down on the send level knob to change the level sent to the auxiliary channel strip.**

   You can also double-click the send knob and enter the value directly.

3. **Add the insert effects you want to the auxiliary track.**

   The effects on the auxiliary track process the signal you route through the send.

Like insert effects, you can click the left side of the send slot to turn it on and off. Click-hold the right side of the slot to choose a different bus and how the fader and pan controls affect the send. Select pre-fader when you want the send effect level to remain unchanged as you lower or raise the channel strip's fader. Select post-fader when you want the send effect level to remain relative to the fader level. Select post-pan (the default) to allow both the fader and the pan knob to determine how the signal is sent to the auxiliary channel strip.

TIP

Auxiliary channel strips in the mixer aren't automatically added to the track list in the main window. If you want to see your auxiliary track in the track list, Control-click the channel strip in the mixer and choose Create Track. The auxiliary channel strip will now have a corresponding track in the tracks area, so you can do things such as reorder the track in the list and edit the track when the mixer isn't open.

Controlling Signal Flow

After your tour of channel strips and insert and send effects, you understand that signal flow is flexible in Logic Pro. You can route audio in many ways for both practical and creative purposes. In this section, you find more ways to control signal flow.

Understanding how insert and send effects work

As you learned previously, insert effects are inserted in a series. Therefore, the order in which you insert the effects can have a big effect on your sound. However, many Logic Pro plug-ins, such as Compressor and Space, have a mix control, so you can adjust the level of dry and wet signal, effectively turning it into a parallel effect.

Send effects are always parallel to the source that's sending the signal. But on the auxiliary channel hosting the effects, the plug-ins are inserted in series. You can also use effects sends on auxiliary channel strips. For example, you can add a distortion or overdrive plug-in to an auxiliary track and send a small amount of your lead vocal to it so it's in the background, adding a little color. Then you can send the distorted auxiliary track to a delay or reverb so that the distorted vocal adds depth and interest to your mix.

TIP

Make your mix as complicated or simple as you want, but don't lose sight of your mixing goal and concept. What are you trying to do to the audio? Is it necessary? Is it enhancing your audio? Experiments are great, but give yourself a time limit. Otherwise, you'll easily spend your creative time tweaking instead of finishing your project. You'll have plenty of opportunities to try out new ideas on your next project.

Using auxiliary channel strips

You just discovered some ways to use auxiliary channel strips creatively. For example, you can also use auxiliary channels as submixes, without any effects, by selecting an auxiliary track as the output on a group of tracks. To create submixes, follow these steps:

1. **Select the tracks you want to group.**

   Drum tracks or background vocal tracks are great examples of tracks you would group in a submix.

2. **Click-hold the output slot on one of the selected tracks and choose an empty bus.**

   An auxiliary channel strip is added to the mixer, and you can control the sound and level of all the tracks as a group from the auxiliary track.

Another important use of auxiliary channel strips is for the efficient use of effects on multiple tracks. For example, there's no need to create a separate reverb effects send on every track that needs reverb, especially because reverbs such as Space Designer and ChromaVerb are processor intensive. Instead, create one reverb effects send and add multiple tracks to the same reverb. This technique also gives your mix a cohesive sound.

Another popular use of an auxiliary channel strip is for parallel compression. Create an effects send on the lead vocal or a group of drums. Add the Compressor plug-in to the auxiliary channel strip and compress it heavily, as described later in the "Using Compressor" section. Then mix the compressed sound with the dry sound, and you'll get more presence and loudness without raising the level of the dry tracks.

TIP

After you have your effects sends set up the way you like, you can add them to a track stack and save them as a patch for later recall. For details, see Chapter 4.

Using multi-output instruments

Multi-output capabilities are built into some Logic Pro software instruments, such as Drum Kit Designer, Studio Horns and Strings, Sampler, and Ultrabeat. Multi-output instruments enable you to route audio within the instrument to its own channel strip. This gives you individual control over sounds and enables you to route sounds to separate outputs. To use a multi-output instrument:

1. **Click the right side of the instrument slot on the channel strip, and choose the Multi-Output option of the instrument on the menu.**

   The mixer channel strip is updated with a plus icon and a minus icon above the solo icon, as shown at the bottom left in Figure 17-3.

2. **Click the plus icon to add instrument outputs to the mixer.**

   New channel strips are added to the right of the first instrument channel strip. Continue clicking the plus icon to add the remaining instrument outputs. The number of outputs you have available depends on the software instrument.

3. **Click the center of the instrument slot to open the instrument interface.**
4. **In the instrument interface, route the sounds to the outputs in the mixer.**

   Each instrument handles routing differently. Some instruments, such as Drum Kit Designer, have sounds hard-wired to specific outputs. Others, such as Ultrabeat, are more flexible and allow you to choose how sounds are routed. Consult the instruments' manual (Help ➪ Logic Pro Instruments) for specific instrument instructions.

FIGURE 17-3: A multi-output instrument.

TIP

Because setting up multi-output instruments can be time-consuming, save your multi-output instrument setups as patches for later recall. For instructions on saving patches, see Chapter 4.

Using the output and master channel strips

Unless you have a channel strip's output set to No Output or an additional output that doesn't return into Logic Pro, all channel strips are summed to your stereo output channel strip and finally to your master output. Here's a description of your output and master channel strips:

- **Output:** Your stereo output behaves similarly to other channel strips, except you can't add send effects to it. You can use this channel strip to adjust the final level of your mix, as described in Chapter 19. The output channel strip has a bounce icon below the meter to export your project to a single file or burn it to a CD or DVD. For details on bouncing your project, see Chapter 20.
- **Master:** Your master output is stripped to the basics. Many professionals never touch the master fader during a project and instead use the output for overall volume control. You can adjust the fader level and even automate the fader, as described in Chapter 18. In addition to the mute icon, you have a dim icon, which allows you to adjust the playback volume according to the dim level, which is often used periodically so people in the studio can talk without stopping playback. To set the dim level, choose Logic Pro ➪ Settings ➪ Audio, click the General tab, and adjust the Dim Level slider.

Adjusting the EQ of Your Tracks

EQs (equalizers) are used to adjust the frequencies of your tracks. You might need to correct audio sources, such as microphones, recording environments, or instrument sounds. You can also use EQ to enhance the tone of a track and make it fit in the context of the entire mix. Several EQs are available, such as graphic EQ, parametric EQ, and low- or high-shelf EQ. In this section, you will learn how to use EQ to adjust the sound of your mix.

TIP

There's no right way to EQ your tracks. Different genres of music handle frequencies differently, and a wide range of acceptable EQ exists even within a genre. Because there's no perfect amount of bass or treble, if it sounds good to you and your listeners, it's good.

Adding Channel EQ

Logic Pro Channel EQ is a plug-in with a special place on the channel strip to display the EQ curve and quickly access the EQ controls. You can add the Channel EQ plug-in to a track in the following ways:

- **»** Double-click the EQ display on the channel strip. The Channel EQ plug-in is added to the first available insert effects slot, and the plug-in interface opens, as shown in Figure 17-4.
- **»** Choose EQ ⇨ Channel EQ from the channel strips effects slot, just as you would add any other insert effect.

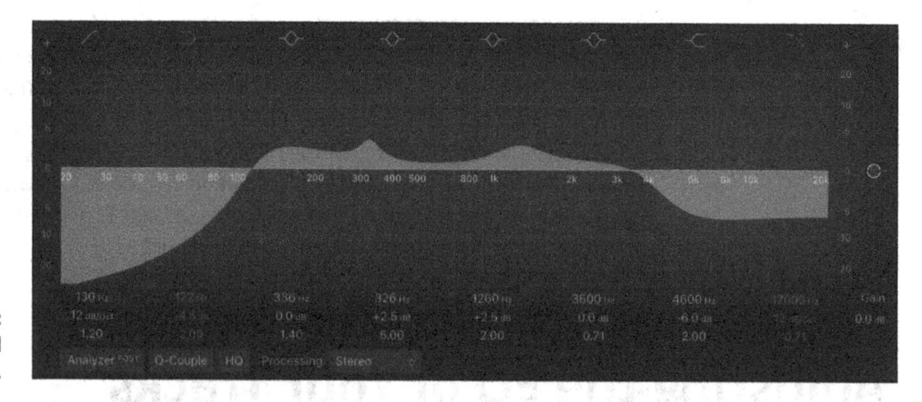

FIGURE 17-4: The Channel EQ plug-in.

Channel EQ is capable of adjusting eight frequency bands simultaneously. You can turn any of the bands on or off by clicking the colored buttons at the top of the interface. Bands 1 and 8 are low-pass and high-pass filters, respectively. Bands 2 and 7 are low-shelving and high-shelving filters, respectively. Bands 3 through 6 are parametric bell filters. You can adjust the boost or cut of each band by dragging the frequency handle in the graphic display, dragging the values of the fields below the graphic display, or double-clicking the fields and entering the value directly. Here's a description of the values:

- **» Frequency:** The top value adjusts the center frequency of the band.
- **» Gain/Slope:** The center value adjusts the level of boost or cut of the frequency band. In the case of the low-pass and high-pass filters, the value adjusts the slope of the filter.
- **» Q factor:** The bottom value adjusts the frequency range that's affected, known as resonance or Q factor. Higher values lead to narrower frequency ranges.

You can adjust the gain of the entire track with the Gain slider on the right side of the interface. Click the Analyzer button below the frequency display to turn the Channel EQ analyzer on or off. Inside the Analyzer button is a Pre/Post button that toggles the analyzer display of the frequency curve before and after the EQ. This feature is useful when you want to see both the frequency content of the track (pre) and how the EQ is affecting the track (post). The Q-Couple button connects the frequency gain with the Q factor so that the resonance band narrows as you increase the boost or cut of the frequency. The HQ button enables oversampling, which is useful when filters become too harsh at the higher end of the frequency spectrum. The Processing pop-up menu allows you to process both sides of a stereo signal, or the left only, right only, mid only, or side only.

Using Match EQ

Logic Pro comes with several EQ plug-ins, and one of the more magical EQs is Match EQ, shown in Figure 17-5. It allows you to grab the frequency spectrum of one audio track and apply it to another. You can use it on individual tracks, groups of tracks, or even your stereo output. Of course, it can't take a folk song and make it sound like a heavy metal song, but it has many practical uses, especially for beginners.

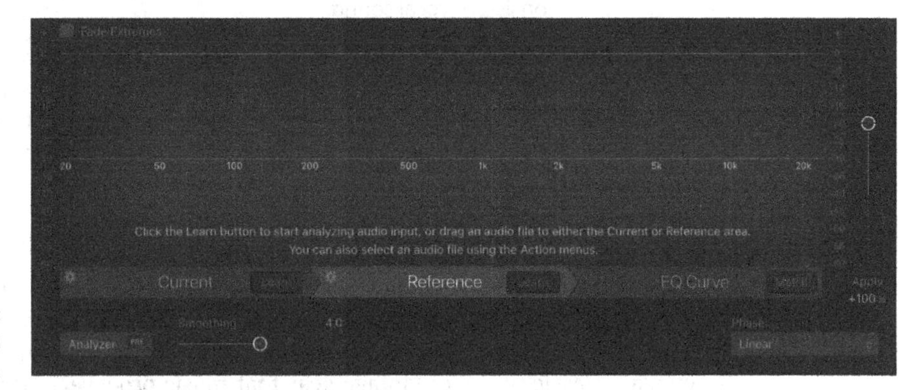

FIGURE 17-5: The Match EQ plug-in.

To match the EQ of one track to another, follow these steps:

1. **Add the Match EQ plug-in on the track to which the EQ is to be applied.**

   The track with the Match EQ plug-in is the track that will be modified by the frequency curve of a different track.

2. **On the Side Chain menu in the plug-in header, select the track to use as a template.**

3. **Click the Reference Learn button at the bottom of the Match EQ main display.**

   Learn mode turns on so you can create a template from the track selected on the Side Chain menu.

4. **Play the project until the frequency curve stabilizes in the graphic display.**

5. **Click the Reference Learn button to turn off learn mode.**

6. **Click the EQ Curve Match button.**

   The new Match EQ curve is added to the graphic display and applied to the current track.

You can adjust the frequency curve by dragging the curve in the graphic display. You can also adjust how much of the curve is applied to the track with the Apply slider on the right side of the interface. Drag the Smoothing slider to smooth the filter curve. Here are a couple more uses for Match EQ:

- **Match the EQ of vocals recorded on a different day.** The voice is a dynamic instrument and often changes from day to day, causing issues if you track a vocal on different days. You can use Match EQ to remove the differences and create a consistent vocal sound.
- **Reverse-match conflicting tracks.** If you have a track that's conflicting with another, drag the Apply slider to a negative value, which will invert the frequency curve and remove the conflicting frequencies. This trick is especially useful on bass guitar and kick drum, which frequently fight for space in the low end.

Logic Pro includes many other EQ plug-ins for you to explore. Channel EQ and Match EQ are transparent plug-ins that don't add a lot of color to your tracks. But if you want to add some vintage vibe to your project, play around with the Vintage EQ Collection located in the EQ menu of your Effects plug-ins. These plug-ins emulate classic pieces of vintage gear that music producers often turn to for an extra level of excitement and analog authenticity.

Adding Depth with Reverb and Delay

You accomplish the illusion of depth in your mix through the use of reverb and delay. Reverbs are designed to simulate acoustic spaces. Aside from using actual acoustic spaces, many different techniques have been used to create reverb effects. Logic Pro can re-create them all.

Mechanical and digital reverbs have distinct sounds and have become classic studio accessories. Plates and spring reverbs are classic mechanical reverbs. A *plate reverb* is a metal plate that vibrates when audio signals are sent into it. A pickup on the plate records the signal to get a shimmery reverb. *Spring reverbs* are common in guitar amps and have a spring that bounces and gets picked up and mixed into the output. An *impulse response* is a digital algorithm created by sending a sound into the space, recording what comes back, and analyzing the difference to replicate it. Impulse responses are realistic and more adjustable than an acoustic space.

If you want an intimate sound, it should be dry and up close, with less chance to bounce around. But if it's a huge sound you want, you might want it to bounce through an arena or echo across mountaintops.

Delays are reproductions of the original signal that are delayed. A tape delay uses electromagnetic tape to record the signal and play it back, adding its own character. Digital delays provide exact copies of the original sound. The delayed signal can also be altered with filters or other processing to add interest.

You can use short delays to create a double-tracking effect as if you had recorded the signal twice. Delays set to around 50 milliseconds can also simulate a small space. Sometimes, delays are preferable to reverbs for simulating acoustic spaces because they take up less room in the mix. For example, you can get the famous slapback echo of Elvis and others with delays between 65 and 150 milliseconds. You can also tempo match multiple delays (multi-tap) at different parts of the beat for interesting rhythmic effects.

The Haas effect, also known as the precedence effect, is a binaural psychoacoustic effect that makes your audio signal sound as if it's coming from outside the speakers. Add a 30-millisecond delay panned opposite to the original signal for ultra-wide stereo signals. This technique is useful when you're running out of room in the stereo spectrum of your mix. Some delays default to note values instead of milliseconds, but you can switch to time-based adjustments by turning off tempo sync in the plug-in.

Using Space Designer

The Space Designer plug-in, shown in Figure 17-6, is a convolution reverb. It *convolves*, or combines, the audio signal with an impulse response to simulate acoustic environments. Space Designer can create any space, such as inside a cabinet, a plastic or cardboard box, or an opera house. Space Designer also has a synthetic impulse response to create unique effects.

FIGURE 17-6: The Space Designer plug-in.

REMEMBER

You can access all of your plug-ins through the channel strip Audio FX slot. Click the effects slot to open the plug-in menu. The plug-in menu is organized by submenus, and you'll find additional reverb plug-ins in the Reverb submenu.

Space Designer is often used on auxiliary tracks and as an effects send because it is processor intensive. That way, you can send several instrument tracks to the same Space Designer plug-in on the auxiliary track.

Space Designer comes with many presets that you can access on the preset menu at the top of the plug-in. You can adjust the presets within the plug-in interface. Here's a description of the Space Designer interface:

- **Sampled IR/Synthesized IR buttons:** Click the Sampled IR button or Synthesized IR button at the top of the interface to switch between the impulse response samples and synthesized impulse response.
- **Display mode bar:** Below the IR buttons is the display mode bar, where you control what is shown in the main display and the parameter bar. You can choose between Volume Envelope, Filter Envelope, and Output EQ views.
- **Main display and parameter bar:** You edit the volume and filter envelopes or EQ curve in the center display and parameter bar. You change the view mode of the main display in the display mode bar at the top of the display.
- **Volume envelope, filter envelope, and output EQ parameters:** You can edit the envelopes and EQ in the main display or numerically in the parameter bar below the display. The parameters you can adjust depend on the view mode of the main display.

» **Impulse response and output parameters:** Below the main display and parameter bar are the impulse response and output parameters. Choose a value from the Quality pop-up menu to adjust the sample rate. The Input slider, to the left of the impulse response parameters, sets how to process stereo or surround signals. The impulse response settings include knobs to adjust the reverb pre-delay, length, and size. The output parameters allow you to adjust the stereo spread and crossover frequency of the impulse responses and the output levels.

Using ChromaVerb

The ChromaVerb plug-in, shown in Figure 17-7, uses a unique approach to create reverb. It emulates the gradual absorption of sound into specific types of rooms. ChromaVerb includes more than a dozen reverb algorithms based on common types of rooms.

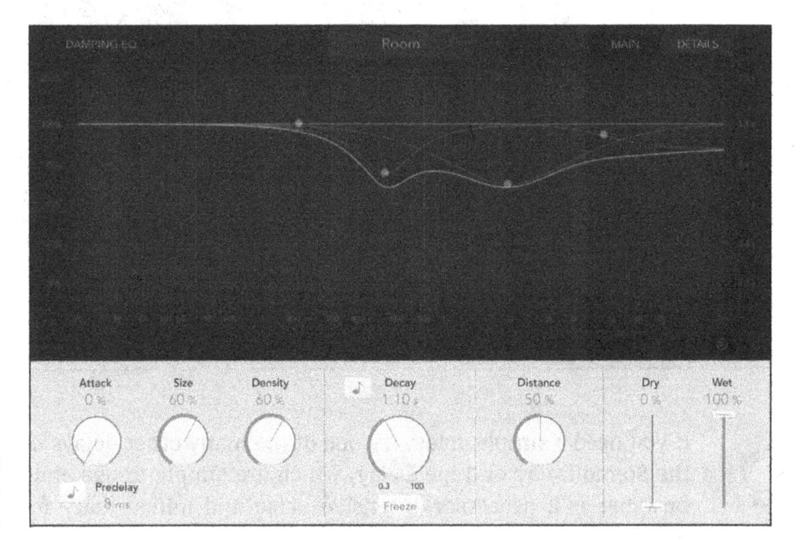

FIGURE 17-7: The ChromaVerb plug-in.

You choose the room type by using the pop-up menu at the top of the ChromaVerb interface. To the right of the pop-up menu are the Main/Details buttons, which you use to edit the reverb settings. The main window gives you access to common reverb parameters; the details window gives you more advanced parameters for shaping the reverb.

TIP

Use less reverb, if at all, on tracks with a lot of low end, such as bass guitar and kick drum tracks. Adding a lot of low end to your reverb can muddy your mix and leave less room for the instruments themselves. Instead, stick to using reverb on instruments that benefit from depth, such as vocals and snare. Not every track in your mix needs to have reverb. And because your listeners can perceive only two or three levels of depth at a time, you can create distinct groups of depth: close (dry), near (a little reverb), and far (the most reverb). Review the "Adding send effects" section in this chapter to learn more about creating reverb groups.

Using Delay Designer

The Delay Designer plug-in, shown in Figure 17-8, is a multi-tap delay with up to 26 individual taps (delays). In addition, each tap can have a distinct sound, such as changes in the volume level, pan, filter, and even the pitch of the delay. It's the most advanced delay that comes installed in Logic Pro.

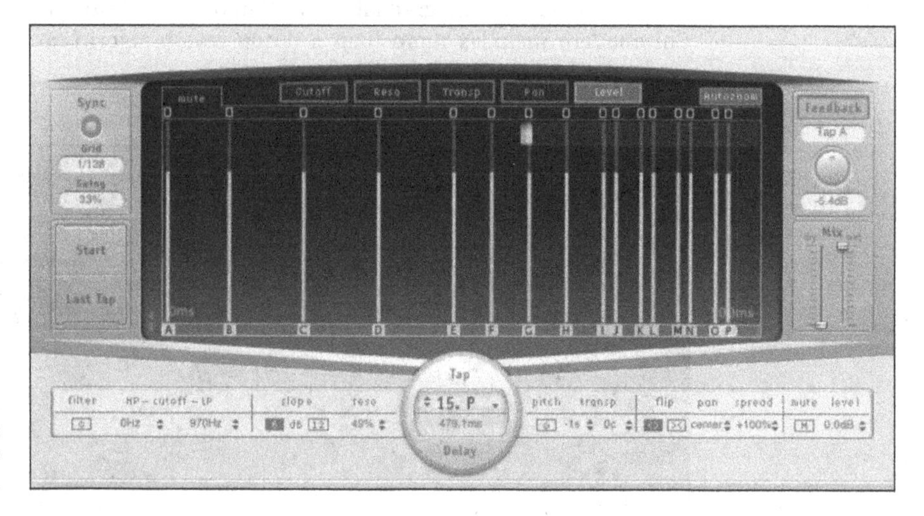

FIGURE 17-8: The Delay Designer plug-in.

TIP

If you need a simple delay, try one of the many other delays in Logic Pro, such as the Stereo Delay or Tape Delay, which are simple to use and sound great. Delay Designer is a processor-intensive delay and unnecessary for most basic delay needs.

Delay Designer comes with presets that include everything from basic to warped delays. The presets are great starting places for designing your own delays. Here's a description of the Delay Designer interface:

- **Sync section and tap pads:** On the left side of the interface, you can create taps in real time with the tap pads. Click the Start button to begin. The Start button changes to the Tap button, which you click every time you want to record a tap. Click the Last Tap button to end. Click the Sync button to sync

Delay Designer to the project tempo. Adjust the sync settings with the Grid resolution and Swing fields.

- **Main display:** In the middle of the interface is the main display, where you adjust the taps. Click the view mode buttons at the top of the display to change the displayed parameters. Click the toggle buttons at the top of each tap to turn tap parameters on or off. Each tap has a bright bar or dot that you can drag to change the level of the selected parameter. At the bottom of each tap is an identification letter that you can drag back and forth through time. You can copy taps by Option-dragging them.
- **Master section:** On the right side of the interface is the master section. The Feedback field sends the output of a single tap back through the effects to make the rhythm continuous. The Mix sliders set the wet/dry mix so you can use the effect as an insert effect as well as a send effect and still control the balance of the wet/dry signal.
- **Tap parameters:** The wide bar at the bottom of the interface controls the selected tap. You can turn the filter on or off, and set the frequency cutoff, resonance, and filter slope. To the right of the Tap Delay circle, you can turn the Pitch parameter on or off and set the transpose level. The Flip parameter sets the pan to its opposite. You can also set the stereo spread, mute, and output level for the selected tap.

TIP

Get the classic delay made famous by The Edge of U2 by using a dotted eighth-note delay. An eighth-note or a quarter-note delay can get lost behind all the other instruments playing on the beat, but a dotted eighth note allows the delay to peek out in the space between the other instruments.

Adding or Removing Dynamics with Compression

Controlling the dynamics of the overall mix as well as individual instruments is a crucial fundamental of mixing. A compressor is your main tool for controlling dynamics as well as for effect. Compressors work well on individual sounds, groups of instruments or sounds, and even the entire mix. There are many types of dynamic control, including limiting, multiband compression, de-essing, gating, envelope shaping, side chaining, and parallel compression. In this section, you learn how to use the dynamics tools of Logic Pro to create powerful and interesting mixes.

Using Compressor

The Logic Pro Compressor plug-in, shown in Figure 17-9, can emulate a wide variety of hardware compressors. Many compressors have become classics in the recording and mixing world, and there's a giant market for software emulations of these compressors. Unfortunately, Logic Pro doesn't advertise its compressor emulations, but savvy users know what they are, and you will too after you read this section.

FIGURE 17-9: The Compressor plug-in.

Like Channel EQ, Compressor has its own meter in the channel strip, right below the EQ display. The meter shows you the amount of gain reduction. You can double-click the meter to open Compressor. If it's not currently inserted on the channel strip, it's added to an open slot. Here's a description of Compressor parameters:

» **Circuit Type:** The Circuit Type menu is where you choose a compressor emulation:

- **Platinum Digital:** This compressor is Logic Pro's original compressor. It's a transparent compressor you can use on any audio source to control dynamics.
- **Studio VCA:** Studio VCA is similar to the Focusrite Red compressor. This compressor is colorful, with low distortion, and has a faster attack than FETs. Use it to protect a track from clipping and on instruments with complex harmonics, such as pianos, harpsichords, or 12-string guitars.

- **Studio FET:** Studio FET is similar to the UREI 1176 Rev E "Blackface" compressor. FET-style compression is warm with a fast attack time. It's often used on individual tracks and groups of tracks for both compression and color.
- **Classic VCA:** Classic VCA is similar to the dbx 160 compressor. It's a clean compressor used for low-distortion dynamic control and fast limiting.
- **Vintage VCA:** Vintage VCA is similar to the SSL compressor. This compressor has a fast response and a distinct sound, and is useful for protecting tracks from clipping. It's often used on bass, kick, and snare. The Vintage VCA is great for providing "glue" on groups of tracks or the entire mix.
- **Vintage FET:** Vintage FET is similar to the UREI 1176 Rev A "Bluestripe" compressor. This compressor can provide a lot of aggressive color to groups of instruments, such as drums, and individual instruments, such as bass and lead vocals.
- **Vintage Opto:** Vintage opto is similar to a Teletronix LA-2A. Electro-optical tube compression is smooth and rich, with a warm tone. It's often used on piano, bass, and vocals, as well as on instrument groups and the entire mix. (I encourage you to search the names of these famous compressors for audio examples while being mindful of the rabbit hole you're entering!)

» **Threshold:** Rotate the Threshold knob to adjust the level at which compression starts.

» **Ratio:** The Ratio knob sets the ratio of gain reduction when the threshold is reached.

» **Make Up:** Use the Make Up knob to raise or lower the gain after the signal has been compressed. You can choose to automatically raise the gain to −12 dB or 0 dB on the Auto Gain menu.

» **Knee:** The Knee knob adjusts how extreme the compression begins as the signal reaches the threshold. Low values compress harder, and higher values compress softer.

» **Attack:** The Attack time knob sets the reaction speed of the compressor.

» **Release:** The Release time knob sets the speed it takes for the compressor to release after the signal falls below the threshold. Click the Auto button to let the compressor choose the release time.

» **Limiter Threshold and button:** Click the Limiter button to turn on the limiter, which limits the signal from going beyond the level set by the limiter threshold.

» **Limiter controls:** You can choose the type of output distortion, which adds color to the compressor in varying degrees. You can adjust the output mix to blend the compressed signal with the original signal for parallel compression. Use the Output Gain knob to control the final signal level. You can also add a side-chain filter.

TIP

You can set Compressor to react to the dynamics in another track in the project. Choose the track that will trigger Compressor on the Side Chain menu at the top of the plug-in interface. Now Compressor will compress the audio only when the track chosen on the Side Chain menu reaches the threshold. This feature is useful when you want to make room for a track in the mix by automatically lowering any track that's competing with it.

Using Limiter

The Limiter plug-in, shown in Figure 17-10, is used when you need to restrict the level from going any higher. Limiters differ from compressors because they have a strict threshold, whereas compressors only reduce the level above the threshold. You might put a limiter on a single track to guarantee that it doesn't clip or on a group of tracks to raise the volume as high as possible before it distorts. Limiters can have a big effect on the sound, so a little goes a long way.

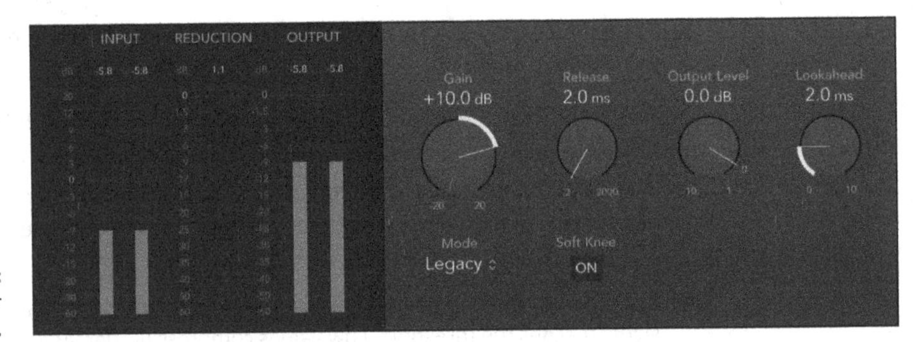

FIGURE 17-10: The Limiter plug-in.

Limiters are often added as the last insert effect in the chain. But sometimes, you can add a limiter before a compressor that's set to a slow attack time to catch any large peaks and avoid the pumping and breathing effect that can come from too much compression. A description of the Limiter parameters follows:

- **Gain:** Rotate the Gain knob to adjust the input level.
- **Release:** Rotate the Release knob to set the time it takes for the compressor to release after the signal falls below the threshold.
- **Output Level:** Rotate the Output Level knob to adjust the overall output level.

- **Lookahead:** Rotate the Lookahead knob to adjust how far in advance the limiter will analyze the signal. Higher levels cause latency, so it's best not to use this setting on instrument or group tracks because it will change the timing. Higher lookahead times work better when Limiter is on the main output.

Using other dynamics tools

Logic Pro comes with many dynamics tools that can make your time mixing easier. Dynamics tools can help you solve problems and be creative at the same time. As a beginner, it's a good idea to use the presets that come with these plug-ins. The preset names are educational, and you'll often find a preset that describes exactly what you want to do. Following are other dynamics tools you can use:

- **De-esser:** Remove hiss from vocal sibilance. This frequency-dependent compressor can lower specific frequency ranges, such as those that cluster around sibilant sounds.
- **Expander:** Expand the dynamic range. For example, you can use an expander to reduce the amount of hi-hat leakage in a snare track by expanding the distance between the main snare sound and the background hi-hat.
- **Noise gate:** Lower the level of sounds below the threshold. The noise gate enables you to remove unwanted room noise, such as amp hump.
- **Enveloper:** Adjust the attack and release of a sound's transients. An enveloper is a tone-shaping tool that works well with instruments having sharp transients, such as drums, picked and plucked sounds, and pianos.

Third-party plug-ins are a great way to expand the capabilities of your music production in Logic Pro. Plug-ins can range from effects to virtual instruments, allowing you to customize your sound with a vast array of tools. When working with third-party plug-ins, be aware of the following pitfalls:

- Always make sure that you download and install plug-ins from reputable sources to ensure their security and compatibility with your system.
- Make sure that the plug-in is compatible with the version of Logic Pro you're using; some older plug-ins may not function properly in newer versions.
- Remember to bounce tracks that use third-party plug-ins, which you learn about in Chapter 20, to ensure that your tracks aren't lost to a plug-in that becomes abandoned or incompatible with a newer version of Logic Pro.

In this chapter, I touched on most of the important Logic Pro plug-ins, but there are others you'll want to explore, such as distortions, filters, multi-effects, and modulation plug-ins. I encourage you to master the plug-ins you have before you spend money on third-party effects. Logic has everything you need to make a great mix. And as your mix nears completion, you'll want to head over to Chapter 18 to learn how to automate your mix so it's dynamic and exciting.

IN THIS CHAPTER

» Planning your automation

» Adding and editing automation

» Automating MIDI events

» Recording live automation

Chapter 18
Automating Your Mix

Imagine The Beatles standing around a mixing board with Sir George Martin, each lending a hand to pull faders up and down without stopping, hoping to get the perfect mix take. That's how mixing was done before mechanical faders brought hardware automation to modern mixing boards.

With software automation, you have even greater and more flexible control over your mix. Automation can help you first stabilize your mix and then design an exciting mix. In this chapter, you learn how to add interest to your mix and automate your own mix blueprint.

To get all that you can from this chapter, you must have Enable Complete Features selected in Advanced Settings. Choose Logic Pro ⇨ Settings ⇨ Advanced and select Enable Complete Features.

Turning Your Mix into a Performance with Automation

Every track, channel strip, and plug-in is capable of being automated. However, automation is best to add after the mix is stable. For example, if you're still arranging or editing your project, having automation on a track can get in the way of your workflow because you have more things to focus on as you edit. And if your mix isn't stable, your mixing affects the automation as well. For these reasons, it's a good idea to use project and track notes to jot down any automation ideas you may want to try later.

After your mix is stable, take a pass through the project, focusing on each of the core mixing concepts described in Chapter 16. Then, automate the levels, panning, EQ changes, effects, and dynamics. If you plan your automation this way, the final result will be an elevated and exciting mix.

Choosing Your Automation Mode

To automate a track, you need to display the track's automation in the tracks area. Choose Mix ⇨ Show Automation or press A. The track headers will display automation parameters, as shown in Figure 18-1. To enable automation for a track, click the left side of the automation button in the track header. Using the automation button, you can choose between two types of automation:

- **Track:** When track-based automation is enabled, you can view and edit automation on the entire track.
- **Region:** When region-based automation is enabled, you can view and edit automation on the track's regions.

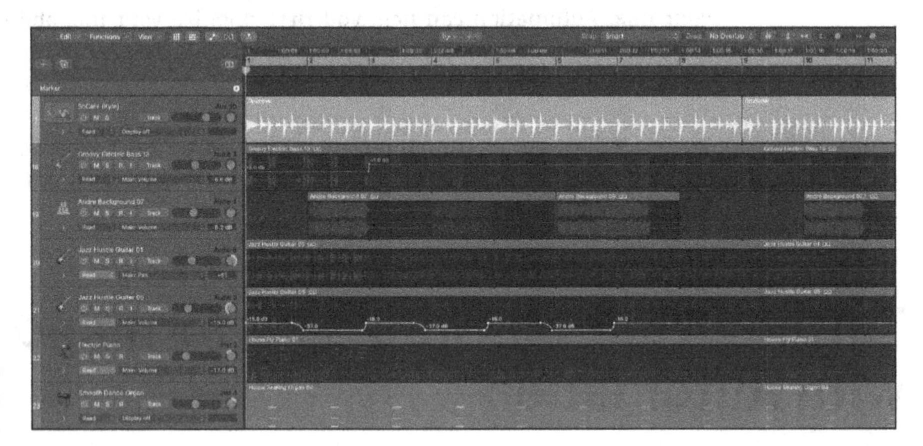

FIGURE 18-1: Track automation in the tracks area.

After you've enabled automation on a track, you can choose between the following automation modes on the menu:

- **Read:** Automation is played back but not recorded.
- **Touch:** Automation is recorded during playback while the parameter is being touched; when the parameter is released, it returns to the previous value. Touch mode is useful when you're automating a temporary change in the parameter that you want to return to the previous value.
- **Latch:** Automation is recorded during playback while the parameter is being touched; after the parameter is released, automation stops at that last touched value. Latch mode is useful when you're automating a parameter to a new value that you want to keep.
- **Write:** Automation is recorded during playback, and existing automation is erased as the playhead passes over it. Write mode is useful when you want to record and erase automation simultaneously.
- **Trim:** Automation values are offset by the amount you move the control.
- **Relative:** A secondary automation curve is added to offset the primary automation curve.

TIP

You can also turn on automation and select the automation mode from the automation mode pop-up menu in the mixer.

Adding Automation to Your Tracks

After you enable automation, adding it to your track is as simple as clicking a location on a region with the pointer tool. A control point is added with the current parameter's value, as shown in Figure 18-2.

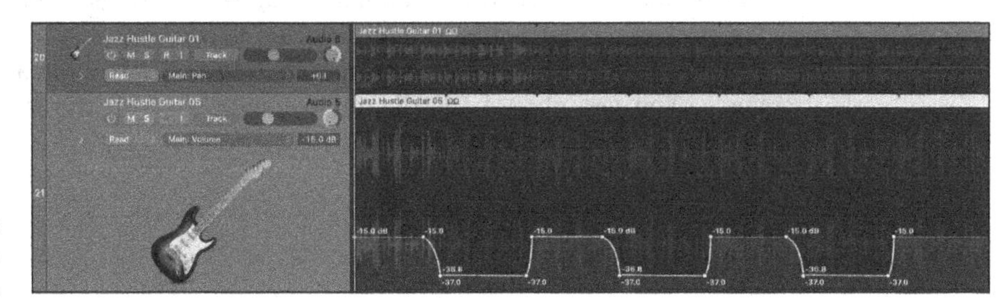

FIGURE 18-2: Track automation control points.

The parameter you automate is chosen in the automation parameter drop-down menu on the track header. You can display multiple automation lanes by clicking the disclosure triangle to the left of the automation parameters. You can quickly add an automation point at the playhead position for the volume, pan, and sends by choosing Mix ➪ Create Automation ➪ Create 1 Automation Point Each for Volume, Pan, Sends.

TIP

If you know that the parameter value you're automating will return to its original value, you'll save time by creating an automation point at the beginning and end of regions. Choose Mix ➪ Create Automation ➪ Create 1 Automation Point at Region Borders.

Adjusting automation points

The fastest way to adjust automation points is to select them with the pointer tool and drag them up, down, left, or right. You can select multiple points by Shift-clicking them. You can delete selected points by pressing Delete.

To create smooth curves between automation points, use the automation curve tool. Drag left, right, up, or down on the automation line with the automation curve tool to create automation curves. You can temporarily switch from the pointer tool to the automation curve tool by holding down Shift-Control while dragging the automation curve.

Moving regions with or without automation

You may end up needing to make an edit to a track or your song arrangement after you've started automating your mix. Moving regions with automation can cause complications if you don't do it right. That's why Logic Pro's default state is to ask you whether you really want to move regions that have automation. You can change this behavior in the automation system settings, shown in Figure 18-3. Choose Logic Pro ➪ Settings ➪ Automation to open the automation settings, and then choose the default behavior from the Move Track Automation with Regions menu. You can choose the default behavior also on the Mix ➪ Move Track Automation with Regions menu.

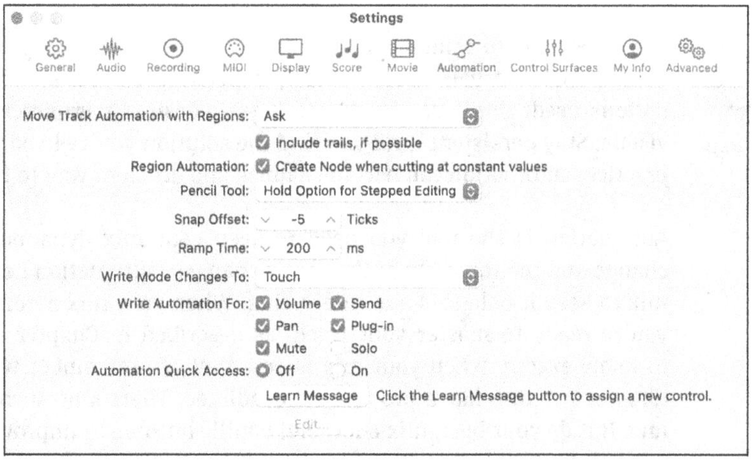

FIGURE 18-3: The automation settings.

Recording Live Automation

The most precise way to automate your mix is by inputting automation data in the tracks area, as you learn in this chapter. But when automation was created on hardware mixing consoles, the automation was performed in real time as the project was playing. Recording live automation is an enjoyable experience because you get to perform along with your track, and perhaps create some magic along the way. Here's how to record live automation:

1. **Choose an automation mode on the tracks you want to automate.**

   You can choose Touch, Latch, or Write, as described in the "Choosing Your Automation Mode" section previously in this chapter.

2. **Play your project and adjust the parameters you want to automate.**

   Automation data is added to the tracks.

3. **When you've finished automating, stop the project and set the track's automation mode to Read.**

TIP

To speed up the automation process, you can combine it with smart controls. You can use the smart controls to record live automation and also automate the smart controls from the tracks area. Because smart controls can control more than one parameter at a time, you can build some dynamic mixes quickly.

REMEMBER

It's important to remember that different tracks will have their own available parameters that you can automate, depending on the insert effects and other options used. Don't get frustrated if your desired parameter isn't immediately visible. Stay persistent until you find the solution you're looking for. With a bit of practice, automation can become a quick and efficient way to fine-tune your mix.

Automation is the tool you need to keep your mix dynamic. As song sections change and the intensity of the music changes, automation helps you adjust your mix to keep it balanced and interesting. When your mix automation is complete, you're ready to master your track, as described in Chapter 19. It may be hard to know exactly when your mix is complete, so remember what you learned in Chapter 1: Think like a pro and set deadlines. There's no such thing as a perfect mix. But do your best, mix a lot, and you'll continue to improve.

IN THIS CHAPTER

» Enhancing the mix with EQ

» Controlling dynamics with compression

» Reaching competitive loudness with limiting

» Using audio references to improve your mix

Chapter 19

Mastering Your Final Track

After a track has been mixed, a mastering stage puts the final polish on the track. The level is matched to reference levels, and if needed, EQ and dynamics processing is applied. In the case of an album project, the mastering stage can include making global refinements, song sequencing, and cataloging.

For professional situations, the best practice is to hire someone to master your track. Someone with a fresh set of ears and nothing but the best equipment can raise the level of your mix and turn it into a golden master. Sometimes, just sending a signal through a high-end compressor or EQ can improve the sound. Other times, a mastering engineer will need to adjust the sound to compensate for a poorly tuned room. Then, if the project warrants it, let go and let your work be improved.

However, not all projects need outside mastering. Sometimes it's not practical to pay for a mastering service. You might be under a deadline that forces you to do it yourself, or you might want to avoid extra expenses when the project budget is low. Logic Pro gives you the tools you need to master your own music. You can even be the fresh set of ears that masters someone else's mix.

In this chapter, you discover how to enhance your mix, boost the volume to competitive levels, and support your mastering choices with reference audio.

Fine-Tuning EQ

Ask mastering engineers if they put EQ before or after compression, and they'll tell you that it depends, leaving you more confused than before. I think it's safe to say both alternatives can work just fine. If you're making only subtle changes, the EQ and compressor shouldn't affect each other too much. If your EQ is inserted before the compressor and you make extreme EQ changes, the compressor's behavior can be affected. In that case, you might want to put the EQ after the compressor. Sometimes, cutting unwanted frequencies before the compressor can help you do a better job of compressing. Beginners should put the EQ before the compressor when they master their track and keep the changes subtle.

TIP

Your mastering effects chain is commonly used on the stereo output channel strip. That way, all the audio in your project will be affected by the plug-ins. However, suppose you've imported multiple audio tracks into your project for mastering, as in the case of an entire album. In that case, you may want separate control over each track's mastering effects chain. In this case, you would put the effects on the individual track's channel strip, not the stereo output. More details are given in the "Matching levels to other recorded material" section later in the chapter.

Using Linear Phase EQ

Linear Phase EQ, shown in Figure 19-1, has the same controls as Channel EQ, which is described in Chapter 17. Linear Phase EQ works best in mastering situations because it keeps the stereo audio in phase even under extreme levels of gain, which requires more processing power. Linear Phase EQ doesn't work well on single tracks because it introduces latency that can affect the timing of the signal relative to the other tracks. The higher processing and latency make this EQ ideal for mastering single stereo tracks.

TIP

Linear Phase EQ shares settings with Channel EQ. You can switch between the two plug-ins, and they will retain the current settings.

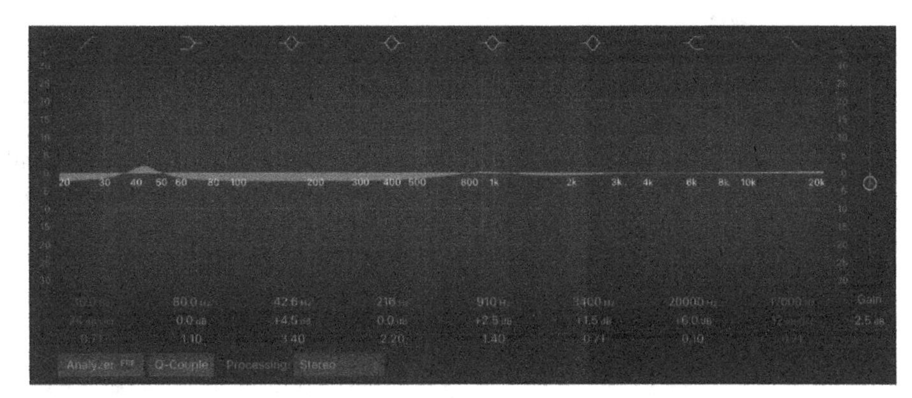

FIGURE 19-1: The Linear Phase EQ plug-in.

Matching a reference mix with Match EQ

Match EQ (described in Chapter 17) is a great tool to help you get a specific sound. You can manipulate your audio only so far after your mix is completed, but the mastering phase is a great time to analyze the frequency content of your mix and compare it to a reference mix. You can even adjust your mix to capture more frequencies similar to the reference mix. To analyze a reference audio file using Match EQ, follow these steps:

1. **Drag the reference audio file onto the Reference Learn button.**

   The audio is analyzed, and a frequency curve is added to the graphic display, as shown in Figure 19-2.

2. **Drag the current audio file that's being mastered onto the Current Learn button.**

   The audio is analyzed, and a frequency curve is added to the graphic display.

3. **Click the EQ Curve Match button to apply a Match EQ curve to the audio.**
4. **Drag the Apply slider to adjust the level of the Match EQ.**

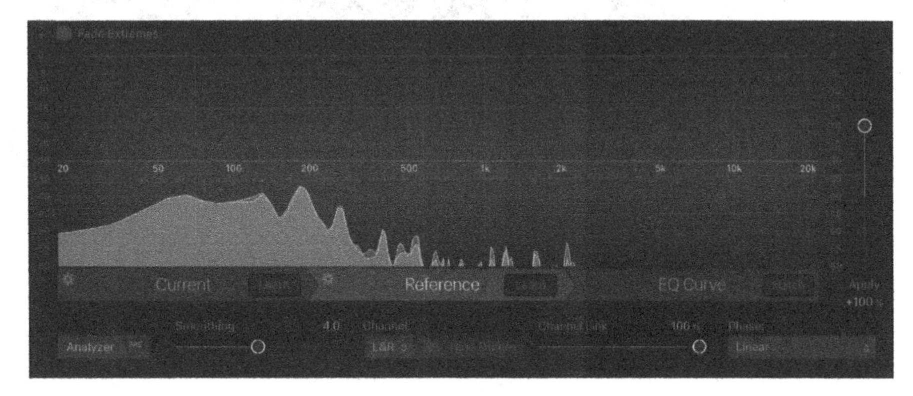

FIGURE 19-2: The Match EQ plug-in.

TIP

Don't expect miracles with Match EQ. I've heard it produce excellent and horrible results. But, at the very least, the graphic display will enlighten you about the difference in EQ between the audio you're mastering and the reference audio. But remember that Match EQ can change the gain of your track, and you may need to make adjustments. And because Match EQ is nondestructive and you can always undo your work, you can feel free to experiment with it.

Adding Multiband Compression

Multiband compression allows you to separate frequency bands and compress them independently. (For the details on compression, see Chapter 17.) Logic Pro has a plug-in called Multipressor, shown in Figure 19-3, which can compress up to four different frequency bands.

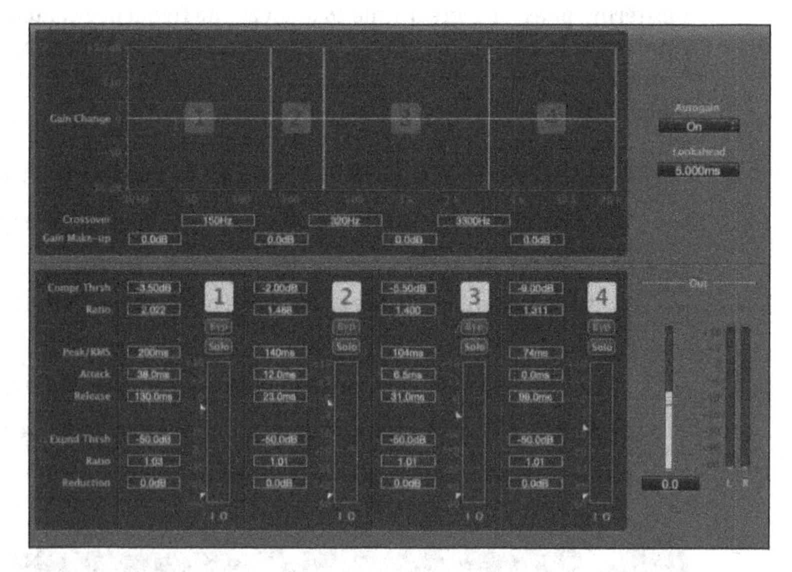

FIGURE 19-3: The Multipressor plug-in.

Multiband compression can produce subtle or extreme results. Your main goal is to raise the overall level without negatively affecting the balance of the mix. You also want the volume level to match the music in your marketplace. Multipressor has many presets that will help you achieve these goals.

Using Multipressor

You can insert Multipressor on a channel strip by clicking an effects slot and choosing Dynamics ➪ Multipressor. Many of its parameters are similar to the

Compressor plug-in described in Chapter 16. Here's a description of the parameters that are unique to the Multipressor plug-in:

- **Gain Change:** Drag the horizontal bars in the graphic display up or down to raise or lower, respectively, the gain of that particular frequency band. You can adjust the gain also by using the Gain Make-Up fields. You raise the gain after a signal has been compressed because, as you reduce the level of the peaks via compression, you can raise the overall level to get the signal louder.
- **Crossover:** Drag the vertical lines in the graphic display left or right to lower or raise, respectively, the crossover frequency. You can also enter values directly in the crossover fields. It might be easier to start with fewer bands in the beginning. You can turn the frequency bands on and off by clicking their number button above each level meter.
- **Expand Threshold, Ratio, and Reduction:** Multipressor has a built-in expander, which helps you keep some of the dynamics you may be losing due to compression. The gain of the signal below the expander threshold is reduced based on the ratio. The expander takes the quiet signals and makes them quieter.
- **Band Bypass and Solo:** Click the bypass and solo buttons above each band's level meter to hear precisely how the compressor is affecting the frequency band. These buttons are excellent ear-training tools. Soloing and bypassing bands helps you identify where instruments have the most and the least frequency energy.

Avoiding a squashed mix

Dynamic mixes need transients. The crack of a snare and the pluck of a string have sharp and fast attacks with fast decays. If you lower the attack and raise the decay, you've compressed the signal. Compressors have the potential to remove all the transients and squash the sound, making it weak and muffled. To preserve the transients, aim for subtle compression during mastering.

Turn It Up

It's easy to get seduced into thinking that louder is better. But there's clearly a limit to the loudness that digital audio can reproduce. So how loud is too loud? A trained ear listens for distortion, the loss of transients and dynamics, and unfavorable effects on the balance. Remember that your goal isn't to get your track louder than everyone else's; it's to get it within the competitive range of your marketplace.

The kind of audio material you're mastering will dictate your loudness needs. A track being delivered and sold on iTunes should match other tracks in the genre. A track delivered to a film or TV music supervisor for background purposes can withstand lower levels and less compression because its job is to not compete with the foreground audio. This section reveals how to maximize and match the level to an ideal reference source.

REMEMBER

The Loudness Meter, which you learn about in Chapter 16, can help you measure the loudness of your entire mix by using the LUFS standard. The goal of LUFS is to normalize the loudness of audio for broadcasting instead of rewarding mixes that are created louder merely for the purpose of gaining attention through volume. Television commercials are the primary culprits; however, *Loudness Wars* refers to a practice in the music industry of making mixes sound louder than their competitors while sacrificing sound quality. Instead, aim to give the listener a consistent and enjoyable listening experience.

Using Adaptive Limiter for maximum loudness

The last effects plug-in to put in your mastering signal chain is Logic Pro's Adaptive Limiter, shown in Figure 19-4. It can raise the overall level of a song and add an analog-sounding saturation to your mix. If your mix is already dirty with lots of distorted instruments, you might go easy on adding any more distortion. But if your mix is relatively clean, a little distortion can excite the mix while smoothing the edges.

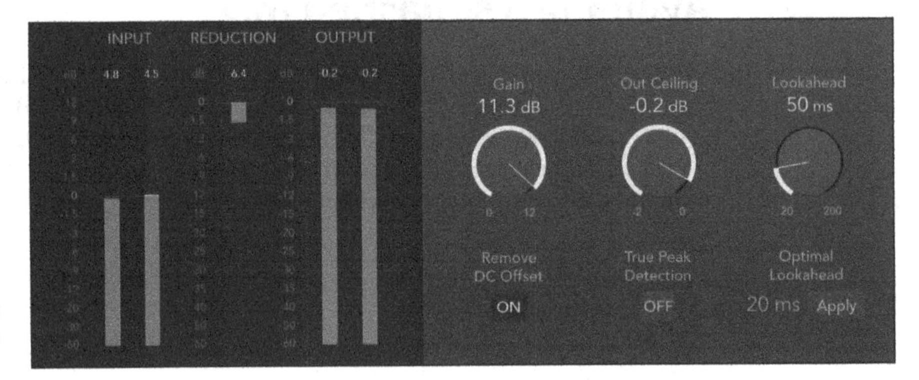

FIGURE 19-4: The Adaptive Limiter plug-in.

Adaptive Limiter is located on the Compressor menu of the effects slot. Here's a description of the Adaptive Limiter controls:

- **Gain:** Use the Gain knob to get your signal the rest of the way to your target level.
- **Out Ceiling:** Set your target peak level with the Out Ceiling knob. For popular music, this level is often just below 0 dBFS, for example, -0.02 dBFS.

Matching levels to other recorded material

To gauge whether you're getting the final level in the same ballpark as your marketplace, it's a good idea to import a reference mix into your project and compare it with your mastered track. (You learn how to import media into your project in Chapter 8.) It's a good idea to keep a playlist of mixes that you know and love for reference as you work. If you're working for a client, ask them to provide reference tracks so you can better achieve the client's vision for the project.

If you've been putting your mastering effects on the stereo output, the plug-ins will affect all the audio you import. You can solve this in the following ways:

- Move the plug-ins on your output to the channel strip of the track you're mastering.
- Move the plug-ins on your output to an auxiliary channel strip, and then route the output of your track to the auxiliary channel strip, as you learn in Chapter 17.

After you've moved your mastering plug-ins from the output, here's how to A/B between your track and the reference track:

1. **Mute the reference track.**

   You only hear your mastered track while the reference track is muted.

2. **Option-click either track's solo icon to A/B between the two tracks.**

   Option-clicking the solo icon turns solo off on all other tracks, allowing you to hear only one track at a time.

TIP

You can use your ears to judge whether the volume levels of your track and the reference track match, or you can measure with an SPL meter. If you've calibrated your listening position with an SPL meter, as you learn in Chapter 16, your ear should be a good judge of the levels between the two tracks. Having a meter on hand will help you build confidence in your ability to judge volume. And don't forget about the Loudness Meter plug-in, which I describe in Chapter 16. It's almost always the last plug-in on my mastering effects chain.

You put the final polish on your track in the mastering phase. Your project is almost complete. Now is the time to render your project and introduce it to an audience. You learn how to bounce and share your project in the next chapter.

IN THIS CHAPTER

» Bouncing full mixes to final audio files

» Backing up and archiving your projects

» Creating alternate mixes and music stems

» Sharing your music

Chapter 20
Bouncing and Sharing Your Music

I think all Logic Pro users anticipate the moment when they can turn their full mix into a format they can share with the world. Seeing a project through to the end takes patience and persistence, and you go through so many emotions during the process that the project ends up holding a lot of personal value. Rendering your full mix to a single audio file (called *bouncing*) is an exciting moment because you make your song real and ready for the world.

You have many reasons to want to bounce your project as soon as possible. You can deliver it to a DJ. You can test it in your car. You can put it on your website. And you can celebrate your creation. In this chapter, you render the final audio file and share it with me so I can help you celebrate being a music producer.

Bouncing Your Project

You usually end up bouncing several times during the course of a project. For example, you might want to listen to your song on the go or share it with someone for feedback. And at some point, you're going to bounce the project because it's complete. In this section, you learn how to render your project into an audio file that you can listen to in software such as Apple's Music app.

TIP

Back up your projects. It's an awful feeling to lose your precious data due to a hard drive crash or, even worse, a stolen computer. You can lessen the pain by backing up your data so you can get back to work.

Apple's Time Machine software, which is installed on your Mac operating system, can get you started backing up. It will back up your entire computer at regular intervals to an external hard drive. In addition, you can back up your data offsite to inexpensive storage solutions such as Amazon S3 or Google Drive. Finally, you can sync your current projects between computers and the cloud using services such as Apple's iCloud. And if you're like me, obsessive at backing up data, you can keep a small external hard drive handy and manually back up your current projects. Then one day, you will be thankful that your data is backed up.

Recording external instruments before you bounce

Before you can bounce your project into a single audio file, or even multiple audio files for archival, you need to record all your external MIDI instruments to audio tracks, as described in Chapter 6. External MIDI instruments are controlled by the MIDI tracks in your project, but the audio of the external instrument needs to be captured in your project on an audio track.

Capturing your software instrument tracks as audio is also a good idea, because you never know when a third-party software instrument will become obsolete. If you want to future-proof your project, bounce all your software instruments to audio tracks. To bounce a software instrument track, follow these steps:

1. **Select a software instrument track.**
2. **Choose File ⇨ Bounce ⇨ Track in Place, or press Control-⌘-B.**

   The Bounce Track In Place dialog opens, as shown in Figure 20-1.
3. **Name the file and select your options.**

   To retain the track's current sound and mix automation, deselect Bypass Effect Plug-ins and select Include Volume/Pan Automation (the default settings). If you want to bounce the raw track, as you would for raw archival, or share with other mix engineers who plan on adding their own effects, do the opposite and select Bypass Effect Plug-ins and deselect Include Volume/Pan Automation.

4. **Click OK.**

   Your software instrument track is bounced to audio and added to a track in the project.

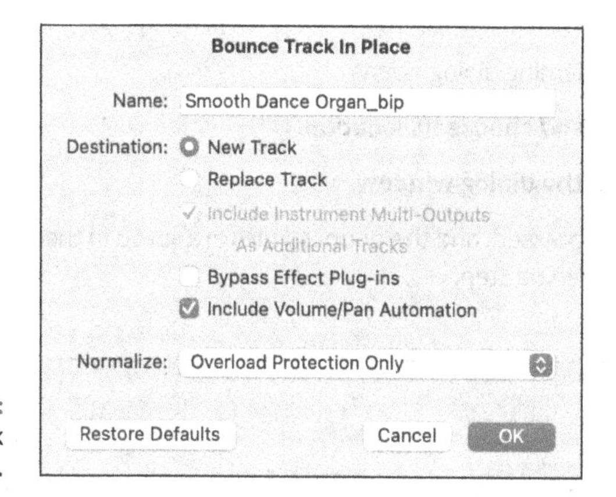

FIGURE 20-1: The Bounce Track in Place dialog.

Bouncing to an audio file

Bouncing is an exciting step. You've been building the house piece by piece, and now you finally get to observe it as a whole. Even though the file will sound the same as your mix in the project, bouncing is an essential final step that symbolizes your commitment to the project. You can bounce your project in the following ways:

1. **Select the length of the section to be bounced by selecting regions, setting the cycle mode, or deselecting regions to use the project end point.**

   If cycle mode is off and a region is not selected, Logic Pro defaults to bouncing the length of the entire project. If cycle mode is on, Logic Pro will use the cycle length to determine the bounce length. If cycle mode is off and a region is selected, the bounce length defaults to the length of the selected region. I like to determine the bounce length by the cycle length. That method allows me to precisely define the start and end points to remove empty space in the beginning and capture any reverb or delay tails at the end.

2. **Start the bounce in either of the following ways:**

   - Click the bounce icon at the bottom of the Stereo Output channel strip.
   - Choose File ➪ Bounce ➪ Project or Section, or press ⌘-B.

The Bounce window opens, as shown in Figure 20-2. You can select the destination file type, file type parameters, and bounce mode, and you can refine the start and end times. Normalizing raises the level of your track to 0 dBFS, so it's best to leave it off.

3. **Click Bounce.**

   The bounce file-naming dialog opens.

4. **Name your file and choose its location.**

5. **Click Bounce in the dialog window.**

   The bounce is processed, and the bounced files are saved in the location you chose in the preceding step.

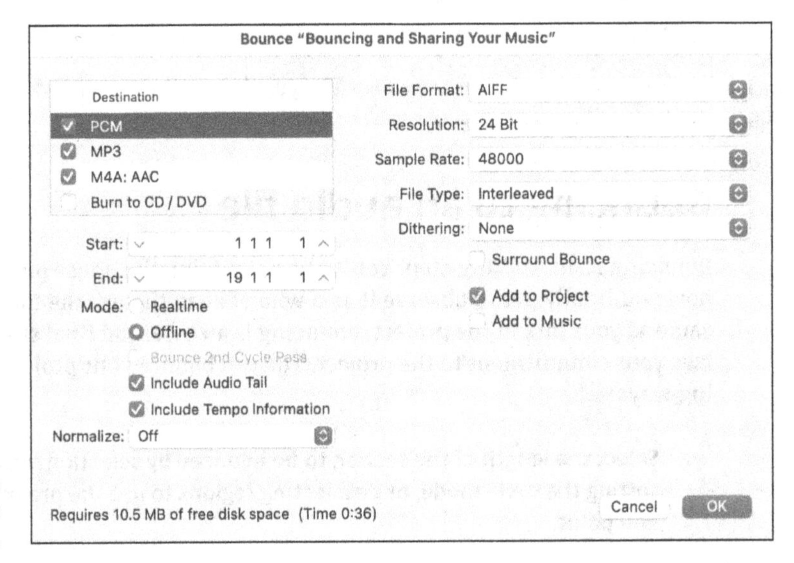

FIGURE 20-2: The Bounce window.

TIP

If you're completely done with the project, it's a good idea to export all your tracks as audio files. Audio files aren't going away any time in the distant future. Third-party plug-ins and software change regularly, so you never know whether your favorite third-party plug-in will work in a later version of Logic Pro. Choose File ⇨ Export ⇨ All Tracks as Audio Files or press Shift-⌘-E. In the dialog that appears, choose the file location, file type, and file settings. Click Save, and every track will have a separate audio file. You could also give the resulting audio to a Pro Tools user, who could easily import your audio and begin mixing immediately.

Creating Stems and Alternate Mixes

Bouncing several different versions of your final mix is a common practice. Clients have different needs, and alternate versions of mixes can cover all your bases. Sometimes, artists need versions without the vocal for performing live to backing tracks. You may even need to bounce groups of tracks separately, creating a stem mix. Whatever your needs, bouncing alternate mixes and stems is easy using Logic Pro.

Bouncing a stem mix

Stem mixes have many uses. You can give a remix artist the stems of the mix for remixing and DJ gigs. Film scores and surround sound projects often use music stems. Mixing and mastering engineers might want stems instead of individual tracks or bounced stereo tracks.

The fastest way to create a stem is to solo the tracks you want to include in the stem and bounce as described previously. Name and catalog your stems clearly. That way, they're easy to recognize, and you won't accidentally leave out individual tracks.

The stems you create will depend on your project and collaborators. For example, a remixer might want separate vocals, drums, and musical instruments. A film score might want to separate the orchestra from electronic instruments or sound effects. If stems are required, you can meet the demand.

TIP

You can also use track stacks to build and organize your stems. Using track stacks makes it easy to solo or mute the instruments in the track stack for bouncing the stems.

Bouncing alternate mixes

When it comes to mixing popular music, you can make several standard alternate mixes. And it's probably best to make them even if no one asks for them because making an alternate mix takes little time, and you may like one of the alternates better. Here's a list of standard alternate mixes:

- **Without vocals:** This mix can be used when a performer needs to sing to a backing track. TV appearances often use live vocals with backing tracks instead of live performers to guarantee great sound. An alternate version of this mix is to leave the background vocals in, known as the TV mix. This type of mix is also helpful if your song is used for karaoke.

- **Vocals only:** Remixers and DJs love this mix because they can provide their own beats and music under the lead vocal track. It can also be mixed with the vocal-free backing track version when film and TV editors need independent control of the vocal level because it coincides with dialog.
- **Vocal up/vocal down:** These mixes involve raising and lowering the overall level of the lead vocal by 1 or 2 dB. The change in level is subtle but often necessary to please the discerning ears of your clients.
- **Without solo:** If the song has a solo instrument or solo section, you might treat it as a lead vocal and provide versions with and without the solo instrument.

Sharing Your Music

You've poured your heart and soul into Logic Pro, and now that you're done, what do you do? Please share it with your friends, including me! I love listening to new artists. There's an audience out there who hasn't discovered your music and wants to hear it. In this section, you discover how to get your music heard.

Sharing your project to Apple's Music app

If you bounced your project files already, you can open them in Apple's Music app from Finder. But you can also bounce and share your project directly with the Music app, so you can sync your project to your iOS devices and play your tracks wherever you go. First, choose File ➪ Share ➪ Song to Music to display the Share to Music window, shown in Figure 20-3. Next, name your project and set the tags, choose the quality level and bounce length, and click Share. The bounce is processed and imported to Music, ready to play and sync.

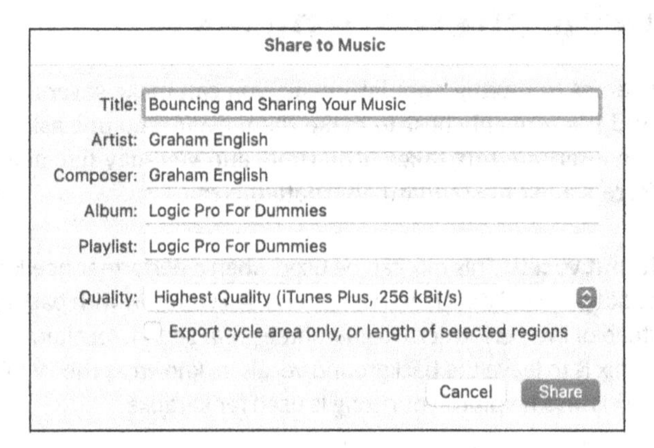

FIGURE 20-3: The Share to Music window.

Sharing your project to SoundCloud

SoundCloud is an online service and social network for posting and sharing audio. The service is free, and you can set up and begin sharing your music quickly. Choose File ➪ Share ➪ Song to SoundCloud. The first time you visit the site, you're asked to enter your username and password, so you must be connected to the internet.

When you're logged in, the Share to SoundCloud window opens, as shown in Figure 20-4. Enter your details and click Share. Your audio is processed on SoundCloud's servers and added to your profile, ready to share in minutes.

FIGURE 20-4: The Share to SoundCloud window.

Additionally, you can share your Logic Pro project using AirDrop, Apple Mail, and GarageBand for iOS. When you share to GarageBand for iOS, you get a single track that contains a bounce of the entire project so you can continue to work on your iOS device. Reopen your project in Logic Pro and watch as the new tracks are seamlessly added to the original project.

I hope you're inspired to make more music and excited to share it with all the eager listeners of the world. And I hope you count me as one of your eager listeners. So stay in touch and share your tracks with me at `https://grahamenglish.com`.

6

The Part of Tens

IN THIS PART . . .

Use an iPad to write songs, play and record software instruments, edit and mix, and navigate Logic Pro.

Speed up your workflow, use key commands and screensets, save time with presets, and troubleshoot Logic Pro.

IN THIS CHAPTER

» Playing and recording software instruments

» Editing and mixing with your iPad

» Navigating and commanding Logic Pro

» Writing songs on your iPad

Chapter 21

Ten Ways to Use an iPad with Logic Pro

Logic Remote is a free app for iOS devices running iOS 11 or higher. With Logic Remote, you can remotely control three Apple music apps on your Mac: Logic Pro, GarageBand, and MainStage. You connect Logic Remote to your music app with Wi-Fi or Bluetooth. Download Logic Remote at `https://logicstudiotraining.com/logicremote`.

When you open Logic Remote, it attempts to connect to the software running on your Mac, so you must have Logic Pro, GarageBand, or MainStage running. Otherwise, Logic Remote will shut down. If you have more than one of these applications running, you're asked to choose the application you want to control. After you're connected, you can control the application using your iPad or iPhone.

In this chapter, you learn how to use your iOS device to play software instruments, edit your tracks, mix your project, and write songs.

Playing Keys

The first thing you should check out in the Logic Remote app is the control bar at the top of the interface, shown in Figure 21-1. The center display shows the playhead location, the track name, and the track number. You can navigate through the track list with the arrows on either side of the display.

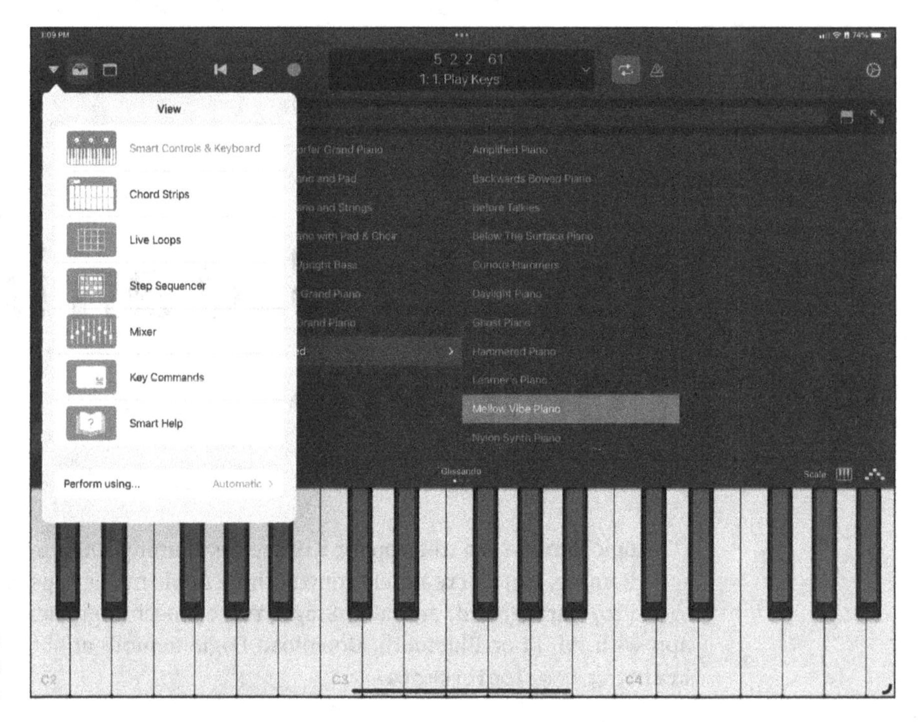

FIGURE 21-1: The Logic Remote library.

Tap the view icon to navigate between different views, including the touch instruments, mixer, smart help, and key commands. The View menu is dynamic and gives you options based on the selected track. For example, drummer tracks have a kit view, while keyboard tracks have a chord strip view.

Tap the library icon to open the patch library. From the Patch Library menu, you can access your entire Logic Pro library and choose new patches for the currently selected track. Logic Remote does its best to recognize the instrument you load and give you the right tools for the job. For example, load a bass guitar patch, and Logic Remote will display a fretboard. Load a drum patch, and Logic Remote will display drum pads. The Logic Remote smart controls give you cool-looking interfaces that are easy to use and fun to play.

Playing Guitar

If you're a guitar player, you may prefer playing all Logic Pro software instruments using a fretboard on your iPad. However, if you're not a guitar player, the fretboard is easy to learn (tap the frets to play notes), and frets can inspire you to play differently. For example, you can drag the strings up or down to bend notes, just like on a guitar. You can also play with chord strips, as shown in Figure 21-2.

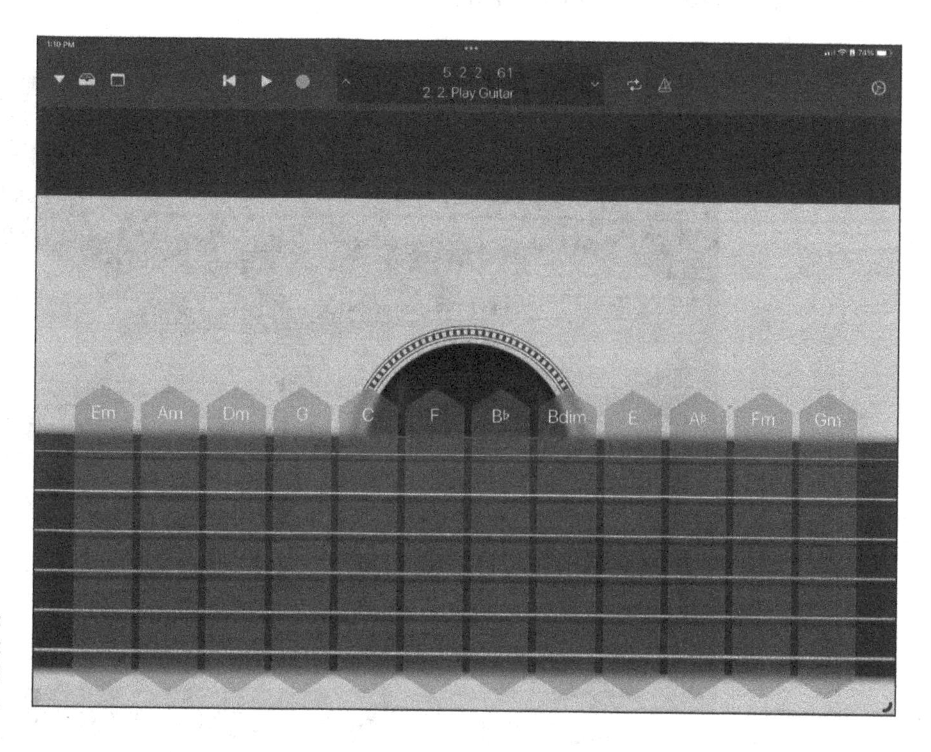

FIGURE 21-2: The acoustic guitar chord strips.

To view the chord strips, tap the view icon and choose the chord strips. Tap the top of the chord strip to play all six strings simultaneously. You can tap any individual string to play it or swipe up and down along the chord strip to simulate strumming. The chord strips can be a wonderful songwriting tool. You can use them to quickly try out different chord progressions as you write. Even a musical virtuoso will find chord strips handy for focusing exclusively on harmony.

Playing Drums

Logic Remote turns your iPad into an easy-to-use beatmaker. With an Ultrabeat track selected, you can view a large screen of 24 drum pads. With a Drum Kit Designer track selected, you can choose the kit view, as shown in Figure 21-3. The drums are touch-sensitive. If you tap a drum with two fingers, it plays repeating notes (great for hi-hat), and as you spread your fingers apart, the pattern gets faster (great for drum rolls).

FIGURE 21-3: The drum kit view.

Both Ultrabeat and Drummer are tightly integrated with Logic Remote. The smart controls take on the instruments' look and feel, helping you adjust quickly as you change between tracks. But you're not restricted to the controls Logic Remote gives you. You're more than welcome to play drums using frets, a keyboard, or even chord strips. And while you're thinking about it, you can use drum pads (but not the drum kit) to play keyboard and guitar parts. You might not get the expressiveness of hardware, but it should help you compose something new and different.

Editing Tracks and Your Arrangement

Editing is often filled with repetitive tasks. When you find yourself choosing the same menus and functions repeatedly, open the key commands view, shown in Figure 21-4, and add a customized key command.

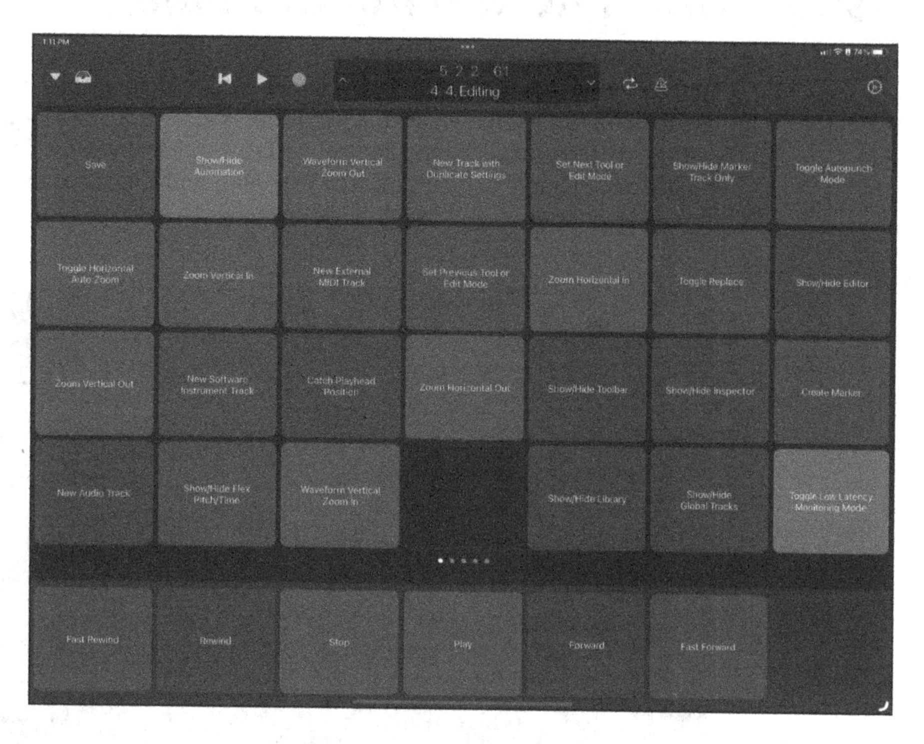

FIGURE 21-4: Logic Remote key commands.

To add a key command, follow these steps:

1. **Tap the view menu and choose the Key Commands view.**

   The key commands view opens.

2. **Tap an empty key command space.**

   The key command search dialog opens.

3. **Browse or search for the key command.**
4. **Tap the key command.**

   The key command is added to the space.

TIP

Several commands to split regions will save you from having to switch to the scissors tool. You can add key commands to set your locators, nudge regions, and go to markers so you can edit your arrangement and your tracks quickly.

Using Your iPad Mixing Console

Tap the view icon and choose the mixer. In my opinion, the mixer, shown in Figure 21-5, is worth the entire price of the iPad. I've spent a lot more money on various MIDI controllers and control surfaces to get the same functionality that I get from a simple and elegant free iPad app. And I never found a solution that was as easy to set up and use as Logic Remote.

FIGURE 21-5: The mixer.

The mixer shows you level meters; icons for automation, record, mute, and solo; pan knobs; faders; and the track names and numbers. You can swipe across the track names or meters to show different groups of faders. To view effects sends in the mixer, tap the Sends 1-4 icon beside the master fader.

TIP

Many audio engineers like to listen to their mix from right outside the control room to test how it sounds from a distance. But not every engineer can listen from outside the room and make changes from an iPad. And with smart controls, you have the most important controls at your fingertips.

Recording Remotely

For many laptop artists, the computer is inseparable from the art. Unfortunately, so is the laptop fan noise. And as you add professional mics and preamps to your studio, they get better at picking up all the imperfections of your recording environment. With Logic Remote, you can leave the computer where it won't impede the quality of your recording.

To begin recording on the selected track, tap the record icon on the transport. Tap the stop icon when you're finished. Tap the cycle mode icon to record multiple takes, as described in Chapter 6.

Commanding Logic Pro

Tap the settings icon to open the edit and track submenus, as shown in Figure 21-6. On this menu, you can undo and redo edits, create and duplicate tracks, adjust the velocity range of the touch instruments, and open Logic Remote Help.

TIP

It's probably a good idea to add a key command to save your project. (See the "Editing Tracks and Your Arrangement" section to find out how to add custom key commands.) In addition to adding a key command to save your project, you can add other project commands such as New Empty Project, Close Project, and Close Project without Saving.

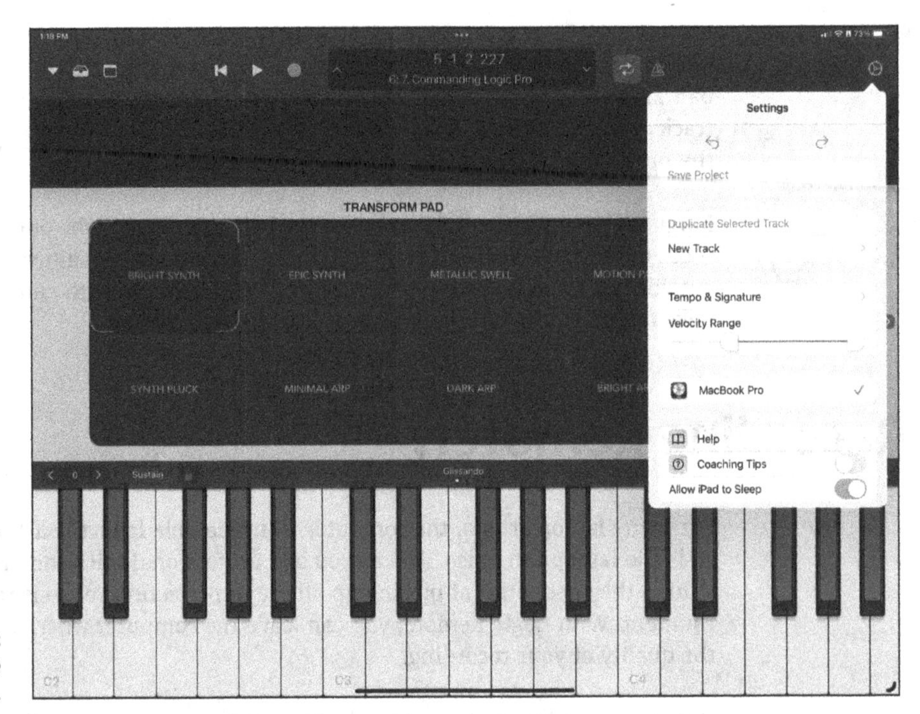

FIGURE 21-6: Logic Remote settings.

Navigating Logic Pro

The control bar display shows you the location of the playhead in musical time or clock time, depending on the general project settings. Tap the control bar display, and a ruler opens below the control bar, as shown in Figure 21-7. Swipe left or right in the ruler or display area to move the playhead.

TIP

The benefit of having a computer keyboard with a numeric keypad is that the numeric keys are assigned to marker numbers. If you're working on a laptop or with a keyboard that doesn't have a numeric keypad, you can create key commands in Logic Remote that go to marker numbers. (See the "Editing Tracks and Your Arrangement" section for information on creating key commands.) If you want to edit a key command, tap it with two fingers to remove or replace the key command or give it a color.

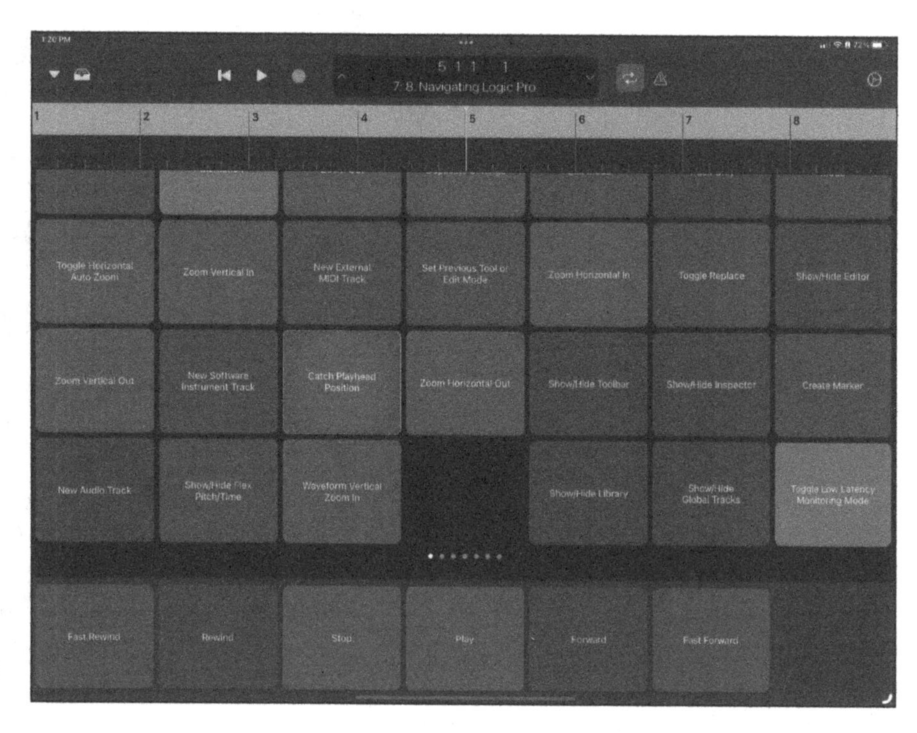

FIGURE 21-7: The ruler.

Adding Tracks with GarageBand for iOS

You can start any project in GarageBand for iOS (iPhone, iPad, or iPod touch), import it to GarageBand on your Mac to work on it a little more, and then import it to Logic Pro to finish the project. The workflow is smooth. One benefit of using all three apps is that you have a version of the project saved on your iOS device, GarageBand projects folder, and Logic Pro projects folder. Redundant backups will save the day.

GarageBand for iOS is a simple app to master. It has a limited feature set but gives you more recording and editing capabilities than The Beatles had in their early days. Plus, you can use its many auto-accompaniment features, such as smart drums, shown in Figure 21-8, to sketch new songs. Tap the settings icon (wrench) to open the GarageBand help system.

You can also share your Logic Pro projects with GarageBand for iOS. This feature allows you to add new tracks remotely using GarageBand on your iOS device. In Logic Pro, go to File ➪ Share ➪ Project to GarageBand for iOS and save the project to iCloud, so you can open it on your iOS device. The GarageBand project will contain a single track containing a reference mix of your Logic Pro project, to which you can add more tracks directly in GarageBand for iOS. New tracks that you add in GarageBand for IOS are added to the original Logic Pro project.

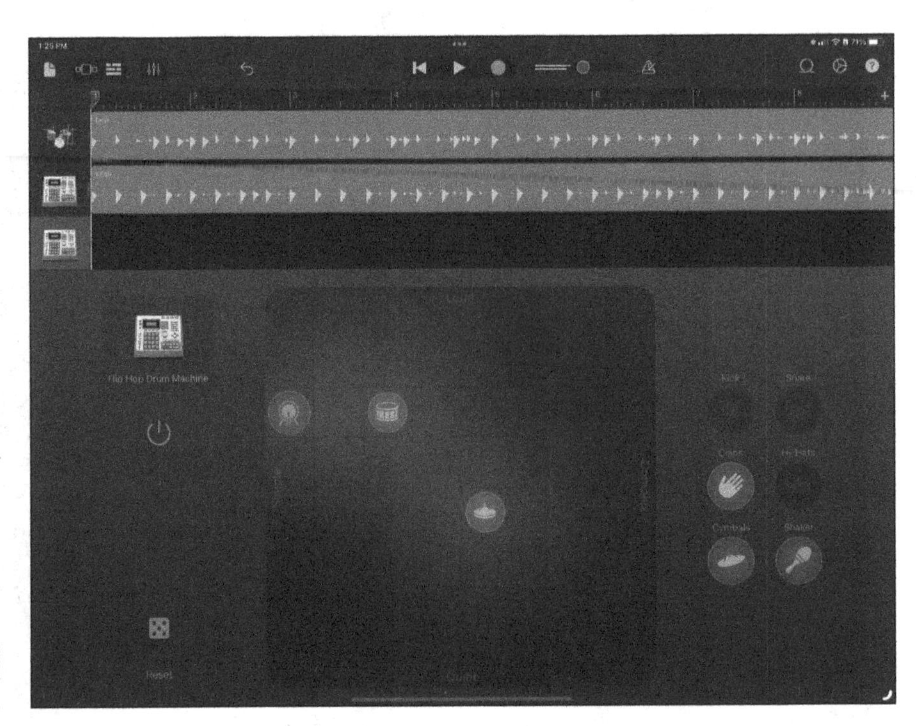

FIGURE 21-8: GarageBand smart drums.

GarageBand for Mac and iOS are both free apps. With your Logic Pro knowledge, I'm confident you'll master GarageBand quickly. To find out more about GarageBand for iOS and Mac, visit the following links:

- `https://logicstudiotraining.com/garagebandios`
- `https://logicstudiotraining.com/garageband`

Importing iPad Audio

GarageBand for iOS can record other music applications on your device through Inter-App Audio. Many iOS apps support the Inter-App Audio protocol. It's a great way to get audio from your third-party synths and drum machines into GarageBand and, eventually, into Logic Pro. In GarageBand for iOS, here's how to record audio from an app with Inter-App Audio capabilities:

1. **Tap the Instruments button at the top of the interface.**

   The touch instrument browser opens, as shown in Figure 21-9.

2. **In the instrument browser, choose Inter-App Audio from the External menu.**

   The screen updates to show you apps capable of inter-app audio.

3. **Choose the app you want to add to your song.**

   An audio track is added to your song using the output of your inter-app audio app. You can now record audio from the chosen app within GarageBand.

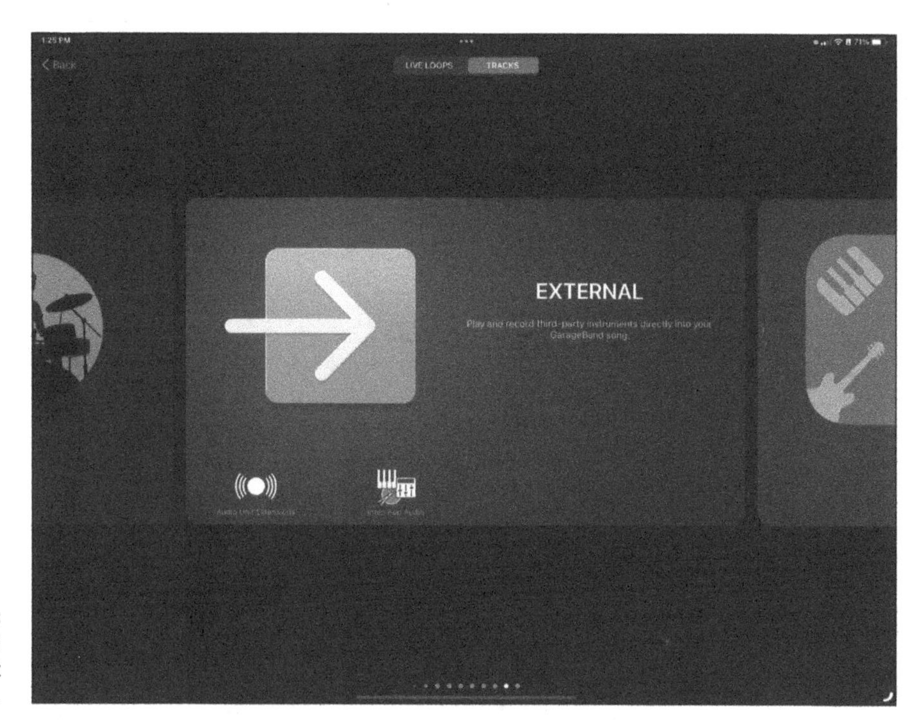

FIGURE 21-9: The GarageBand touch instrument browser.

TIP

You can also use third-party audio effects via Inter-App Audio. Some fantastic-sounding guitar amp simulators can be used as an effect on your audio tracks. Recording apps into GarageBand for iOS and then importing projects to your computer is a simple and effective way to get the most from your iPad, GarageBand, and Logic Pro.

IN THIS CHAPTER

» Getting organized and productive

» Using key commands and screensets

» Saving time with patches and presets

» Troubleshooting and mastering Logic Pro

Chapter 22
Ten Tips to Speed Your Workflow

Finishing a Logic Pro project is like completing a puzzle: You have to fit all the pieces together to see the final picture. Some parts of the game are more fun than others, so you need a good reason to see it through, such as the enjoyment of listening to your project come together. A better motivator is to make every moment that you create with Logic Pro a purposeful moment.

Start by giving your Logic Pro project deadlines and clear outcomes. If you don't have a song to work on, create a learning project. I create projects to learn tools, workflows, plug-ins, and software instruments, in addition to songwriting and music producing. You'll speed up your chops with experience, especially if your purpose is clear.

In this chapter, I offer practical tips to help you speed up your workflow and make more music.

Keep Detailed Notes

I start almost every project by opening the project notes (Option-⌘-P). I write a brief sentence about my project goal and sign my projects, adding my URL and copyright notice. That simple sentence alone will save you time as you search through projects for stuff. For example, sometimes I need to find a piece of music or an effects chain to use in a current project, and the search is much easier when each project has a short description in the notes.

Project notes are also an ideal place to list all your deliverables. In Chapter 16, I list the fundamentals of mixing; I like to collect all that information in the project notes. I list reference tracks, effects ideas, track groups, bar numbers, lyrics, and anything else I can dump out of my brain. I also use project notes as a to-do list to track what I've done and what's left to do.

Along with the ultimate archival tool, scratch paper, use track notes to store specific information about the track, such as instrument frequency ranges and effects chains. I also keep a change log of significant edits and patches or channel strip settings. Sometimes, I'll change the patch or channel strip setting of a track, but I always write down the name of the current patch. Keeping notes saves you from having to commit anything to memory, so you have more mental energy for the music.

Use Key Commands

You can use more than 1,200 key commands in Logic Pro — many more than you'll ever need. But many Logic Pro functions can be achieved only by key command, so one key command you should memorize is Option-K, which opens the Key Commands window.

If there's anything you want to do in your project, you can search through menus to find it, or you can quickly open the Key Commands window and search. The second way is usually faster. Open the Key Commands window regularly and search for commands related to your task.

TIP

Keep one key combination for ad hoc key command assignments. When you find a key command that doesn't have an assignment, you don't have to find a permanent home for it right away. Instead, you can assign it to your temp key command and use it immediately. For example, my temp key command is Control-Option-⌘-K, and I can remember it quickly because it's similar to Option-K.

Use Screensets

Screensets are snapshots of your current screen layout. I use them all the time. To access your screensets, press any number key on your keyboard from 1 to 9. Hold Control for the first digit of double-digit screensets. I reserve screenset 9 for project notes and 8 for track notes, the first screensets I create. I also lock my screensets. That way, I can move them around but get them back to their locked position by pressing the screenset number.

If you've downloaded the templates I've shared throughout this book at `https://logicstudiotraining.com/templates`, you've seen my detailed screensets. Each task deserves a screenset. I have a project of audio editing screensets, MIDI editing screensets, and mixing screensets, and I regularly import them to my current project, as described in Chapter 2. So if you have my templates, you can import them as well.

TIP

As with key commands, keep an ad hoc screenset that you can set up and duplicate from the screenset menu. I usually reserve screenset 1 for my ad hoc workspace. As soon as I have my workspace set up to complete a particular task, I duplicate the screenset, name it, and lock it so I can recall it later.

Save Track Stack Patches and Channel Strip Settings

Track stacks are one of my favorite Logic Pro features. They're easy to pass back and forth between projects and free your precious time. For example, you can save groups of tracks, effects chains, instrument sounds, and so much more. I provide an entire orchestra patch in Chapter 12 that you can add to any project. Track stacks and patches are massive timesavers.

You can also save project templates to save time, but merging templates isn't as easy as loading a few patches. Open the library (Y) and load several patches, and you have a complete band and orchestra with all the audio routing and effects you can imagine.

I like to save channel strip settings almost as much as patches. Adding a standard set of plug-ins to a channel strip is easy from the mixer. Open the mixer (X) and save the channel strip settings from the Setting button at the top of the channel strip. When I get a good sound going, I save it as both a channel strip setting and a patch in the library. They make great starting places for future projects.

Choose a Tool and Master It

If you spend an entire week focusing on a new tool while working on your projects, you'll know its strengths and weaknesses and exactly when to use it. Press T to open the tool menu. Press T twice to select the pointer tool, the most common tool.

Master the pointer, pencil, marquee, scissors, and zoom tools. You'll use them a lot. Editors can be tools too. Spend a week using the step editor in your projects, and you'll master it. Spend a week working on flex time, score editing, smart controls, and other powerful features, and you'll dominate Logic Pro.

Choose a Tool and Ignore It

Logic Pro is so deep that you'll probably never touch some of its features. For example, you might never need to open the score editor if you don't read music. You might never mix your own music. Whatever your situation, feel free to ignore what you don't need.

Many new Logic Pro users come to me wanting to learn it front to back. I always question their motivation because I've never known a music producer who uses Logic Pro that way. The goal isn't to master Logic Pro; the goal is to have Logic Pro help you master making music.

Use the Fastest Way, Not the Right Way

I've wasted a lot of time trying and failing to do something that I knew should be possible. So instead of beating your head against the desk as I do, if something isn't working, ask yourself whether there's another way to accomplish the same goal. Sometimes what seems like the right way isn't always the fastest way.

A long time ago, I tried to create a template that was perfectly connected to my studio environment. I added external MIDI instruments for all my synths, audio tracks for all my inputs, software instruments I always used, and a detailed MIDI controller setup to control every knob and fader. It should have worked, but it never truly did. And I didn't make a whole lot of music during that experiment.

There's no such thing as perfect software. But there is such a thing as software used well. So my motto is: A proper project is a finished project.

Establish a Troubleshooting Strategy

It's not often that I have to troubleshoot Logic Pro. But when I do, I use the following strategy:

1. **Is the problem with the hardware?**

   Test all your hardware, including instruments, cables, audio interface, speakers, and anything else that could be connected to your system. Check the computer's audio system preferences.

2. **Is the problem with the software?**

   Test different projects and new projects. Check the Logic Pro settings and the I/O buffer size.

3. **Is the problem with the project?**

   Import parts of the project into a new project. Gradually add more parts of the project.

4. **Is the problem with a project component?**

   Test your third-party effects and software instrument plug-ins.

I almost always find a solution by searching in this order. If I don't, I search online and visit the Apple discussion forums. Set a time limit for troubleshooting before you ask for help online. Then set a time limit for your online search before you call Apple or visit a local Apple store.

Save and Back Up Frequently

In Chapter 20, I plead with you to back up your work. One of the greatest productivity tips of all time is don't lose your data. Digital storage is getting cheaper and cheaper, and you can find many free storage services online. Back up your data.

I've had hard drives fail. I've dropped hard drives. I've deleted things by mistake. But I don't lose time. I have onsite and offsite backups, plus I sync my core projects. I call this process abundance by redundancy. Backing up doesn't take much time to set up and is easy to automate. Search for *cloud storage and backup* or something similar, and you'll be presented with competitive offers.

I don't like to brag about my backup system, but I do think it is great. I can vouch for its effectiveness, and I've most certainly needed it.

Don't Lose Sight of the Music

Sometimes you have to make things more complicated before making them simple. For example, when you have a song you're going to record, you have to break it into parts and rebuild it so you can share it. Likewise, when you compose, you essentially break your ideas apart and rebuild them. Logic Pro is a tool to help you compose, record, mix, and produce your own music.

Songs are waiting to be written. Singers are waiting for you to record them. Bands are waiting for you to mix them. Listeners are waiting for you to produce music for them. Don't wait to make music. You have Logic Pro. You're ready.

Index

B

C

D

E

F

G

H

I

K

L

M

N

O

P

Q

R

S

T

U

V

W

X

Z

About the Author

Graham English owns logicstudiotraining.com, the popular website for learning Logic Pro and music production. Graham is an experienced writer, instructor, and entertainer and has been a music and audio professional for over 35 years. You can catch Graham gigging around the New England area and online at `https://grahamenglish.com`.

Author's Acknowledgments

I am forever grateful for the love and support of my partner, Katie, and our beautiful children, Desmond and Violet. Their constant encouragement and belief in me is a shining star.

I want to thank my coach, John Chancellor, for keeping me accountable and continually encouraging me while I wrote this book. Writing a book is no easy feat, so I am thankful that I have a strong supporting mentor and friend in John.

Thanks to Steven Hayes for keeping this book successful throughout the years. This third edition of *Logic Pro For Dummies* is further proof that Steven's vision and support are valuable resources to Logic Pro users worldwide.

I also want to thank Aaron Black for making the first edition of *Logic Pro For Dummies* happen. Your impact on my life is immeasurable, and I am forever grateful.

Susan Pink has always been a true leader in making *Logic Pro For Dummies* better. Her ability to improve everything has inspired me to strive for success. With her guidance, expertise, and encouragement, I've always felt deeply supported. A massive round of applause is certainly deserved. Thank you, Susan!

Ryan Williams deserves a huge thank-you for all his hard work in reviewing and contributing to this book's success. *Logic Pro For Dummies* would not be the same without him. His expertise and knowledge were essential in helping me write the best book possible. Ryan is truly an invaluable asset. I am both privileged and humbled to have been given the opportunity to write a *Dummies* book. *Dummies* and Wiley have been setting the bar for quality in how-to books, working hard to deliver the best possible value for readers. Their dedication is truly inspiring, and I can only hope that it's reflected in my own work as an author. My sincere thanks go out to *Dummies* and Wiley for entrusting me with this project. I've always felt honored to be part of their mission of excellence.

Finally, I'd like to thank all the enthusiastic and passionate Logic Pro users. You've created an enormous community that loves to support one another. You ask a lot of questions, you give a lot of answers, and you make a lot of music. I hope I can help you make more music, and I hope you share it with other *Logic Pro For Dummies* readers at `https://logicstudiotraining.com`.

Publisher's Acknowledgments

Executive Editor: Steve Hayes

Project and Copy Editor: Susan Pink

Technical Editor: Ryan Williams

Proofreader: Debbye Butler

Production Editor: Mohammed Zafar Ali

Cover Image: © SeventyFour/Shutterstock